PROCESSING AND CHARACTERIZATION OF MULTICOMPONENT POLYMER SYSTEMS

New Insights

PROCESSING AND CHARACTERIZATION OF MULTICOMPONENT POLYMER SYSTEMS

New Insights

Edited by

Jose James

Sabu Thomas

Nandakumar Kalarikkal

Yang Weimin

Kaushik Pal

AAP | APPLE ACADEMIC PRESS

Apple Academic Press Inc.
3333 Mistwell Crescent
Oakville, ON L6L 0A2
Canada USA

Apple Academic Press Inc.
1265 Goldenrod Circle NE
Palm Bay, Florida 32905
USA

Exclusive worldwide distribution by CRC Press, a member of Taylor & Francis Group

International Standard Book Number-13: 978-1-77188-724-3 (Hardcover)
International Standard Book Number-13: 978-0-429-46979-4 (eBook)

CIP data on file with Canada Library and Archives

Library of Congress Cataloging-in-Publication Data

Names: James, Jose (Chemist), editor | Sabu Thomas, editor | Nandakumar Kalarikkal, editor | Yang Weimin, editor | Kaushik Pal, editor

Title: Processing and characterization of multicomponent polymer systems : new insights / editors, Jose James [and four others].

Description: Toronto : Apple Academic Press, 2019. | Includes bibliographical references and index.

Identifiers: LCCN 2019007174 (print) | LCCN 2019008241 (ebook) | ISBN 9780429469794 (ebook) | ISBN 9781771887243 (hardcover : alk. paper)

Subjects: LCSH: Polymers in medicine. | Biomedical materials. | Elastomers. | Medical instruments and apparatus.

Classification: LCC R857.P6 (ebook) | LCC R857.P6 P765 2019 (print) | DDC 610.28/4--dc23

LC record available at https://lccn.loc.gov/2019007174

Apple Academic Press also publishes its books in a variety of electronic formats. Some content that appears in print may not be available in electronic format. For information about Apple Academic Press products, visit our website at **www.appleacademicpress.com** and the CRC Press website at **www.crcpress.com**

ABOUT THE EDITORS

Jose James

Jose James is an Assistant Professor in the Research and Post-Graduate Department of Chemistry at St. Joseph's College in Moolamattom, Idukki, Kerala, India. Mr. James has recently submitted his PhD thesis under the faculty development program of the Indian University Grants Commission, under the supervision of Prof. (Dr.) Sabu Thomas (Pro-Vice Chancellor and Professor of Polymer Science and Engineering in School of Chemical Science, Mahatma Gandhi University, Kottayam) and Dr. George V. Thomas (Principal and Associate Professor, Department of Chemistry, St. Joseph's College, Moolamattom, India). He has extensive research experience in the field of interpenetrating polymeric networks (IPNs) and has two publications dealing with diffusion of IPN in *Polymer* (Elsevier) and on viscoelastic behavior of IPNs in the *New Journal of Chemistry* (RSC). He has attended more than 20 international conferences with one publication in IOP conference proceedings and has presented many research papers. In addition, he has co-authored three chapters published in international books. He has four years of research experience in polymer chemistry and nanotechnology at the laboratories of the School of Chemical Science, Mahatma Gandhi University, Kottayam, Kerala, India.

Sabu Thomas, PhD

Sabu Thomas is currently the Vice Chancellor of Mahatma Gandhi University, Kottayam, Kerala, India. He was the Founder Director of the International and Inter University Center for Nanoscience and Nano-technology and is also a full professor of Polymer Science and Engineering in the School of Chemical Science of the same university. He is a fellow of many professional bodies. Prof. Thomas has co-authored many papers in international peer-reviewed journals in the area of polymer processing. He has organized more than 50 international conferences. Prof. Thomas's research group is in specialized areas of polymers, which include polymer blends, fiber-filled polymer composites and their morphological characterization, aging, and degradation, pervaporation phenomena, sorption, and diffusion, interpenetrating polymer systems, recyclability and reuse of waste plastics and rubbers, elastomeric cross-linking, and dual porous nanocomposite scaffolds for tissue engineering. Prof. Thomas's research group has extensive exchange programs with different industries, research, and academic institutions all over the world and is performing world-class collaborative research in various fields. The professor's center is equipped with various sophisticated instruments and has established state-of-the-art experimental facilities that cater to the needs of researchers within the country and aboard. He has more than 930 publications with 34,227 citations, 55 books, 30 international awards and three patents to his credit. His H index is 84. He is a reviewer for many international journals. Prof. Thomas has attained the 5th position in the list of Most Productive Researchers in India in 2008-17.

Nandakumar Kalarikkal, PhD

Nandakumar Kalarikkal is an Associate Professor and Director at the School of Pure and Applied Physics and Director of International and Inter University Center for Nanoscience and Nanotechnology of Mahatma Gandhi University, Kottayam, Kerala, India. His research activities involve applications of nanostructured materials, laser plasma, phase transitions, etc. He is the recipient of research fellowships and associateships from prestigious government organizations, such as the Department of Science and Technology and Council of Scientific and Industrial Research of the Government of India. He has an active collaboration with national and international scientific institutions in India, South Africa, Slovenia, Canada, France, Germany, Malaysia, Australia, and the US. He has more than 130 publications in peer-reviewed journals. He has also co-edited nine books of scientific interest and co-authored many book chapters. He has 920 citations, and 4 patents to his credit. His H Index is 18. He is a reviewer for many international journals.

Yang Weimin, PhD

Yang Weimin is the Taishan Scholar Professor of Qingdao University of Science and Technology in China. He is a Professor of the College of Mechanical and Electrical Engineering and Director of the Department of International Exchanges and Cooperation, Beijing University of Chemical Technology, Beijing, China. In addition, he is a fellow of many professional organizations. Dr. Weimin has authored many papers in international peer-reviewed journals in the area of polymer processing. He has contributed to a number of books as author and editor and acts as a reviewer to many international journals. In addition, he is a consultant to many polymer equipment manufacturers. He has also received numerous awards for his work in polymer processing. His interests include polymer processing and CAD/CAE/CAM of polymer processing.

Kaushik Pal, PhD

Kaushik Pal is a Research Professor in the Department of Nanotechnology and is a Founding Director of Center for Multidisciplinary Research Excellence (CMRE) at Bharath University, Chennai, India. His research areas are high-yield novel nanomaterials/polymer/grapheme-assembled liquid crystals, novel electro-optic switchable device for nanoelectronics, static memory devices, transparent electrodes, super capacitors, and smart, flexible display performances. His novel research findings have resulted over 50 publications that have been published in many international peer-reviewed journals and reputed magazines. He has published eight book chapters as well. Dr. Pal has been cited 20 times, and and his current h-index is 7. In recent years, he has been serving on the editorial boards of over 30 international journals. Dr. Pal has received several significant awards, including the prestigious Marie-Curie Experienced Research Fellowship (Postdoctoral) honor from Aristotle University of Thessaloniki, Greece (offered by the European Commission) and has been selected as Visiting Scientist for the project BK-21, Brain Korea Postdoctoral Fellowship (Visiting Scientist).

CONTENTS

Contributors .... *xi*

Abbreviations .... *xv*

Preface .... *xix*

PART I: Bio-Composites and Nanocomposites 1

1. Tailoring and Assessing Ethylene Vinyl Acetate (EVA)/Organoclay Nanocomposites for Biomedical Applications 3

Azlin Fazlina Osman, Tuty Fareyhynn M. Fitri, Asna Rasyidah Abdul Hamid, Fatimah Hashim, Md Rakibuddin, and Abdulkader M. Alakrach

2. Gold (Silver)-Polymer Nanocomposites for Biosensing and Microfluidic Applications 25

Simona Badilescu and Muthukumaran Packirisamy

3. Growth of Mesenchymal Stem Cells on Surface-Treated 2D Poly(Glycerol-Sebacate) Bio-Elastomers of Varying Stiffness 43

Raju B. Maliger, Peter J. Halley, Justin J. Cooper-White, and Donna Dinnes

4. Enhancement of Impact Strength in Jute Fiber-Reinforced Polypropylene-Spent Coffee Ground Composites 63

N. Jaya Chitra

5. Processing and Mechanical Characterization of Short Banana-Jute Hybrid Fiber-Reinforced Polyester Composites 77

Siva Bhaskara Rao Devireddy and Sandhyarani Biswas

6. Graphene and Graphene-Based Polymer Nanocomposites: The New Wonder Materials of the Nanoworld 101

Ujjal K. Sur

7. Preparation and Characterization of ZnO Doped Cts/PEG-Ag Nano-Composite for Impacting Photocatalytic Response 115

T. Kokila, P. S. Ramesh, D. Geetha, and S. Kavitha

8. Development of High-Performance *In-Situ* Polypropylene/Nylon 6 Microfibrillar Composites 135

B. D. S. Deeraj, K. Jayanarayanan, and Kuruvilla Joseph

9. **Effect of Co-Precipitation on the Physico-Mechanical and Dielectric Properties of Nickel Ferrite Embedded Natural Rubber Composites.......... 155**
Sankar S. Menon, Radhu Krishna, Lida Wilson, Sreedha Sambhudevan, Balakrishnan Shankar, Anshida Mayeen, and Nandakumar Kalarikkal

PART II: Interpenetrating Polymeric Networks and Nanostructured Materials.......... 177

10. **Effect of Initiating Systems on the Viscoelastic Behavior of SBR-PMMA Interpenetrating Polymer Networks.......... 179**
Jose James, George V. Thomas, and Sabu Thomas

11. **Influence of Cationic, Anionic, and Nonionic Surfactants on Hydrothermal Synthesis of Nano CuS: Structural, Morphological, and Capacitance Behavior.......... 191**
D. Geetha, P. S. Ramesh, and Surekha Podili

12. **An Experimental Study on the Influence of Polymeric Nano-TIO_2 in Cement and Concrete for Its Dispersion, Structural Characterization, Mechanical Properties, and Its Performance Under Aggressive Environment.......... 213**
Mainak Ghosal and Arun Kr. Chakraborty

PART III: Theoretical Protocols for Polymers and Clusters.......... 235

13. **Cooperative Molecular and Macroscopic Dynamics of Amorphous Polymers: A New Analytical View and Its Potential.......... 237**
José Joaquim C. Cruz Pinto and José Reinas S. André

14. **Vibrational Modes of the Van der Waals-London Silicon Clusters Si_2 and Si_3 in the Gas-Phase.......... 261**
Srećko Botrić and Ivan Zulim

PART IV: Special Topics in Polymer Processing and Polymer Coating.......... 275

15. **Initial Flow Analysis in Crucible-Nozzle During Planar Flow Melt Spinning Process.......... 277**
Sowjanya Madireddi

16. **Corrosion Protective Methyl Substituted Polyaniline Coatings on Oxidizable Metals.......... 301**
Vandana P. Shinde and Pradip Patil

Index.......... 331

CONTRIBUTORS

Abdulkader M. Alakrach
Center of Excellence Geopolymer and Green Technology (CE GeoGTech), School of Materials Engineering, University Malaysia Perlis, Arau, Perlis 02600, Malaysia

José Reinas S. André
PhD, Coordination Professor, UTC-Engineering, and Technology, UDI-Research Unit for Inland Development, Guarda Polytechnic Institute, Technology, and Management School, Guarda, Portugal, E-mail: jandre@ipg.pt

Simona Badilescu
Optical-Bio Microsystems Laboratory, Micro-Nano-Bio Integration Center, Department of Mechanical and Industrial Engineering, Concordia University, Montreal, Canada, H3G 1M8

Sandhyarani Biswas
Department of Mechanical Engineering, National Institute of Technology, Rourkela, Odisha, 769008, India

Srećko Botrić
Faculty of Electrical Engineering, Mechanical Engineering and Naval Architecture, Department of Mathematics and Physics, University of Split, Ruđera Boškovića 32, HR–21000 Split, Croatia, E-mail: srecko.botric@fesb.hr

Arun Kr. Chakraborty
Research Scholar, Associate Professor, Department of Civil Engineering, Indian Institute of Engineering Science and Technology, Shibpur, Howrah 711103, India

N. Jaya Chitra
Department of Chemical Engineering, Dr. MGR Educational and Research Institute (University), Maduravoyal, Chennai–95, India, E-mails: jayakarun7696@gmail.com; hod-chem@drmgrdu.ac.in

Justin J. Cooper-White
AIBN/School of Chemical Engineering, University of Queensland, St. Lucia, Brisbane, Queensland–4072, Australia

B. D. S. Deeraj
Department of Chemistry, Indian Institute of Space Science and Technology, Kerala–695547, India

Siva Bhaskara Rao Devireddy
Department of Mechanical Engineering, St. Ann's College of Engineering and Technology, Chirala, Andhra Pradesh, 523187, India, E-mail: sivabhaskararao@gmail.com

Donna Dinnes
AIBN/School of Chemical Engineering, University of Queensland, St. Lucia, Brisbane, Queensland–4072, Australia

Tuty Fareyhynn M. Fitri
Center of Excellence Geopolymer and Green Technology (CE GeoGTech), School of Materials Engineering, University Malaysia Perlis, Arau, Perlis 02600, Malaysia

D. Geetha
Department of Physics (DDE Wing), Annamalai University, Annamalai Nagar – 608002, Tamilnadu, India, E-mail: geeramphyau@gmail.com

Mainak Ghosal
Adjunct Assistant Professor, JIS College of Engineering, Kalyani, Nadia, W. Bengal, and Research Scholar, Associate Professor, Department of Civil Engineering, Indian Institute of Engineering Science and Technology, Shibpur, Howrah 711103, India

Peter J. Halley
AIBN/School of Chemical Engineering, University of Queensland, St. Lucia, Brisbane, Queensland–4072, Australia

Asna Rasyidah Abdul Hamid
Center of Excellence Geopolymer and Green Technology (CE GeoGTech), School of Materials Engineering, University Malaysia Perlis, Arau, Perlis 02600, Malaysia

Fatimah Hashim
School of Fundamental Science, University Malaysia Terengganu, 21030 Kuala Terengganu, Terengganu, Malaysia

Jose James
Research and Post-Graduate Department of Chemistry, St: Joseph's College, Moolamattom and International, and Interuniversity Center for Nanoscience and Nanotechnology, and School of Chemical Sciences, Mahatma Gandhi University, Kottayam, 686560, Kerala, India. josejameskadathala@gmail.com

K. Jayanarayanan
Department of Chemical Engineering and Materials Science, Amrita University, Coimbatore, India

Kuruvilla Joseph
Department of Chemistry, Indian Institute of Space Science and Technology, Kerala–695547, India

Nandakumar Kalarikkal
School of Pure and Applied Physics, Mahatma Gandhi University, Kottayam, Kerala, India

S. Kavitha
Department of Physics, Annamalai University, Annamalai Nagar 608002, Tamilnadu, India

T. Kokila
Department of Physics, Vivekanandha College of Arts and Sciences for Women (Autonomous), Elayampalayam, Namakkal–637205, Tamilnadu, India

Radhu Krishna
Department of Chemistry, Amrita School of Arts and Sciences, Amritapuri, Amrita Vishwa Vidyapeetham, Amrita University, India

Sowjanya Madireddi
CVR College of Engineering, Hyderabad, Telangana – 501510, India, E-mail: madireddisowjanya@gmail.com

Raju B. Maliger
AIBN/School of Chemical Engineering, University of Queensland, St. Lucia, Brisbane, Queensland–4072, Australia, Tel.: +61450425510, E-mail: r.maliger@gmail.com

Anshida Mayeen
School of Pure and Applied Physics, Mahatma Gandhi University, Kottayam, Kerala

Sankar S. Menon
Department of Chemistry, Amrita School of Arts and Sciences, Amritapuri, Amrita Vishwa Vidyapeetham, Amrita University, India

Azlin Fazlina Osman
Center of Excellence Geopolymer and Green Technology (CE GeoGTech),
School of Materials Engineering, University Malaysia Perlis, Arau, Perlis 02600, Malaysia,
E-mail: azlin@unimap.edu.my

Muthukumaran Packirisamy
Optical-Bio Microsystems Laboratory, Micro-Nano-Bio Integration Center,
Department of Mechanical and Industrial Engineering, Concordia University, Montreal, Canada H3G 1M8

Pradip Patil
Department of Physics, North Maharashtra University, Jalgaon 425001, Maharashtra, India

José Joaquim C. Cruz Pinto
PhD, Professor, Retired Full Professor from the University of Aveiro/CICECO,
Department of Chemistry, 3810–193 Aveiro, Portugal, E-mail: jj.cruz.pinto@ua.pt

Surekha Podili
Department of Physics, Annamalai University, Annamalai Nagar – 608002

Md Rakibuddin
School of Materials Science and Engineering, Yeungnam University, 280 Daehak-ro, Gyeongsan, South Korea

P. S. Ramesh
Department of Physics (DDE Wings), Annamalai University, Annamalai Nagar 608002, Chidambaram, Tamilnadu, India, E-mail: psrddephyau@gmail.com

Sreedha Sambhudevan
Department of Chemistry, Amrita School of Arts and Sciences, Amritapuri,
Amrita Vishwa Vidyapeetham, Amrita University, India

Balakrishnan Shankar
Department of Mechanical Engineering, Amrita School of Engineering, Amritapuri,
Amrita Vishwa Vidyapeetham, Amrita University, India

Vandana P. Shinde
Department of Physics, North Maharashtra University, Jalgaon 425001, Maharashtra, India

Ujjal K. Sur
Department of Chemistry, Behala College, University of Calcutta, Kolkata–60, India,
E-mail: uksur99@yahoo.co.in

George V. Thomas
Research and Post-Graduate Department of Chemistry, St: Joseph's College, Moolamattom, Kerala, India

Sabu Thomas
International and Interuniversity Center for Nanoscience and Nanotechnology, School of Chemical Sciences, Mahatma Gandhi University, Kottayam, 686560, Kerala, India

Lida Wilson
Department of Chemistry, Amrita School of Arts and Sciences, Amritapuri, Amrita Vishwa Vidyapeetham, Amrita University, India

Ivan Zulim
Faculty of Electrical Engineering, Mechanical Engineering and Naval Architecture, Department of Electronics and Computing, University of Split, Ruđera Boškovića 32, HR–21000 Split, Croatia, E-mail: zulim@fesb.hr

ABBREVIATIONS

ABCDQPA	nozzle domain
AFM	atomic force microscopy
$AgNO_3$	silver nitrate
AIBN	azobisisobutyronitrile
BCGFB	slit
BPO	benzoyl peroxide
BRNS	Board of Research in Nuclear Sciences
BSA	bovine serum albumin
CB	carbon black
CBC	carpet backing cloth
CH_3COOH	glacial acetic acid
CH_4	methane
CM	coupling model
CNFs	carbon nanofibers
CNTs	carbon nanotubes
CO_2	carbon dioxide
CPCs	conducting polymer composites
CPs	conducting polymers
CR	corrosion rate
CS	coconut sheath
CSTMD	cooperative segmental theory of materials dynamics
CTAB	cetyltrimethylammonium bromide
CTM	compression testing machine
CTMD	clustering theory of materials dynamics
Cu	copper
CV	cyclic voltammetry
CVD	chemical vapor deposition
CVs	cyclic voltammograms
DABCO	1,4-diazabicyclo[2,2,2]-octane
DAE	Department of Atomic Energy
DCP	dicumyl peroxide
DFT	density functional theory
DLS	dynamic light scattering

DMA	dynamic mechanical analysis
DMEM	Dulbecco's modified Eagle medium
DMF	dimethylformamide
DMRL	Defense Metallurgical Research Laboratory
DMSO	dimethyl sulfoxide
DMTA	dynamic mechanical thermal analysis
DVB	divinylbenzene
ECP	electrochemical polymerization
EDAC	1-ethyl–3-(3-dimethyl aminopropyl) carbodiimide
EDLCs	electric double layer capacitors
EDS	energy dispersive spectroscopic
EDX	energy dispersive X-ray
EDTA	ethylene diamine tetraacetic acid
EDXA	energy dispersive X-ray analysis
EDXMA	energy dispersive X-ray microanalysis
EG	ethylene glycol
EG	expanded graphite
EIS	electrochemical impedance spectroscopy
EMT	effective-medium theory
ES	emarldine salt
EVA	ethyl vinyl acetate
EVA	ethylene vinyl acetate
FBR	foreign body reaction
FBS	fetal bovine serum
FRGS	fundamental research grant scheme
FSI	fluid-structure interaction
FTIR	Fourier transform infrared
FWHM	full width at half maximum
GCE	glassy carbon electrode
GO	graphene oxide
H_2	hydrogen
HA	hyaluronic acid
HCl	hydrochloric acid
HDPE	high-density polyethylene
IIST	Indian Institute of Space Science and Technology
IPNs	interpenetrating polymer networks
IR	infrared band
IR	infrared spectral range
ISRL	Indian Synthetic Rubber Limited

ITO	indium-tin oxide
LCDs	liquid-crystal displays
LCPs	liquid crystalline polymers
LCS	low carbon steel
LDH	layered double hydroxide
LDPE	low-density polyethylene
LEDs	light-emitting diodes
LSPR	localized surface Plasmon resonance
MCT	mode coupling theory
MES	morpholino (ethane-sulfonic acid)
MFBs	microfibrillar blends
MFCs	microfibrillar composites
MMA	methyl methacrylate
MMT	montmorillonite
MS	multiple sclerosis
MSC	mesenchymal stem cells
MW	microwave
NaOH	sodium hydroxide
NHS	N-hydroxysuccinimide
NP	nanoparticles
NSL	nanosphere lithography
OMMT	organically modified montmorillonite
OPC	Ordinary Portland Cement
PANI	polyaniline
PB	pernigraniline base
PBT	poly (butylene terephthalate)
PC	polycarbonate
PCE	polycarboxylate ether
PDMS	poly(dimethylsiloxane)
PE(%)	percentage protection efficiency
PE	polyethylene
PEG	ploy (ethylene glycol)
PE-g-MA	polyethylene grafted maleic anhydride
PEI	polyethyleneimine
PET	poly (ethylene terephthalate)
PF	phenol formaldehyde
PFA	paraformaldehyde
PFMS	planar flow melt spinning
PMMA	poly [methyl methacrylate]

POT	poly(o-toluidine)
PP	polypropylene
PQ	domain inlet
PQDCBAP	nozzle
PS	polystyrene
PVA	poly(vinyl alcohol)
PVC	polyvinyl chloride
PVP	polyvinylpyrrolidone
QHE	quantum Hall effect
RA	rheumatoid arthritis
RB	round bottom flask
RFCs	rubber ferrite composites
RMS	root mean square
ROS	reactive oxygen species
S	sisal
SA	sodium hyaluronate
SAED	selected-area electron diffraction
SBR	styrene butadiene rubber
SCG	spent coffee ground
SDBS	sodium dodecyl benzene sulphonate
SDS-PAGE	sodium dodecyl sulfate-polyacrylamide gel electrophoresis
SEM	scanning electron microscope
SLE	systemic lupus erythematosus
SO^{2}_{4-}	sulfate
SS	stainless steel
TEM	transmission electron microscopy
TPU	thermoplastic polyurethane
TST	transition state theory
UGC	University Grants Commission
UV	ultraviolet
VA	vinyl acetate
VOF	volume of flow
VOF	volume of fluid technique
XPS	X-ray photoelectron spectroscopy
XRD	X-ray diffraction
XRD	X-ray diffractometer
$Zn(NO_3)_2.6H_2O$	zinc nitrate

PREFACE

Recent years have witnessed the sheer growth of macromolecular concepts and nanotechnology-based innovations in polymer science. This book is a collection of contributions from materials science experts across the globe. The fabrication and characterization of polymeric systems still exist as the evergreen domain in materials science. The quality measurements of newly designed polymeric products demand systematic and unexplored characterization protocols. As an outcome of this, research on this has attracted the considerable attention of the scientific community all over the world. Polymer science and technology has now emerged as a specific stream taught in universities and practiced in industries for the attainment of a new products with specific features. We hope that our new book, *Processing and Characterization of Multicomponent Polymer Systems: New Insights*, will surely be an asset for scientists, engineers, and budding researchers working in the area of polymer science and nanotechnology.

This book contains 16 chapters and are categorized under four major streams, which are "Bio-Composites and Nanocomposites," "Interpenetrating Polymeric Networks and Nanostructured Materials," "Theoretical Protocols for Polymers and Clusters," and "Special Topics in Polymer Processing and Polymer Coating."

Under the stream "Bio-composites and Nanocomposites," nine chapters are included. Chapter 1 presents the incorporation of soft, flexible, and biocompatible polymers such as polyurethane and ethyl vinyl acetate with nanofiller (organically modified montmorillonite) to enhance their performance as insulation materials in implantable biomedical devices. Chapter 2 presents the *in-situ* synthesis of Au-PDMS and Ag-PDMS nanocomposites, both at the macro scale and inside the channel of a microfluidic chip and their application for plasmonic detection of biological entities in cancer research. Chapter 3 presents research on poly(glycerol-sebacate), a bioelastomer as a potential material for tissue engineering applications. This chapter deals with the effect of poly(glycerol-sebacate) substrate stiffness and surface treatment methods on the morphology and lineage of mesenchymal stem cells (MSC).

Chapter 4 deals with processing of a Jute sandwich composite through compression molding. In that work, coffee grounds were used as a matrix and braided jute fiber as a reinforcing material. The high-impact strength of the composites makes it applicable for automobile and machinery parts. Chapter 5 deals with the mechanical characteristics of short randomly oriented banana-jute hybrid fiber-reinforced polyester composites and were studied as a function of the overall fiber loading and different weight ratios of banana and jute fibers. In Chapter 6, a brief overview on various aspects of graphene—such as synthesis, functionalization, self-assembly, and some of its amazing properties along with its various applications ranging from sensors to energy storage devices—have been illustrated. Chapter 7 narrates the investigation on ZnO-doped Cts/PEG–Ag nanomaterials, which have been synthesized by reduction method at 80°C. The XRD spectrum shows that all the samples are facing center cubic structure. The ZnO doped Cts/PEG–Ag nanocomposite showed excellent photocatalytic behavior and underlying mechanical properties. Chapter 8 deals with the synthesis of microfibrillar *in-situ* composites of polypropylene/nylon 6 blends in a twin screw extruder. The storage modulus of the composite was found to be superior to the individual polymers throughout the temperature range of analysis. Chapter 9 presents spinel structured nickel ferrite and its fabrication using a co-precipitation method. Natural rubber composites were prepared with different loadings of nickel ferrite. Dielectric measurements show that permittivity decreases with an increase in frequency and increases with an increase in ferrite loading.

Three chapters were included under the stream on "Interpenetrating Polymeric Networks and Nanostructured Materials." Chapter 10 deals with the synthesis of a series of interpenetrating polymer networks (IPNs) based on styrene butadiene rubber and poly [methyl methacrylate] by a sequential polymerization technique. The effects of two initiating systems (benzoyl peroxide and azobisisobutyronitrile) in the polymerization of MMA during the fabrication of IPN were studied in detail. The viscoelastic properties of these IPNs were investigated in the temperature range of –80 to 200°C and at a frequency of 1 Hz. Chapter 11 deals with the synthesis and formation mechanism of CuS nanostructures by a simple hydrothermal route using organic surfactants anionic, cationic, and non- ionic surfactants as templates and thiourea as the sulfur source at 130°C. Chapter 12 aims to investigate the long-term effect of M-40 Grade concrete made with nano-TIO_2 when exposed under aggressive chemical attacks. The results corroborated the fact that though strength gains for cement mortar and M-40 grade concrete made with 1% nano-TiO_2 is minor in the short term, their long-term durability property enhances multifold in a natural atmosphere and also when exposed to chemical attacks.

The stream "Theoretical Protocols for Polymers and Clusters" consists of two chapters. Chapter 13 addresses the cooperative nature of the molecular and macroscopic dynamics of amorphous-phase polymers as the result of truly many-body processes. A new theory of clustering of specifiable primitive relaxors is proposed. Chapter 14 is an odd chapter, and it indicates the possibility for the Van der Waals-London clusters Si_2 and Si_3 to exist as the vibrational structures. By applying the Lagrange's formalism of classical mechanics to small oscillations of weakly bound clusters Si_2 and Si_3 proved that their harmonic frequencies are in the infrared spectral range (IR).

The stream "Special Topics in Polymer Processing and Polymer Coating" consists of two chapters. Chapter 15 deals with a numerical model that is used to investigate the flow patterns in the crucible-nozzle for a few designs. A volume of fluid technique with fluid-structure interaction has been employed along with conservation equations and a temperature-dependent viscosity equation. Chapter 16 narrates poly(o-toluidine) (POT) coatings and were electrodeposited as corrosion protective coatings on low carbon steel (LCS), 304 stainless steel (SS) and copper (Cu) substrates. POT coatings on metal substrates have been carried out under cyclic voltammetric conditions in an aqueous sodium salicylate solution.

Polymer science and technology has emerged as a hot topic from the postgraduate level to the levels of graduation and diversified industrial courses in polytechnics and university colleges. Keeping young researchers in mind, this book is edited to highlight the latest innovations and principles behind these findings, specific to nanostructured polymeric materials and polymer nanocomposites. This book is devoted to novel architectures at the nano-level with an emphasis on new synthesis and characterization methods. Chapterwise bibliographies have been included for further research in these topics. This survey is an outstanding resource reference for anyone involved in the field of polymer materials design for advanced technologies.

We appreciate the efforts and enthusiasm of all the authors for writing their chapters in spite of their busy schedules. Furthermore, without the tireless efforts of the Apple Academic Press, this book would not have been possible. As editors, we express our sincere gratitude to the creative environments of the International and Inter University Center for Nanoscience and Nanotechnology (IIUCNN), Mahatma Gandhi University, Kerala, India, and to our colleagues for their encouragement and support.

PART I
Bio-Composites and Nanocomposites

CHAPTER 1

TAILORING AND ASSESSING ETHYLENE VINYL ACETATE (EVA)/ ORGANOCLAY NANOCOMPOSITES FOR BIOMEDICAL APPLICATIONS

AZLIN FAZLINA OSMAN[1], TUTY FAREYHYNN M. FITRI[1], ASNA RASYIDAH ABDUL HAMID[1], FATIMAH HASHIM[2], MD RAKIBUDDIN[3], and ABDULKADER M. ALAKRACH[1]

[1]*Center of Excellence Geopolymer and Green Technology (CE GeoGTech), School of Materials Engineering, University Malaysia Perlis, Arau, Perlis 02600, Malaysia, E-mail: azlin@unimap.edu.my*

[2]*School of Fundamental Science, University Malaysia Terengganu, 21030 Kuala Terengganu, Terengganu, Malaysia*

[3]*School of Materials Science and Engineering, Yeungnam University, 280 Daehak-ro, Gyeongsan, South Korea*

ABSTRACT

The growth in the medical industry has encouraged the innovation and development of implantable medical devices. New biomedical materials are being investigated, including those elastomeric materials required for mimicking human soft tissue. Soft, flexible, and biocompatible polymers such as thermoplastic polyurethane (TPU) and ethyl vinyl acetate (EVA) have been incorporated with nanofiller to enhance their performance as biomaterials. In this chapter, we review current researches on EVA nanocomposites incorporating 'pre-dispersed' organically modified montmorillonite for use as insulation materials in the implantable biomedical devices. Besides of having a low cost, the engineered nanoclay has low cytotoxicity level, can provide toughening, plasticizing, thermal, and biostabilizing effects to host polymer, making the resulting polymer nanocomposite tougher, more flexible, durable, and thermally stable. Here, we also highlight the advantages of EVA-based nanocomposites and their potential to replace non-recyclable silicone elastomer for implant applications. Optimizing the nanofiller dispersion and interfacial interactions between the nanofillers and the host copolymer are critical to achieving the desired nanocomposite performance as implantable materials.

1.1 INTRODUCTION

For several decades, the use of cross-linked silicone elastomers can be capaciously seen in the biomedical field, including implantable devices [1–3]. As a biomaterial, silicones possess useful properties such as biocompatible, biostable, low modulus and hardness, encouraging their applications for medical implantable devices such as a pacemaker, breast, facial, and cochlear implants [1–5]. However, there are some limitations of silicone elastomers which are restricting further advance in the new generation of devices. The high surface tack and inherently poor mechanical properties of these biomaterials, especially in relation to tensile and tear strength can be the absolute drawbacks and limit their application in long-term implantable medical device [1, 4–6].

With regards to the above-mentioned problems, our research intends to develop a new biocompatible material with excellent flexibility, toughness, and thermal stability as a future candidate for biomedical applications. Previous researches proved that polymeric materials can be tailored to meet specific property requirements by the incorporation of organically modified nanoclays such as montmorillonite (MMT), fluoromica, and hectorite [7–9]. This combination of polymers and organoclays resulted in a new form of materials called 'polymer nanocomposites' which possess various advantages over the neat polymer such as the improvement in mechanical and barrier properties, biocompatibility, biostability, and also thermal stability [7–9]. EVA is a type of copolymer which is composed of long ethylene and vinyl acetate (VA) monomers [10]. EVA possesses thermoplastic characteristics and therefore can be processed using conventional industrial techniques such as extrusion and injection molding. The properties of this copolymer can be tailored by varying its VA composition [10, 11]. More flexible and rubbery EVA can be obtained with the higher composition of

VA. The enhancement in EVA performance as the biomedical material can be achieved by adding a small amount of nanoclay as reinforcing filler [9]. However, poor dispersion of nanoclay in the host polymer structure inhibits optimized nanoscale reinforcement in the produced nanocomposite material. The exfoliation and dispersion of the organo-MMT nanofiller are vital to ensure the improvement in thermal, mechanical, and barrier properties of the end nanocomposite product [6–10, 12]. It has been reported by many researchers that poorly dispersed nanoparticles could degrade the mechanical properties of host polymers [6–13]. Even though it is highly important to produce well dispersed and exfoliated layered nanofiller in the host polymer, many researches proved that fully exfoliated nanoclay structure is not easily achieved, even the nanoclay has been organically modified prior to melting compounded with the polymers [7–9, 12–15]. Therefore, improving the processing method or route could be the best way to optimize the organoclay delamination (exfoliation) and dispersion in the host polymer. The breaking or loosening of the large tactoids should be achieved prior to melt compounding with the host polymer. This is to ensure better nanofiller dispersing ability inside the polymer matrix. We have, therefore, introduced the pre-dispersing process of the organoclay in a liquid medium prior to melt compounding with polymers. This process has successfully weakened the tactoid bonding and further facilitated the exfoliation and dispersion of the nanoclay during the melt compounding process. Besides, this pre-dispersing procedure was also being investigated as a new processing approach to plasticize this biomedical plastic, while improving its toughness, thermal stability, biostability, and biocompatibility. The scientific concept used to tailor the EVA properties was based on the manipulation of the nanoscale interactions between the EVA (host polymer) and the organo-MMT (nanofiller). This chapter describes the potential of this new and simple technique in developing the EVA nanocomposite system as new biomedical material.

1.2 BIOMATERIALS FOR IMPLANTABLE MEDICAL COMPONENTS

Biomaterials are materials of natural or synthetic origin that are suitable for close contact with living tissues of a human body, especially as part of a medical device or implant. American National Institute of Health has defined biomaterial as "any substance or combination of substances, other than drugs, synthetic or natural in origin, which can be used for any period of time, which augments or replaces partially or totally any tissue, organ or function of the body, in order to maintain or improve the quality of life of the individual" [16]. The term of 'implants' is applied to the biomaterials which are utilized for devices in dental, ophthalmic instrument, orthopedics, and surgical [17]. There are several types of implant, for example, sutures, bone plates, joint replacements, ligaments, vascular grafts, heart valves, intraocular lenses and dental implants and medical devices such as pacemakers, biosensors, artificial hearts and blood tubes. These devices are widely used to replace and/or restore the function of traumatized or degenerated tissues or organs, to assist in healing, to improve function and to correct abnormalities [1, 2].

Polymers are widely used in biomedical implant applications due to their low cost, versatility, flexibility, ease of processing and shaping [3, 18]. Polymeric material such as silicone elastomer is utilized as an encapsulant or insulator for the implantable electronic devices such as pacemakers and electrode array of a cochlear implant. The encapsulant acts as an insulating layer to protect the living tissues from the conductive parts and to prevent the inclusion of body fluid into the device, to avoid corrosion and device malfunction [19]. However, to serve as an effective insulation material for the implantable devices, there are two main requirements; which are biocompatibility and biostability. Both need to be met in order to avoid complication and human body suffering from adverse effects of implantation [7]. Furthermore, the materials should possess adequate mechanical properties and can be easily processed and shaped into the various design, thickness, and size [6, 18]. To obtain a comfortable implantable device and reducing the risk of pronounced Foreign Body Reaction (FBR) and adverse effects, the final package can be fabricated biomimetic, soft, and stretchable [19]. Even though the medical industry is continuously growing in size and technology, there is a limited number of flexible, tough, and biostable materials currently being used in the biomedical application, particularly as insulation material of the electrically active implantable device. The existing materials being employed such as silicone elastomers have some drawbacks in terms of mechanical properties and biostability. For instance, silicon elastomer possesses relatively low tensile strength and toughness and high surface tack [20–24]. These suggest the need to introduce and investigate new biomaterial as a candidate for implantable biomedical device application.

1.3 NANOCOMPOSITE WITH POLYOLEFIN COPOLYMER MATRIX

Polyolefin's such as polyethylene (PE), polypropylene (PE), and higher poly (α-olefins) and olefin copolymers has been composed as the most widely used group of commodity thermoplastics. There are produced by chain polymerization of chosen monomers or monomer combination. Furthermore, a wide range of applications can be tailor-made by polyolefin's-based materials which consist of rigid thermoplastics to high-performance elastomers [25, 26]. However, due to their brittleness on exposure to severe states, there is a requirement to ameliorate the competitiveness of polyolefin's in engineering applications, such as by improving their impact strength at low temperature, stability, stiffness, impact resistance, strength, and heat distortion temperature without relinquishing their processability [26]. One of the ways is through copolymerization among olefin monomers or with other types co-monomers. For instance, polypropylene can be tailored to be tougher, high impact and flexible when copolymerized with ethylene [27]. The combination of ethylene and vinyl acetate monomers resulting in more flexible and stress crack resistance copolymer named ethylene vinyl acetate (EVA) [10]. In spite of these advantages, the copolymers somehow may exhibit some limitations such as lack of durability, biostability, and thermal stability as compared to their homopolymer counterpart [28]. Therefore, several attempts have been done to further improve the properties and performance of these copolymeric materials, including the incorporation of nanofillers to form the so-called 'copolymer nanocomposites' [29]. The work by Verma and Choudhary [30] has proved that the incorporation of multiwall carbon nanotube (MWCNT) can improve the crystallinity and melting temperature of the PP random copolymer. EVA copolymer was incorporated with 95 wt% $Mg(OH)^2$ and SiO^2 nanofiller to improve the mechanical and flame retardancy of the host copolymer. As a result, an efficient halogen-free flame retardant nanocomposite was successfully produced [31]. George and co-workers have investigated the effects of $Al(OH)^3$ nanofiller on the crystallization, thermal, mechanical, flammability, and optical properties of EVA/carbon monoxide system. They found out that by adding an appropriate amount of $Al(OH)^3$ nanofiller, the flammability and UV resistance of EVA/carbon monoxide-based paints could be improved for possible outdoor applications [32]. Previous researches also proved that copolymeric materials can be tailored to meet specific property requirements by the incorporation of organically modified nanoclays such as montmorillonite (MMT), fluoromica, and hectorite. This combination of polymers and organoclays resulted in a new form of materials called polymer nanocomposites, which possess various advantages over the neat polymer such as the improvement in mechanical and barrier properties, biocompatibility, biostability, flame retardancy and also thermal stability [7–9, 33–37]. While a large body of research concerning polymer-organoclay nanocomposites exists, the number of studies specifically devoted to ethyl vinyl acetate (EVA)-nanoclay nanocomposite is relatively small. Our group is currently investigating the potential of EVA-nanoclay nanocomposite for biomedical applications, specifically as soft, flexible, tough insulation material for electrically active implantable devices [9, 33–37].

1.3.1 ETHYL VINYL ACETATE (EVA)/NANOCLAY NANOCOMPOSITES: ETHYL VINYL ACETATE COPOLYMER (EVA) AS MATRIX MATERIAL

EVA is a type of copolymer which is composed of long chains of ethylene (non-polar) and randomly distributed vinyl acetate (polar) monomers where the rubber-like properties of EVA are attributed to the high amorphous phase (the polar vinyl acetate). The basic chemical structure of EVA is shown in Figure 1.1, PE, and VA is usually formed by free radical polymerization with different acetate contents [10, 11]. Recently, the number of researches on EVA copolymer has kept increasing, further revealing its potential for various applications [10, 11, 25, 29, 32, 37]. EVA possess thermoplastic characteristics, which means it can be easily molded and processed by a conventional industrial method such as calendaring, injection, extrusion, blow molding and rotational molding [10, 11]. The main advantage of the EVA copolymer is the possibility to obtain a wide range of properties by varying the VA content in its composition (3–48%). Therefore, it is possible to broaden their applications from rigid plastic to the rubber-like/elastic products [10, 11]. EVA with low VA content (3–5%) is being contemplated as modified low-density polyethylene. In conventional film applications, flexible EVA is used for meat packaging as stretch films and also for cling-wrap. Besides, EVA copolymers with higher VA contents are used as wax coatings, wax additives, elastomers, and hot melt adhesives. A preferable grade of hydrolyzed EVA, usually nominated as EVOH, presents superior barrier properties with respect to gas (oxygen) permeability [38].

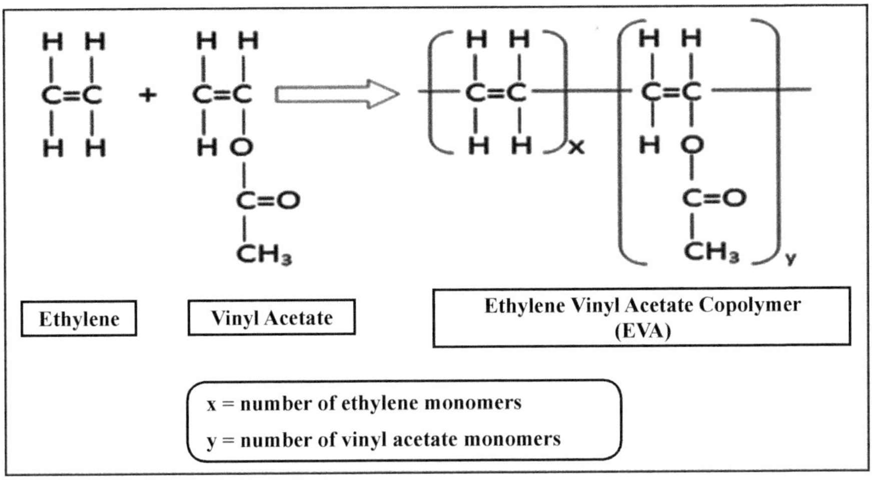

FIGURE 1.1 Chemical structure of EVA.

1.3.2 STRUCTURE AND PROPERTIES OF EVA

EVA copolymer is composed of monomer units which are connected via free-radical addition polymerization through the double bonds of the two monomers which are ethylene and vinyl acetate (VA) [39]. EVA is synthesized by the copolymerization of ethylene and vinyl acetate with bulk polymerization process and also by high pressure and temperature. It is a synthetic random copolymer of hydrophobic ethylene (non-polar) and hydrophilic vinyl acetate (polar) monomers [10, 40]. Besides, based on the initial concentration of vinyl acetate and molecular weight, EVA comprises a wide range of properties. The crystallinity of EVA is controlled by the vinyl acetate ratio. Lower crystallinity, better clarity and flexibility, higher polarity and better adhesion strength of EVA will be produced by increasing the amount of VA [40].

EVA behaves like thermoplastics and it is soft and flexible and can behave like elastomeric materials. In addition, EVA has a good clarity and gloss, barrier properties, stress-cracking resistance, low-temperature toughness, hot-melt adhesion, stability to UV radiation and waterproof properties [41]. Those products made from EVA have no plasticizers, unlike PVC containers. However, to make tubing, fitments, and connection that are similar to PVC products but without plasticizer are still remain a challenge when EVA plastic is used as a replacement. As opposed to plasticized PVC, the EVA-based medical devices and packaging were not widely commercialized. This is because, there is a lack of marketing incentive to make the switch and therefore, it remains as an "on-the-shelf" option [42].

1.3.3 EVA NANOCOMPOSITES

EVA nanocomposites are a new class of materials which consist of nanofiller and elastomeric material that leverage the benefits of engineered plastics. The inclusion of the small amount of nanometer-sized particles has produced a significant effect to the final properties of the EVA in which impressive improvement in its physical and mechanical performance was obtained [9, 33–37, 43].

Among all types of nanofiller, nanoclay received high attention due to their abundance, tailorable surface chemistry, and good reinforcement capability. EVA copolymer is considered as a unique matrix polymer for clay-based nanocomposites due to their strong interaction forces with the clay. Hence, due to an interaction of polar –OH groups that exist on the clay surface with a polar functional group of EVA, their properties can be improved by allowing strong intermolecular interactions between both constituents [44]. However, the studies on EVA/clay-based nanocomposite were relatively small when compared to other types of polymer nanocomposite systems. Most of the research found in the literature focused on the investigation of the EVA-nanoclay nanocomposites for flame retardant applications and packaging films. For instance, Elisabetta et al. [45] studied the effect of nanoclay on properties of EVA nanocomposites as a new material for film packaging. They found that the tensile modulus of EVA has increased gradually with the addition of 10 wt% of nanoclay [45]. The works by Ugel et al. [43] also studied the potential of EVA-clay based nanocomposite for film packaging application. They demonstrate that EVA nanocomposite containing 2.5 wt% of hydrotalcite nanofiller exhibit an increased in tensile strength, elastic modulus, and elongation at break by 70%, 61%, and 10%, respectively when compared with the neat EVA.

The exploration and intensive study on EVA nanocomposite are needed to make this versatile material adaptable for more advanced applications such as biomedical. It is believed that with careful formulation and processing of the EVA nanocomposite and optimization of EVA-nanofiller interactions, the target properties for biomedical material can be achieved. Therefore, our current project attempts to develop a better understanding of structure-processing-property relationships of this particular nanocomposite system.

1.3.4 LAYERED SILICATES (NANOCLAYS) AS NANOFILLERS

Nanoclays are minerals with a high-aspect ratio and usually with the thickness in the nanometer range and at least one dimension of the particle is in nano-size [46]. The length and width of these platelets are in the micron range, with aspect ratios between 300:1 to 1500:1 [47]. The nanoscale interactions between the molecular chains of the host polymer and the high aspect ratio nano-clay platelets provide a large opportunity for reinforcing effects [9]. A reinforcement effect can be formed from the immobilization of a segment in the polymer chains.

Nanoclays can be characterized by their purity and cation-exchange capacity [46]. Nanoclay platelets with a plate-like structure and a thickness of <1 nm are optimal for reinforcement due to the high aspect ratio structure. There are various nanoclays that act as a nanofillers of commercial interest nowadays, such as hydrotalcite, montmorillonite (MMT), mica fluoride, hectorite, and octasilicate. MMT is the most common and widely used nanoclay in polymers [48]. MMT is hydrophilic and belongs to the family of smectite clays with a unique feature of existing in agglomerates of nanoparticles platelets held together by attractive forces [46–48]. Therefore, to render the surface of the MMT into 'hydrophobic' thus encourage its compatibility with the hydrophobic host polymer, surface modification of the clay with organic surfactant can be done [46–48]. There are two main purposes of MMT surface modification: (1) to tailor the surface chemistry of the MMT from hydrophilic to hydrophobic, in order to enhance compatibility of this nanofiller with the host polymer having hydrophobic chains; (2) to increase the basal spacing of the MMT nanoplatelets, in order to disperse them well inside the host polymer matrix. The most widely used organic surfactants are those based on alkyl ammonium or alkyl phosphonium actions. Their functions are to decrease the surface energy of the inorganic MMT nanofiller and also improving the wetting characteristics within the polymer matrix, thus resulted in larger interlayer spacing [46, 48]. Furthermore, organic cations from long-chain alkyl ammonium salts also have bulky hydrocarbon tails to occupy the intergallery space of the MMT, in which can further facilitate silicate delamination. Hence, the resulting "organo-MMT' can be more effectively dispersed in the polymer matrix and to obtain a high degree of nanofiller-matrix interaction [9, 35]. Our technical paper illustrates the interactions between the organic surfactant (dimethyl dialkyl (C14-C18) amine) with the nanoplatelets of the MMT, in which both have distinct polarity [35]. The organic surfactant possesses hydrophobic characteristic due to its long alkyl chains, whereas the surface of MMT nanoplatelets is hydrophilic. Due to a strong electrostatic interaction, the positively-charged head group of the organic surfactant is connected on the negatively-charged MMT nanoplatelet surface; thereby render the MMT nanoplatelets 'physically hydrophobic.' This organically modified MMT (organo-MMT) was used as nanofiller in the EVA copolymer matrix. Since the EVA copolymer consists of hydrophobic polyethylene (PE) and hydrophilic (PVA) molecular chains, the existence of the organo-MMT in the polymer structure would result in polar and non-polar interactions between these components. Besides, the distribution and arrangement of the MMT nanoplatelets and the organic surfactant and their interaction with the host polymer could also be affected by pre-dispersing of the organo-MMT in the hydrophilic (water) and hydrophobic (toluene) as a liquid medium [35].

It is no doubt that the nano-sized clay used to reinforce the polymer and form 'polymer nanocomposite' can result in remarkable enhancement of mechanical, thermal, and barrier properties at very low nanofiller volume fractions, with small (if any) decrease in flexibility. Our studies have proved that the incorporation of low content organo-MMT nanofillers into the EVA copolymer matrix can bring about significant improvement in elongation at break, toughness, and biostability [9, 35, 37]. For example, when 5wt% of organo-MMT was added into the EVA copolymer matrix by twin-screw extruder, the increment of tensile strength, elongation at break and toughness was achieved by 41.8%, 33.7% and 101%, respectively [35]. Nevertheless, the key to successful nanoclay reinforcement in the polymer matrix is its exfoliation and dispersion degree in the respective polymer matrix. Therefore, the choice of the nanoclay type, surface modification, and dispersing technique are crucial in determining the dispersion quality and final nanocomposite properties [35, 46, 48].

1.4 EVA AS BIOMATERIALS

The growth in human populations and the needs to extend an average individual health-span lead to the development of new generation medical devices, medical diagnostic technologies, and drug delivery systems. In the production of the components for medical equipment, the variation of materials used has also expanded. Many devices for use in medicine such as tubing, catheters, probes, packaging for drugs and ointments, nursing aids, and also surgical instruments are now being made from polymeric materials [49, 50]. This is due to the flexibility, ease of shaping and processing of the polymeric materials as compared to metal and ceramic materials. However, new medical device designs continue to reduce in size and thickness, new materials that exhibit improved strength, biostability, and toughness, while maintaining their flexibility and biocompatibility are also required. The medical industries are still keeping an effort to develop these new, sophisticated, and ideal biomedical plastics which are having the above-mentioned characteristics. Ongoing research and invention are needed in order to overcome the limited number of existing biostable, biocompatible, flexible, and tough materials that offer versatility, exceptional performance, and meet an industrially relevant manufacturing process.

The biocompatibility and flexibility of the biomedical plastics are much in concern if the materials are to be used for the implantable device. Biocompatible, soft, and flexible materials are needed for close contact with human tissue in order to avoid irritation and tissue damage. Apart from biocompatibility, the success of a material to be used as a biomaterial in medical devices is often related to the ability and ease of the material to be formed into complicated shapes [51]. Generally, synthetic polymers like EVA offer more advantages than natural materials because they can be tailored to give a broad range of properties and foreseeable lot-to-lot uniformity. Besides, the more reliable source of raw materials that free from concern of immunogenicity can be obtained to form synthetic polymers, nowadays [51].

EVA copolymer was used in medical equipment, packaging, and pharmaceutical applications for the past three decades [3, 4, 52]. EVA is suitable to be used in biomedical applications because it consists of random structure, which offers high weather and ozone resistance and exceptional mechanical properties [3, 52]. It possesses thermoplastic characteristics, therefore can be easily processed using conventional industrial methods such as extrusion and injection molding. Lastly, it is flexible and can be easily shaped into a complex design.

1.5 EVA/CLAY NANOCOMPOSITE FOR BIOMEDICAL APPLICATIONS: TAILORING EVA NANOCOMPOSITES PROPERTIES (STRENGTH/TOUGHNESS/FLEXIBILITY/THERMAL STABILITY/BIOCOMPATIBILITY) THROUGH MANIPULATION OF NANOSCALE INTERACTIONS

While there are several papers reported on the production of EVA copolymer blend and composites for biomedical applications, the first attempt of producing the biomedical EVA nanocomposites using organoclay and non-blend (single) EVA system was done by our research group [9]. In this study, we have used EVA copolymer with 15 wt% VA content and organo-MMT nanofiller to form the EVA nanocomposite with enhanced ambient and in vitro mechanical performance with respect to the neat EVA. The improvement in biostability and biocompatibility of the EVA was achieved by adding 1 wt% organo-MMT. The results can be associated with the formation of the more tortuous path by the inclusion of the exfoliated and well-dispersed organo-MMT nanofiller, thereby resulted in more restricted permeation of the oxidants and water molecules [9]. However, when higher loading of nanofiller was used (5 wt% organo-MMT) in EVA matrix, the biostability decreased due to the presence of large tactoids. This situation occurred due to a higher degree of collision between the nanoclay platelets when a higher concentration of the organo-MMT was used, hence resulted in platelet agglomeration and worsening of the nanocomposite mechanical properties and biostability [9].

In our second paper, we show that the pre-dispersing of the organo-MMT prior to melt compounding with EVA matrix can significantly improve the nanofiller exfoliation and dispersion in EVA [35]. We have obtained a large improvement in EVA when high organo-MMT loading 5 wt% was added [35]. The increment of tensile strength, elongation at break and toughness was achieved by 41.8%, 33.7%, and 101%, respectively. These were attributed to the improved quality of organo-MMT exfoliation and dispersion in the EVA matrix upon the pre-dispersing procedure. This shows that the pre-dispersing method can be a good option to improve the quality of organo-MMT exfoliation and dispersion, especially when high loading of nanofiller is used to reinforce the EVA matrix.

In our more recent study [37], the pre-dispersing method was applied to enhance the organo-MMT dispersion and exfoliation, to assure the enhancement in mechanical and thermal properties and biostability of the

EVA copolymer when 5 wt% organo-MMT was added. Additionally, the pre-dispersing technique was also studied as a new approach to plasticize and compatibilize the PE and VA chains segments in the EVA copolymer, without compromising its other beneficial properties such as tensile strength, toughness, thermal stability, and biocompatibility. As opposed to the previous works, [9, 35] this EVA copolymer comprised of 18 wt% VA in its composition. It is commercially known as COSMOTHENE EVA H2181 and supplied by the Polyolefin Company (Singapore) Pte. Ltd. Organically modified montmorillonite (OMMT), which contains 35–45 wt.% dimethyl daily (C_{14}-C_{18}) amine as an organic surfactant was used as nanofiller. The pre-dispersing process was done by ultrasonication, using water as a medium. The organo-MMTs were prepared as 20 wt% solution in distilled water and the suspension was stirred at room temperature by using an ultrasonic probe. Two ultrasonication times were used, which were 2 minutes and 5 minutes. The 'pre-dispersed' organo-MMT was dried and used as nanofiller in the EVA copolymer. The nanocomposite samples were prepared by melt mixing EVA copolymer with the constant amount of ratio for organo-MMT nanofiller which is 5 wt%. The morphological characterization of the organo-MMT nanofillers was done through transmission electron microscopy (TEM) and presented in Figure 1.2. It can be seen that the degree of organo-MMT tactoid exfoliation is greater in the pre-dispersed organo-MMT as compared to its existing form (non-pre-dispersed). The result was further supported by the FESEM images and XRD data that confirmed the existence of more exfoliated and uniform silicate platelets in the pre-dispersed organo-MMT (Figure 1.3).

The tensile strength and toughness of the neat EVA and EVA nanocomposites are presented in Figure 1.4. The tensile strength values of EVA nanocomposite incorporating organo-MMT and organo-MMT

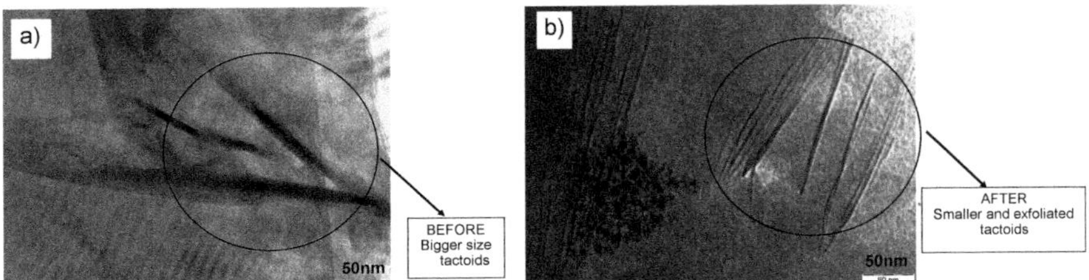

FIGURE 1.2 TEM images of (a) non-pre-dispersed organo-MMT (b) pre-dispersed organo-MMT by 2 minutes of ultrasonication.

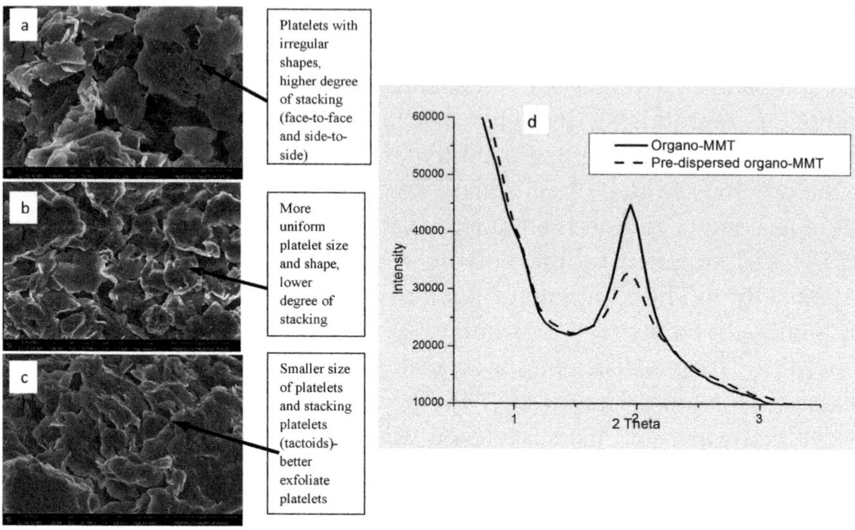

FIGURE 1.3 FESEM images on non-pre-dispersed (b) pre-dispersed in water medium by magnetic stirring (c) pre-dispersed in water medium by ultrasonication d) XRD pattern between the non-pre-dispersed and pre-dispersed organo-MMT.

ultrasonicated for 2 minutes show statistically no significant difference with that of neat EVA. However, the tensile strength value was seen to reduce when 5 minutes ultrasonicated organo-MMT was added into the EVA. The toughness value was seen to enhance with the incorporation of 5wt% organo-MMT. Interestingly, the value was further enhanced when 2 minutes-ultrasonicated organo-MMT was employed as nanofiller. The tensile toughness of the EVA was successfully increased from 109MPa to 126MPa. However, when longer ultrasonication time used to pre-disperse the organo-MMT (5 minutes), the decrease in tensile toughness of the EVA can be seen. These suggest that the pre-dispersing process of the organo-MMT nanofiller might affect the mechanical behavior of the resulting EVA nanocomposite. The optimum ultrasonication time can assist in nanofiller exfoliation and dispersion inside the EVA matrix. The well dispersed and exfoliated organo-MMT can interact more with the EVA chains. The nanoplatelets are more mobile and preferentially align in the direction of the stress for more effective stress transfer. As a result, the EVA copolymer can achieve greater strength and toughness. Unfortunately, the toughening effect might be lost when a longer period of ultrasonication time was used to pre-disperse the organo-MMT. This is because, the excessive 'salvation

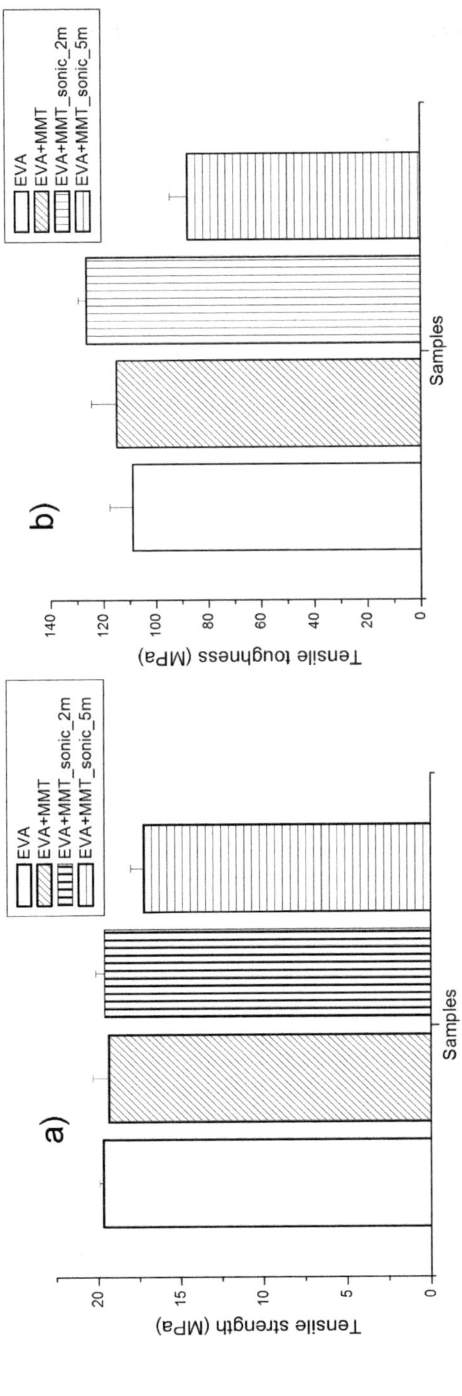

FIGURE 1.4 Mean values of (a) tensile strength (b) tensile toughness of EVA and EVA nanocomposites.

power' of the ultrasonication process can disrupt the organization of the surfactant used to modify the MMT surface, thereby cause imbalance interactions between the nanofiller and the copolymer chains.

The TGA curves of the neat EVA and EVA nanocomposites are shown in Figure 1.5. The incorporation of the 5wt% organo-MMT resulted in enhanced EVA thermal stability. Apparently, the ultrasonication of the organo-MMT for 2 minutes resulted in more pronounced thermal stability enhancement of the host copolymer. As mentioned earlier, the pre-dispersing process can produce better dispersed and exfoliated organo-MMT. Therefore, the EVA-nanofiller interactions can be improved to provide greater shielding upon the thermal degradation process. The enhanced nanoplatelets-EVA chains interactions resulted in more restricted polymer chain mobility, thus require a higher temperature to break the bonds. These can be seen through much higher maximum degradation temperature and higher residue content obtained in the EVA nanocomposite with the pre-dispersed organo-MMT (2 minutes ultrasonication).

The benefit of applying the pre-dispersing process on the organo-MMT was also revealed through the cytotoxicity test. We have done the cytotoxicity study on the organo-MMT and pre-dispersed organo-MMT nanofillers in the in vitro environment to allow assessment on their biocompatibility under the controlled condition as showed in Figure 1.6 The cytotoxicity activities of the nanofillers were monitored on the cultured cell line NIH/3T3 murine embryonic fibroblasts. We found that the IC50 value of

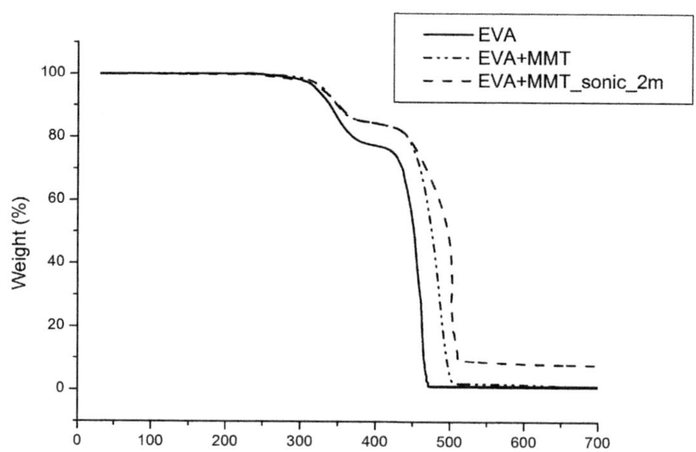

FIGURE 1.5 Thermogravimetric curves of the EVA and EVA nanocomposites.

the organo-MMT was increased from ~31 μg/mL to ~46 μg/mL when the organo-MMT was pre-dispersed by ultrasonication in a water medium. This higher cell viability in the pre-dispersed organo-MMT sample shows that the more exfoliated and dispersed organo-MMT can possibly possess reduced cytotoxicity level. This is because, when existing in their pristine highly stacked form, organo-MMT particles have more tendency to form intracellular reactive oxygen species (ROS) which can damage the cell membrane through localized oxidative stress [37].

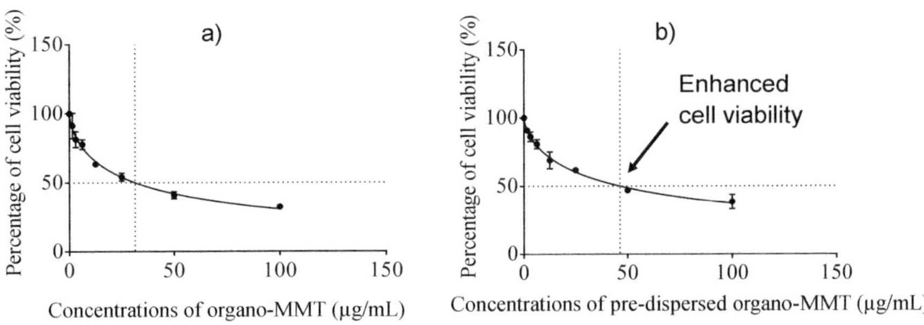

FIGURE 1.6 Cytotoxicity data on normal mouse fibroblast NIH/3T3 cell lines after 72 hrs treatment assessed based on MTT assay for IC50 values. (a) organo-MMT: ~31 μg/mL, (b) pre-dispersed organo-MMT: ~46 μg/mL.

1.6 CONCLUSION AND FUTURE WORKS

Our findings proved that EVA copolymers have the potential to be developed as biomedical materials as their flexibility, thermal properties, biostability, and biocompatibility can be further enhanced by the incorporation of organically modified montmorillonite. Several medical or pharmaceutical devices and packaging applications such as thermoformed trays, containers, tubes, needle covers, clamshells, blister pack and to the greater extent, insulation material for the electrically active medical device could be potentially produced by using well-formulated EVA-MMT nanocomposites. While the biocompatibility, safety, and cost have been the main concern in the biomedical industry, new processing approach towards the development of the plasticized biomedical polymer without or with minimal content of additives such as plasticizer should also be introduced. Biomedical plastics with fewer ingredients are better for safety, processing, and cost considerations.

As summarized in Figure 1.7, our experimental data showed that the application of the pre-dispersing process of the organo-MMT can produce better dispersed, distributed, and more biocompatible nanofiller in the EVA/organo-MMT nanocomposite system. Furthermore, the EVA copolymer mechanical, thermal, and barrier properties can be greatly improved with the incorporation of the pre-dispersed nanofiller without sacrificing its intrinsic property (flexibility and resilience). This soft, flexible characteristic is important to be maintained for close contact with human soft tissue. The clever introduction of nanofillers into this multiphase copolymer morphology can ensure the success of obtaining the desired EVA nanocomposite property profile. In conjunction with this, the level of interplay between the nanofiller-PE-PVA has to be characterized and understood, so that the correct nanofiller type, loading, processing parameters and etc. can be chosen to produce an optimized nanocomposite system. The degree of organo-MMT exfoliation affects the mechanical and thermal performance of the resulting EVA nanocomposite. The pre-dispersed organo-MMT which possesses a higher degree of exfoliation can result in greater tensile strength, toughness, thermal stability and biostability of the EVA copolymer. These improvements are beneficial for EVA nanocomposite intended for use in various applications including biomedical.

To this end, we are currently investigating the use of destabilized hybrid silicate nanofillers (with hydrophobic and hydrophilic characteristic) to further enhance the biostability of the EVA nanocomposites. Furthermore,

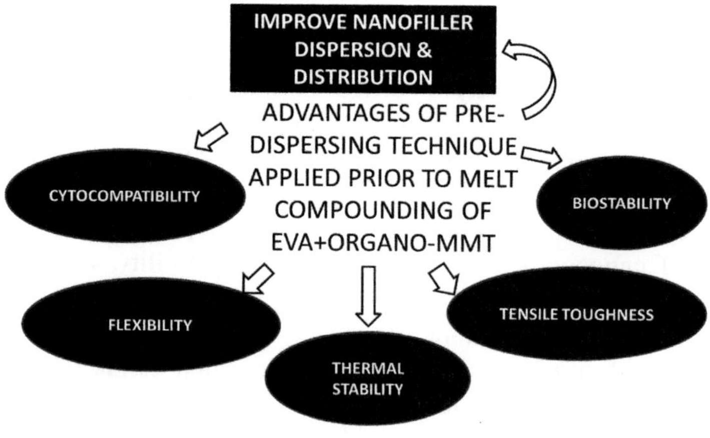

FIGURE 1.7 Advantages of the pre-dispersing process of the organo-MMT which applied prior to melting compound with the EVA copolymer matrix.

we are also developing an in-depth understanding of the structure-property relations of these materials, most importantly the interplay between the EVA co-monomer chains (hydrophobic PE and hydrophilic PVA) with the dual polarity hybrid nanofillers. The purpose was to further enhance molecular interactions between the multiphase copolymer with the nano-filler. In this way, some limitation of the EVA copolymer such as its high biodegradation rate and low dimensional stability upon long-term in vivo application could possibly be improved. Last but not least, it is hoped that the EVA/clay based nanocomposite will reach a great prospect in the future biomedical industry to compete with other biomaterials such as silicone elastomer and polyurethane.

ACKNOWLEDGMENT

The authors would like to acknowledge the Ministry of Higher Education, Malaysia for funding the research through Fundamental Research Grant Scheme (FRGS) (No: 9003–00473).

KEYWORDS

- **biomaterials**
- **ethylene vinyl acetate**
- **nanoclays**
- **nanocomposites**
- **pre-dispersing technique**

REFERENCES

1. Park, J., & Lakes, R. S., (2007). *Biomaterials: An Introduction*, Springer-Verlag, New York.
2. Bhat, S. V., (2013). *Biomaterials, Chapter 2*. In: Bergmann, C. P., & Stumpf, A., (ed.), *Dental Ceramics: Microstructure, Properties, and Degradation*, Springer: Netherlands, p. 9.
3. Ortensia, I. P., Curcio, M., & Puoci, F., (2015). Chapter 1: Polymer chemistry and synthetic polymers. In: Puoci, F., (ed.), *Advanced Polymers in Medicine* (Vol. 1, pp. 1–31), Springer International Publishing: Switzerland.

4. Ratner, B. D., Hoffman, A. S., Schoen, F. J., & Lemons, J. E., (2004). *Biomaterials Science: An Introduction to Materials in Medicine*. Elsevier Academic Press: London.
5. Szycher, M., (2012). *Szycher's Handbook of Polyurethanes* (2nd edn.). CRC Press: Boca Raton.
6. Osman, A. F., (2013). Biomedical thermoplastic polyurethane nanocomposites: Structure-property relationships. PhD Dissertation, Australian Institute For Bioengineering and Nanotechnology (AIBN), The University of Queensland, Australia.
7. Osman, A. F., Edwards, G. A., Schiller, T. L., Andriani, Y., Jack, K. S., Morrow, I. C., & Halley, P. J., (2012). Structure–property relationships in biomedical thermoplastic polyurethane nanocomposites. *Macromol.*, *45*, 198–210.
8. Osman, A. F., Andriani, Y., Edwards, G. A., Schiller, T. L., Jack, K. S., Morrow, I. C., & Halley, P. J., (2012). Engineered nanofillers: Impact on the morphology and properties of biomedical thermoplastic polyurethane nanocomposites. *RSC Adv.*, *2*, 9151–9164.
9. Osman, A. F., Alakrach, A. M., Kalo, H., Azmi, W. N., & Hashim, F., (2015). *In vitro* biostability and biocompatibility of ethyl vinyl acetate (EVA) nanocomposites for biomedical applications. *RSC Adv.*, *5*(40), 31485–31495.
10. Fink, J. K., (2010). *In Handbook of Engineering and Specialty Thermoplastics: Polyolefins and Styrenics* (pp. 187–209). Wiley: Hoboken USA.
11. Peacock, A. J., (2000). *Handbook of Polyethylene: Structures, Properties, and Applications*, CRC Press: New York.
12. Andriani, Y., Morrow, I. C., Taran, E., Edwards, G. A., Schiller, T. I., Osman, A. F., & Martin, D. J., (2013). *In vitro* biostability of poly(dimethylsiloxane/hexamethylene oxide)-based polyurethane/layered silicate nanocomposites. *Acta Biomater.*, *9*(9), 8308–8317.
13. Agubra, V. A., Owuor, P. S., & Hosur, M. V., (2013). Influence of nanoclay dispersion methods on the mechanical behavior of E-glass/epoxy nanocomposites. *Nanomater.*, *3*, 550–563.
14. Gopakumar, T. G., Lee, J. A., Kontopoulou, M., & Parent, J. S., (2002). Influence of clay exfoliation on the physical properties of montmorillonite/polyethylene composites. *Polym. J.*, *43*, 5483–5491.
15. Soulestin, J., Rashmi, B. J., Bourbigot, S., Lacrampe, M. F., & Krawczak. P., (2012). Mechanical and optical properties of polyamide 6/clay nanocomposite cast films: Influence of the degree of exfoliation. *Macromol. Mater. Eng.*, *297*, 444–454.
16. Bergmann, C. P., & Stumpf, A., (2013). *Dental Ceramics: Microstructure, Properties, and Degradation*. Springer Heidelberg: New York.
17. Williams, D., (1999). *The Williams Dictionary of Biomaterials* (1st edn.). Liverpool University Press: Liverpool.
18. Agrawal, C. M., (1998). Reconstructing the human body using biomaterials. *JOM.*, *50*(1), 31–35.
19. Beeck, M. O., Jarboui, A., Cauwe, M., Declercq, H., Uytterhoeven, G., Cornelissen, M., et al., (2013). Improved chip & component encapsulation by dedicated diffusion barriers to reduce corrosion sensitivity in biological and humid environments. In: *Microelectronics Packaging Conference (EMPC), 2013 European* (pp. 1–6). IEEE. Grenoble, France.
20. Rahimi, A., & Mashak A., (2013). Review of rubbers in medicine: Natural, silicone, and polyurethane rubbers. *Plast. Rubber Compos.*, *42*(6), 223–230.

21. Anderson, J. M., Hiltner, A., Wiggins, M. J., Schubert, M. A., Collier, T. O., Kao, W. J., & Mathur, A. B., (1998). Recent advances in biomedical polyurethane biostability and biodegradation. *Polym. Int.*, *46*(3), 163–171.
22. Martin, D. J., Osman, A. F., Andriani, Y., & Edwards, G. A., (2012). Thermoplastic polyurethane (TPU)-based polymer nanocomposites. In: Gao, F., (ed.), *Advances in Polymer Nanocomposites – Types and Applications* (pp. 321–350). Elsevier Woodhead Publishing: Cambridge.
23. Khan, W., Muntimadugu, E., Jaffe, M., & Domb, A. J., (2014). Implantable medical devices. In: Domb, A. J., & Khan, W., (ed.), *Focal Controlled Drug Delivery* (pp. 33–59). Springer: US.
24. Maytin, M., & Epstein, L. M (2011). Insulation. In: Ellenbogen, K. A., (ed.), *Implantable Devices: Design, Manufacturing, and Malfunction. An Issue of Cardiac Electrophysiology Clinics* (Vol. 201, p. 347). Elsevier Health Sciences: UK.
25. Ghosh, P., (2011). *Polymer Science and Technology: Plastics, Rubbers, Blends, and Composites* (3rd edn.). McGraw-Hill: New York.
26. Chrissopoulou, K., & Anastasiadis, S. H., (2011). Polyolefin/layered silicate nanocomposites with a functional compatibilizer. *Eur. Polym. J.*, *47*(4), 600–613.
27. Maier, C., & Calafut, T., (1998). *Polypropylene: The Definitive User's Guide and Databook* (1st edn.). Plastic Design Library: New York.
28. Harper, C. A., & Petrie, E. M., (2003). In: *Plastics Materials and Processes: A Concise Encyclopedia*, Wiley: Hoboken, NJ, USA.
29. Feldman, D., (2016). Polyolefin, olefin copolymers, and polyolefin polyblend nanocomposites. *J. Macromol. Sci. Part A.*, *53*(10), 651–658.
30. Verma, P., & Choudhary, V., (2015). Polypropylene random copolymer/MWCNT nanocomposites: Isothermal crystallization kinetics, structural, and morphological interpretations. *J. Appl. Polym. Sci.*, *132*(13), 41734–41747.
31. Pang, H., Wang, X., Zhu, X., Tian, P., & Ning, G., (2015). Nanoengineering of brucite@ SiO_2 for enhanced mechanical properties and flame retardant behaviors. *Polym. Degrad. Stab.*, *120*, 410–418.
32. George, G., Mahendran, A., & Anandhan, S., (2014). Use of nano-ATH as a multifunctional additive for poly (ethylene-*co*-vinyl acetate-*co*-carbon monoxide). *Polym. Bull.*, *71*(8), 2081–2102.
33. Osman, A. F., Alakrach, A., Kalo, H., Dahham, O. S., & Bakri, A. M. A., (2015). In vitro mechanical properties of metallocene linear low-density polyethylene (mLLDPE) nanocomposites incorporating montmorillonite (MMT). *Appl. Mech. Mater.*, *754–755*, 24–28.
34. Osman, A. F., Hong, T. W., & Alakrach, A., (2015). The effects of melt compounding method on the ambient and in vitro mechanical properties of EVA/MMT nanocomposites. *Appl. Mech. Mater.*, *789–790*, 75–79.
35. Osman, A. F., Kalo, H., Hassan, M. S., Hong, T. W., & Azmi, F., (2016). Pre-dispersing of montmorillonite nanofiller: Impact on morphology and performance of melt compounded ethyl vinyl acetate nanocomposites. *J. Appl. Polym. Sci.*, *133*, 43204–43219.
36. Osman, A. F., Hamid, A. R. A., Rakibuddin, M., Khung, G. W., Ananthakrishnan, R., Ghani, S. A., & Mustafa, Z., (2017). Hybrid silicate nanofillers: Impact on morphology and performance of EVA copolymer upon in vitro physiological fluid exposure. *J. Appl. Polym. Sci.*, *133*, 44640–44655.

37. Osman, A. F., Fitri, T. F. M., Rakibuddin, M., Hashim, F., Johari, S. A. T. T., Ananthakrishnan, R., & Ramli, R., (2017) Pre-dispersed organo-montmorillonite (organo-MMT) nanofiller: Morphology, cytocompatibility, and impact on flexibility, toughness, and biostability of biomedical ethyl vinyl acetate (EVA) copolymer. *Mater. Sci. Eng. C.*, *74*, 194–206.
38. Ghosh, P., (2011). *Polymer Science And Technology* (3rd edn.). Tata McGraw-Hill Education Private Limited: New York.
39. Kaminsky, W., (2013). *Polyolefins: 50 Years After Ziegler and Natta II: Polyolefins by Metallocenes and Other Single-Site Catalysts*, Springer-Verlag: Berlin Heidelberg.
40. Lee, H. M., Park, B. J., Choi, H. J., & Gupta, R. K., (2007). Preparation and rheological characteristics of ethylene vinyl acetate copolymer/organoclay nanocomposites. *J. Macromol. Sci. B: Phys.*, *4*(6), 261–273.
41. Chang, M. K., Hwanga, S. S., & Liu, S. P., (2014). Flame retardancy and thermal stability of ethylene-vinyl acetate copolymer nanocomposites with alumina trihydrate and montmorillonite. *Ind. Eng. Chem. Res.*, *20*(4), 1596–1601.
42. Czuba, L., (1999). *Opportunities for PVC Replacement in Medical Solution Containers.* Medical Device & Diagnostic Industry magazine, pp. 1–4.
43. Ugel, E., Giuliano, G., & Modesti, M., (2011). Poly(ethylene-co-vinyl acetate)/clay nanocomposites: Effect of clay nature and compatibilising agents on morphological thermal and mechanical properties. *Soft Nanoscience Letters.*, *1*, 105–119.
44. Joseph, S., & Focke, W. W., (2010). Poly(ethylene-vinyl co-vinyl acetate)/clay nanocomposites: mechanical, morphology, and thermal behavior. *Polym. Compos.*, *32*(2), 252–258.
45. Elisabetta, U., Gaetano, G., & Michele, M., (2011). Poly (ethylene-co-vinyl acetate)/ clay nanocomposites: Effect of clay nature and compatibilising agents on morphological thermal and mechanical properties. *Soft Nanoscience Letters*, *01*(04), 105–119.
46. Paiva, D. L. B., Morales, A. R., & Díaz, F. R. V., (2008). Organoclays: Properties, preparation, and applications. *J. of Appl. Clay Sci.*, *42*, 8–24.
47. Suresh, R., Borkar, S. N., Sawant, V. A., Shende, V. S., & Dimble, S. K., (2010). Nanoclay drug delivery system. *Int. J. Pharm. Sci. Nanotech.*, *3*(2), 901–905.
48. Kotal, M., & Bhowmick, A. K., (2015). Polymer nanocomposites from modified clays: Recent advances and challenges. *Prog. Polym. Sci.*, *51*, 127–187.
49. Tuzhilkin, I. M., & Rylov, E. E., (1974). Future development of the production of plastic articles for medical purposes. *J. Biomed. Eng.*, *4*, 3–7.
50. Lloyd, A., (2004). Alternatives to PVC. *Mater. Today*, *7*(2), 19.
51. Teoh, S. H., (2004). *Engineering Materials For Biomedical Applications*, World Scientific Publishing: London.
52. Reyes, J. D., (2014). *Innovative Uses of Ethylene Vinyl Acetate Polymers for Advancing Healthcare*. In the plastics SPE polyolefins conference on medical devices, Cleveland, USA.

CHAPTER 2

GOLD (SILVER)-POLYMER NANOCOMPOSITES FOR BIOSENSING AND MICROFLUIDIC APPLICATIONS

SIMONA BADILESCU and MUTHUKUMARAN PACKIRISAMY

Optical-Bio Microsystems Laboratory, Micro-Nano-Bio Integration Center, Department of Mechanical and Industrial Engineering, Concordia University, Montreal, H3G 1M8, Canada

ABSTRACT

Nanometal-polymer composite films are hybrid materials with inorganic nanoparticles immobilized and uniformly dispersed into a polymer matrix. Nanoparticles such as gold and silver, used to enhance the optical properties of the polymers, are called 'optically effective additives' as they lead to new functionalities of the polymer-based materials. On the other hand, metal nanoparticles such as Au or Ag, significantly contribute with their distinctive optical properties. In order to emphasize their major contribution to the overall optical properties of the hybrid material, they are generally called "optically effective additives" [1]. Au (Ag)-polymer and copolymer composite films are suitable for applications such as waveguides, color filters, thermochromic materials etc. The strong plasmon band of Au and Ag nanoparticles (Localized Surface Plasmon Resonance) in the visible spectrum, that originates from the excitation of plasmons by the incident light, makes noble metal-polymer nanocomposites particularly adequate for sensing and biosensing applications. Furthermore, an association of Au and Ag nanoparticles of various shapes with poly(dimethylsiloxane) (PDMS), allows the use of nanocomposite materials for microfluidic biosensing as well.

In this presentation, we will talk about the *in-situ* synthesis of Au-PDMS and Ag-PDMS nanocomposites both at the macro scale and inside the channel of a microfluidic chip. The *in-situ* synthesis of nanoparticles involves the reduction of gold ions in a gold precursor solution by the curing agent of the polymer. The kinetics of the reaction has been investigated and the effect of annealing temperature on the morphology of nanoparticles has been studied by using imaging and spectroscopic methods. The nanocomposite has been successfully used for sensing of antibody-antigen interactions, allowing the detection of various important proteins. Nanocomposites of metals integrated into other polymers such as PMMA, PVA, and PS have been synthesized as well by using UV, microwave, and thermal reduction methods.

The effect of the particle's shape on the properties of nanocomposites and their sensing abilities has also been studied by synthesizing nano-star particles and integrating them into microchannels. Because of the biocompatibility and non-toxicity of gold and silver nanoparticles, their nanocomposites can be used for plasmonic detection of biological entities in cancer research.

2.1 INTRODUCTION

Gold nanoparticles are particularly suited for sensing applications because of their strong plasmon band in the visible spectrum, arising from the excitation of plasmons by the incident light. The Plasmon band is shifted to longer wavelengths when the dielectric constant (refractive index) in the vicinity of the particle changes, and this fact is used for sensing applications (Figure 2.1).

All the nanocomposites, containing Au and Ag, display in the visible spectrum, the strong Localized Surface Plasmon Resonance (LSPR) band that characterizes the noble metals and, for this reason, UV-Visible spectroscopy became a powerful tool for the study, not only of metal nanostructures but of nanocomposites as well.

There are two general approaches to the preparation of nanocomposite materials. Either, a suitable metal precursor is dissolved in the polymer solution and the ions are subsequently reduced to nanoparticles chemically, so no-, or photochemically or, the corresponding monomer is polymerized around the metal nanoparticles.

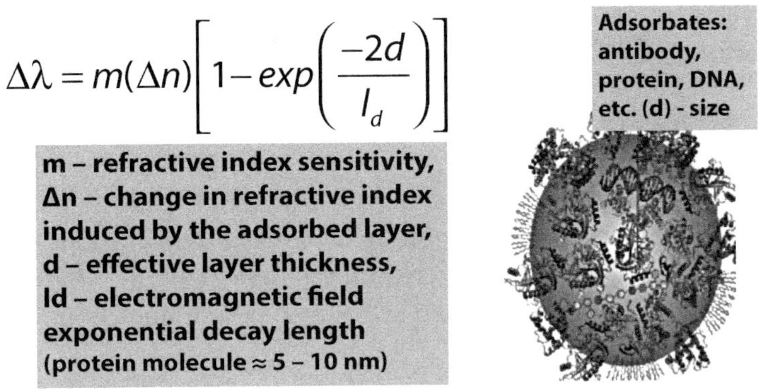

FIGURE 2.1 The dependency of the shift of the plasmon band on the refractive index sensitivity and the effective layer thickness of the layer adsorbed on the gold nanoparticle. Some of the parameters can be manipulated with the purpose of enhancing the sensitivity.

Nanocomposites can also be prepared, either by *in-situ* methods or by incorporating pre-made nanoparticles into a polymer matrix, by using a common solvent.

In addition, physical methods such as chemical vapor deposition, ion implantation, and thermolysis have been successfully used for the synthesis of noble metal nanocomposites (Figure 2.2).

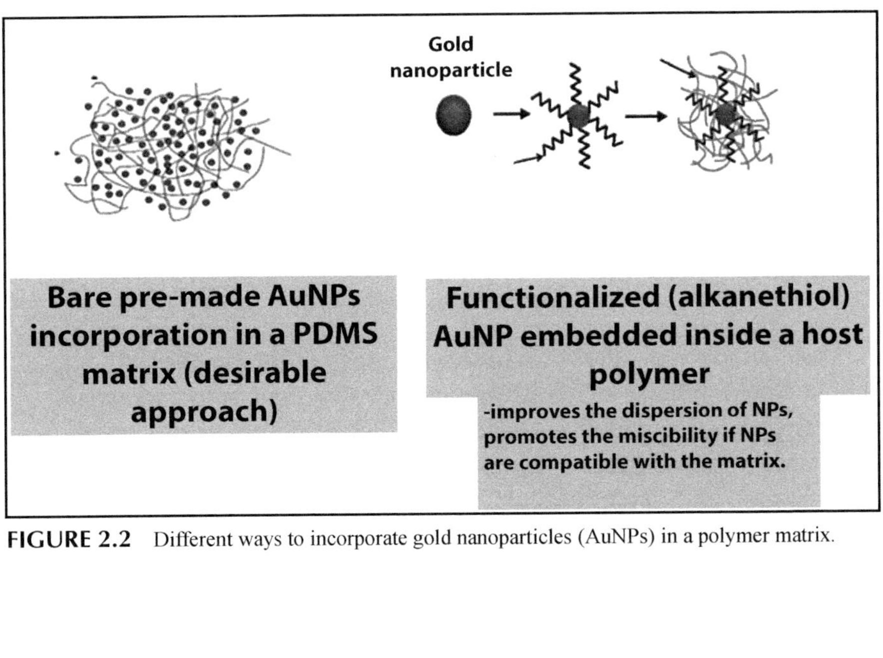

FIGURE 2.2 Different ways to incorporate gold nanoparticles (AuNPs) in a polymer matrix.

2.2 GOLD-POLYMER SENSING PLATFORMS: FABRICATION AND SENSING PROPERTIES

2.2.1 GOLD-PDMS NANOCOMPOSITE SENSING PLATFORMS: MICROFLUIDIC SENSING

Due to the wide utilization of gold nanoparticles with various shapes in biosensing, biological labeling, etc., their association with PDMS, in the form of nanocomposite materials, opens new possibilities in microfluidic biosensing (Figure 2.3).

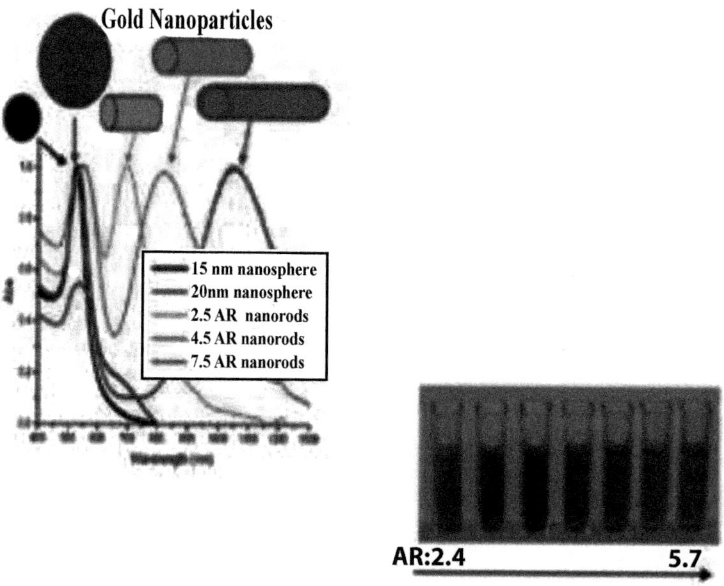

FIGURE 2.3 (See color insert.) LSPR plasmon band of nanoparticles with different shapes. It can be seen that the LSPOR band corresponding to different shapes is centered at different wavelengths.

Gold and silver—PDMS nanocomposites are of considerable interest due to the simplicity of their preparation, low cost, good transparency, oxidative stability, and non-toxicity to cells. In addition, PDMS, the host polymer, has a low glass transition temperature (*T*g), excellent flexibility, high thermal and oxidative stability, good hemo- and biocompatibility. A major drawback of PDMS is the strong hydrophobicity, and therefore, the inertness to biological molecules. To render the PDMS surface hydrophilic, functional groups can be introduced by plasma- or, corona discharge treatments, to promote the filling of microchannels with aqueous solutions and to facilitate PDMS microchip bonding as well. Freshly modified surfaces show good wettability but the effect is not stable and the hydrophobicity is regained over the time.

We have prepared Au-polymer nanocomposites by *in-situ* methods and obtained well-dispersed Au nanoparticles. However, the bio-sensing properties of the as-prepared nanocomposite films have been found to be poor. The sensitivity of the Ag—PDMS platforms can be increased by functionalization, followed by a post-synthesis thermal treatment at moderate temperatures, a treatment that induces important morphological changes. Generally, the biosensing properties of nanocomposites depend on the conditions of their preparation. They are determinant for the distribution of the metal particles in the polymer matrix (Figure 2.4).

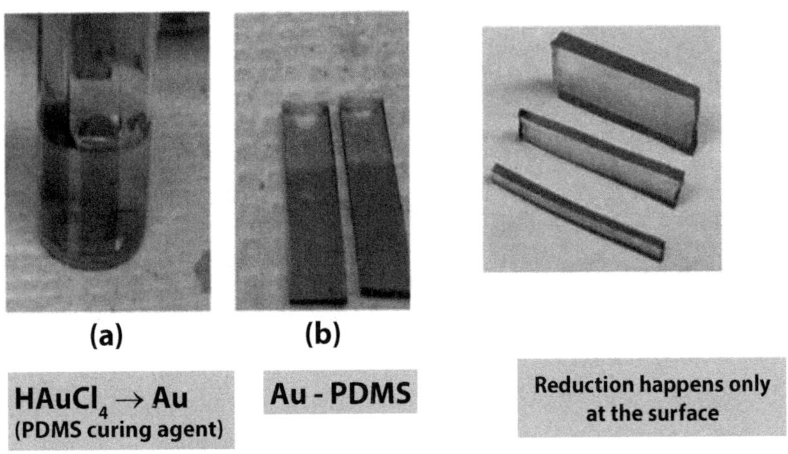

FIGURE 2.4 (See color insert.) *In-situ* synthesis of Au – PDMS nanocomposites. The composite is prepared by immersing a PDMS sample in the solution of chloroauric acid. The change of color shows the presence of AuNPs on the surface of the polymer.

Once the nanocomposite is formed, the samples are annealed at around 500°C. As shown in the SEM images in Figure 2.5, during the annealing process, the morphology of the gold structure is changed gradually and, in the end, an island structure is formed. Gold islands have been proved to be more sensitive to changes in the surrounding environment than the original gold NPs and, for this reason, annealing was used as a general strategy in all our biosensing experiments. The corresponding spectra show that after annealing, the Au plasmon band became narrower and thus more adequate for sensing experiments. In some cases, sensing experiments were performed in microfluidic devices as well. Further, a microfluidic device having the gold nanoparticles in the microchannel was designed, fabricated using the reduction of by sodium citrate, following Turkevich's method. Briefly, 75 mL of a chloroauric acid solution containing around45 μg/mL gold is heated and 5 mL of 1% sodium citrate solution is added to the boiling solution. After the color of the solution turns to purple, the solution is further boiled for 15 min and then left to cool to room temperature. Gold multilayers were deposited on freshly cleaned substrates. Glass substrates were cleaned with soap and de-ionized water, then rinsed with acetone, dried, and finally rinsed with 2-propanol. Before the deposition process, the glass substrates were kept in an oven at 100°C for 1 h. Glass substrates were immersed with an angle (~30 deg) in vials containing the gold colloid solutions (3 to 5 mL) and kept in the oven at temperatures between 60°C and 80°C, until the whole amount of gold was transferred to the substrate, that is 1 to 3 days.

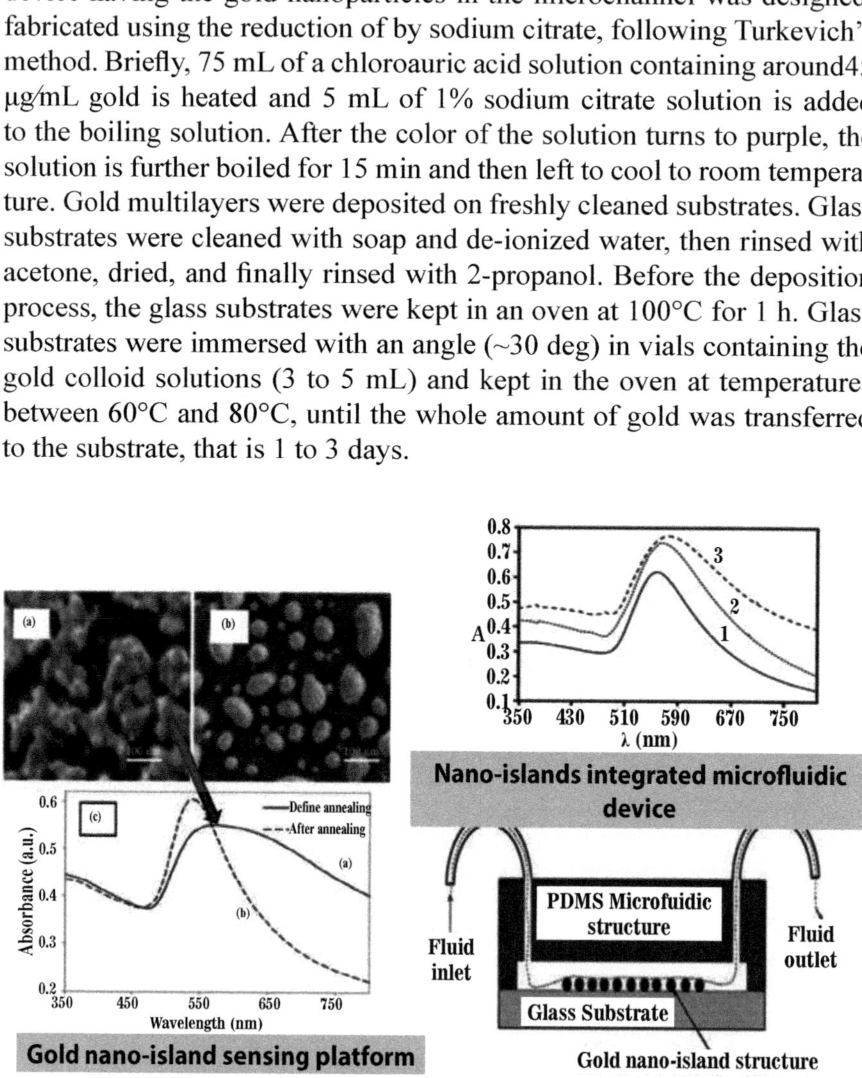

FIGURE 2.5 Tuning the morphology of gold by annealing. The Au plasmon band, around 520 nm is very sensitive to changes in the refractive index of the environment and it can be used for sensing experiments on substrates or in a microfluidic environment.

When sensing is performed in a microfluidic device, the gold NPs, formed by *in-situ* synthesis, show a very narrow size distribution, compared to the wider distribution at a "macro" level, that is, when the synthesis is carried out on a PDMS substrate. As shown in Figure 2.6, in a microfluidic environment, the band is much narrower. That means that when performing a sensing experiment (see Figure 2.7), at each step, the experimentally observed shift ($\Delta\lambda$) will be larger than at macro level and the refractive index sensitivity will be enhanced.

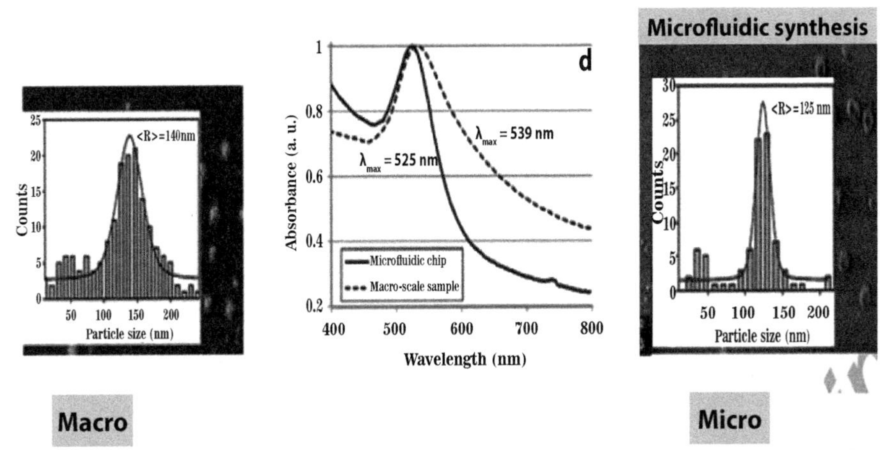

FIGURE 2.6 Au plasmon band and size distribution of particles at the macro level and in a microfluidic device. As seen, microfluidic synthesis results in more uniform AuNPs.

The microfluidic device used for in-situ synthesis is shown in Figure 2.7.

The sensing platform, prepared by convective assembly and annealing, was tested for the detection of bovine somatotropin (bST) through antigen-antibody interactions. A calibration curve that correlates the shift of Au-LSPR band with the concentration of antigen is established in the range of 5 to 1000 ng/mL and the detection limit of the polypeptide is determined. The mechanism of sensing by using the annealed gold nanostructures was also discussed.

The microfluidic detection of the bovine growth hormone has been performed within the detection limit of 5 nm /mL (Figure 2.8). In the case of the microfluidic sensing, the different compounds involved in the sensing process, are pumped through the channel, by using a very low flow rate.

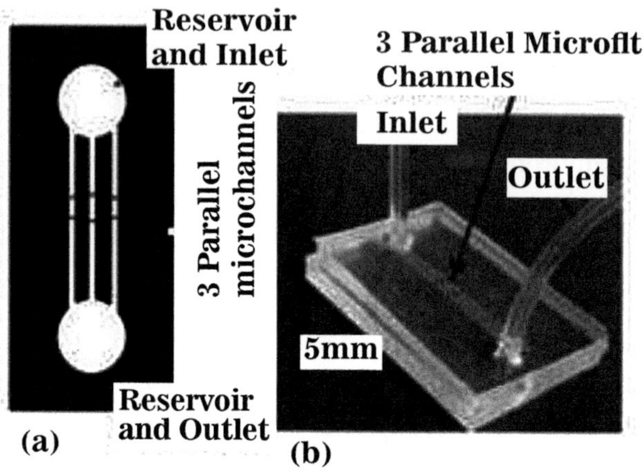

FIGURE 2.7 Microfluidic device used for the microfluidic synthesis. To increase the amount of AuNPs, a three-channel device was fabricated.

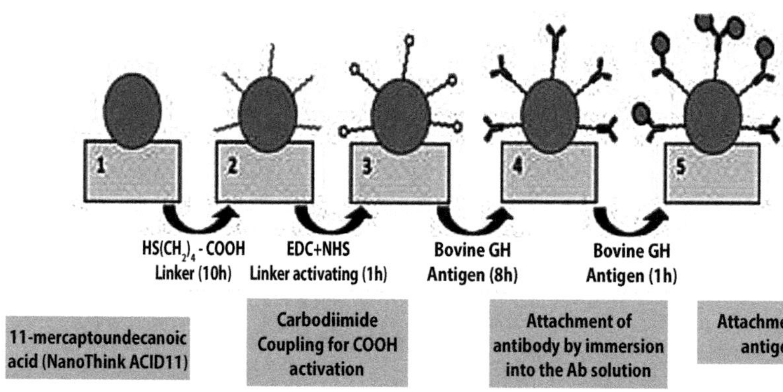

FIGURE 2.8 A general schematic of a sensing protocol, using gold NPs or Au nano-islands. The bonding of a biological entity is done through a linker system. In the case illustrated in the figure, a bovine growth hormone is detected by using the antigen-antibody interaction.

2.2.2 AU-PMMA NANOCOMPOSITES: PREPARATION AND SENSING PERFORMANCE

Different methods have been previously used to prepare Au (Ag)-PMMA nanocomposites, among them, the reduction of Au ions by sodium borohydride, ultrasonic reduction within an emulsion containing the monomer.

This method resulted in silver nanoparticles of about 20 nm. The reduction of metal ions can also be done by the solvent or photochemically. The optical properties of Au-PMMA nanocomposites depend significantly on the size of Au nanoparticles which in turn, depends on the irradiation time. UV irradiation of the PMMA film containing uniformly distributed $AuCl_4^-$ ions has also been used for photopatterning (Figure 2.9).

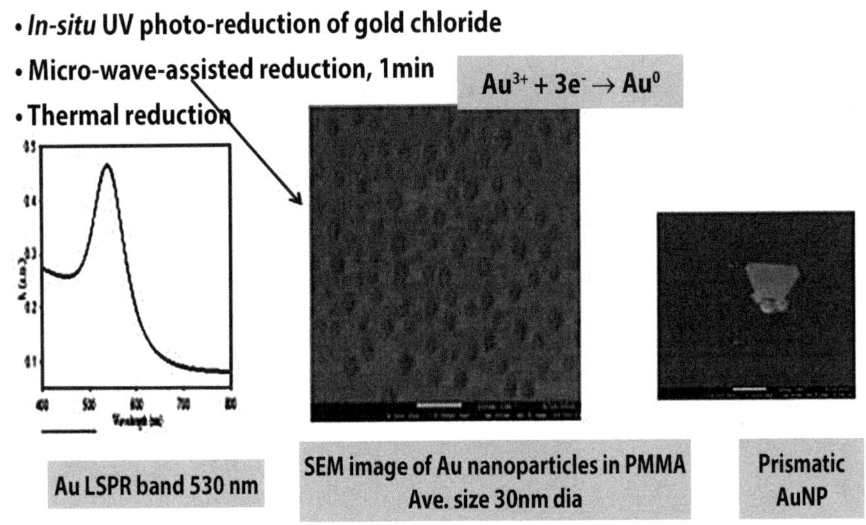

FIGURE 2.9 Microwave-assisted synthesis of Au-PMMA nanocomposite. As shown by the SEM image, AuNP (30 nm) are uniformly distributed in the PMMA matrix. In addition to spherical particles, some prismatic particles can be also seen.

Microwave (MW)-assisted coating of PMMA beads by silver nanoparticles in solution was found to be a new and efficient technique and the method was investigated by our group. The main advantages of MW are the uniform heating and the short thermal induction period that results in a homogeneous nucleation.

To prepare Au-PMMA for bio-sensing applications, three methods were investigated: the UV photochemical and MW-assisted reduction, as well as the thermal reduction method. In order to have a good sensitivity towards the bio-entities, nanocomposites with a low content of Au were synthesized. The samples were prepared in acetone, a common solvent for PMMA and the Au precursor. The mixture was spin-coated on glass and, for performing the reduction to Au nanoparticles, were irradiated (UV or MW) As far as we are aware, until now, microwave-assisted reactions have been carried out and studied only in solutions by using water because of its high dipole moment and high value of the dielectric loss. In our work, a spin-coated Au-PMMA film, containing only traces of acetone, was exposed to microwaves. The sensitivity of non-spherical gold nanoparticles towards bio-molecules was investigated as well (Figure 2.10).

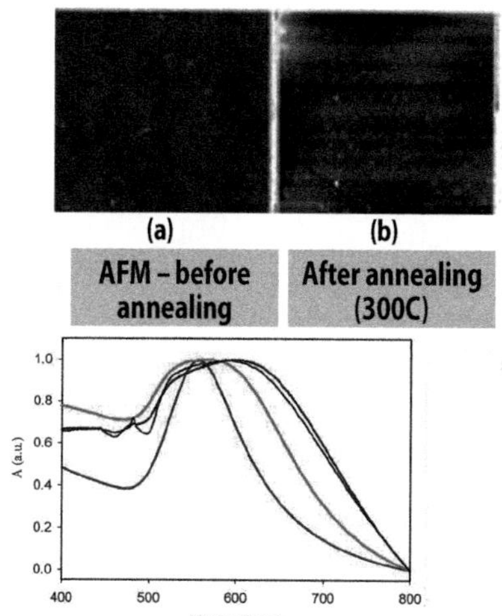

Method of fabrication		$ (nm/RIU)
UV-Irradiation	Non-annealed	77
	Annealed	98
Thermal reduction	Non-annealed	73
	Annealed	98
MW-irradiation	Non-annealed	68
	Annealed	107

FIGURE 2.10 Refractive index sensitivity of the Au-PMMA platforms prepared by different methods.

Au-PMMA nanocomposites have been prepared by UV and microwave irradiation, as well as by thermal reduction for sensing applications. The synthesis was carried out by irradiation of a spin-coated film containing a mixture of gold precursor ($HAuCl_4$) and poly(methyl methacrylate) dissolved in acetone. It is shown that the microwave-synthesized composite showed a uniform distribution of spherical nanoparticles, while, as a result of UV irradiation, a large number of triangular particles can also be seen. As-prepared nanocomposites lacked sensitivity toward the surrounding environment but annealing at 300°C, temperature well above the glass transition temperature, brought about an important change in the structure and properties of the nanocomposite. It is thought that the interactions between the gold and the polymer chains are weakened in the rubbery state of the nanocomposite, and, the mobility of Au nanoparticles is increased. The results show that, in the annealed samples, the surface density of gold nanoparticles is highly increased and the nanocomposites show an enhanced sensitivity toward the surrounding molecules. The results indicate that, by using an adequate post-synthesis heat treatment, gold-polymer nanocomposites can be successfully used as sensing platforms.

2.3 THERMAL MANIPULATION OF GOLD NANOCOMPOSITES FOR MICROFLUIDIC PLATFORM OPTIMIZATION

Gold nano-island structures on polymer substrates were fabricated by the thermal convection method described below. The multilayer films (3-D assembled gold nanostructures) are fabricated from gold nanoparticles by convective assembly from the evaporating meniscus of an aqueous suspension. This method was used by our group for the fabrication and characterization of three-dimensional (3-D) gold nanostructures used as sensing platforms for the detection of different bio-molecules and biological entities. The evaporation from the liquid meniscus at the substrate-solution interface induces the flow of the colloidal suspension. The convective assembly of particles is due to the interaction between particles and particles and surfaces will lead to the formation of ordered arrangements on the substrate. Because of the strong attraction forces between the gold nanoparticles (NP) during the deposition, gold aggregates are formed on the substrates as shown in Figure 2.11.

The main parameters deciding the rate of self-assembly are the evaporation from the liquid meniscus, that is the temperature, the particle concentration, and the particle diameter. The high density of gold nanoparticles in the multilayers and the excellent transparency of the samples allow the use of an ultraviolet (UV)-visible spectrophotometer for the measurement of plasmon bands. The method is simple and can be carried out in any laboratory, without any special apparatus or expertise. The gold aggregates, formed by thermal convection are further annealed to improve the sensibility of the platform. 3-DAu nanostructures fabricated by this method show two plasmon bands, one around 520 belonging to plasmon resonance of isolated gold nanoparticles and one at longer wavelengths (600 to 700 nm) due to a collective surface plasmon oscillation, arising from the coupling of the individual plasmon resonances of closely spaced particles. The intensity of this band is proportional to the degree of aggregation. Multilayers, with different particle packing densities and degrees of aggregation, can be prepared by using a simple oven.

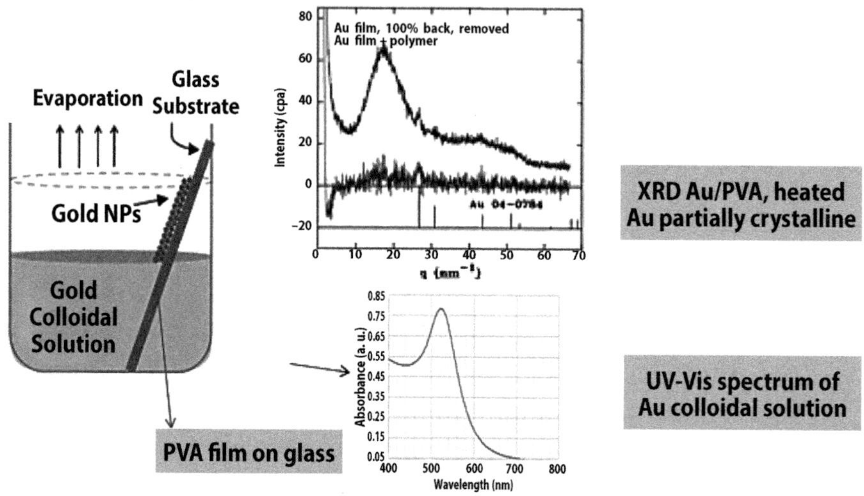

FIGURE 2.11 Schematic sketch of the convective assembly process of gold nanoparticles.

Surface gold–polymer nanocomposites are prepared by using this method for the deposition of gold colloids onto the surface of four polymer films: poly (vinyl alcohol) (PVA), SU–82, poly(styrene) (PS), and poly(dimethylsiloxane) (PDMS. The nanocomposite films are, subsequently, subjected to an incremental heating (in the range of 80–200°C) to increase the plasmonic sensitivity of the platforms as shown in Figure 2.10. Gold-polymer surface nanocomposites have been prepared by depositing pre-synthesized gold colloids on the surface of polymer films, by using the thermal convection method. Subsequently, the films hosting the gold aggregates on their surface were heated incrementally at temperatures in the range of 80 to 200°C and the morphology and spectral properties of the nanocomposite were investigated (Figure 2.12).

During this process, small aggregates are formed and uniformly dispersed on the surface of the film. However, because of the softening of the polymer, a small fraction of nanoparticles may sink into the surface layer of the polymer.

It is well known that the decay length of the electromagnetic field for Localized Surface Plasmon Resonance (LSPR) sensors depends on the size, shape, and composition of nanoparticles and it is generally not more than 5–15 nm, depending on the size of gold nano-islands. However, if the gold nanoparticles sink deeper into the polymer, the biomolecules may be outside of the sensing volume of the particle. In addition, the nanoparticles may be entangled in the polymer network and their mobility may be restricted if the interaction between the particles and the polymer chains is too strong. It is important to be able to control the spatial distribution of nanoparticles to be able to perform a successful sensing experiment. We have explored some ways to increase the mobility of Au nanoparticles and make them accessible to the surrounding environment containing the biomolecule of interest. Among them is the thermal post-synthesis treatment of Au-PDMS nanocomposite that has been proved to be an adequate method to improve the distribution of Au nanoparticles in polymers.

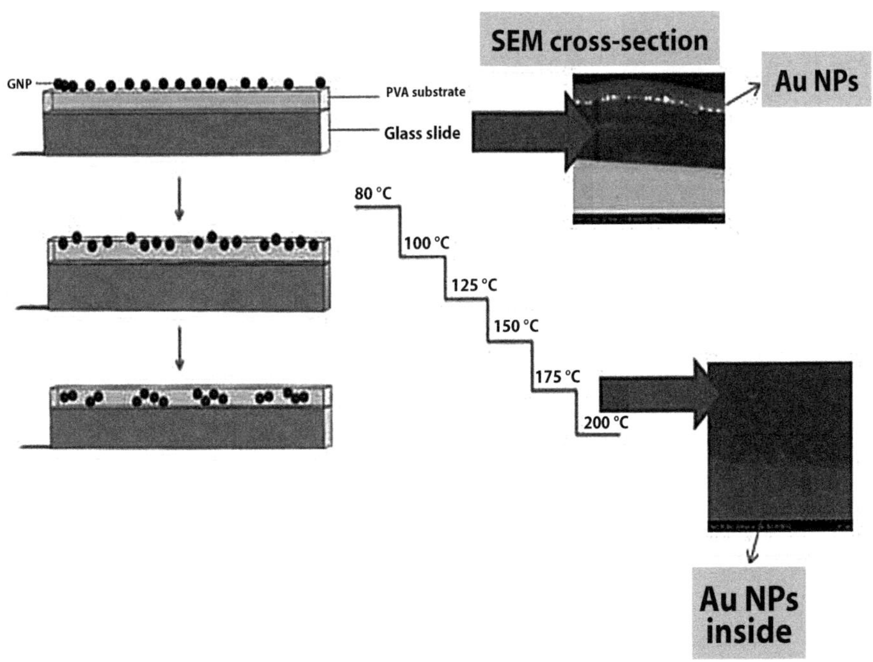

FIGURE 2.12 Thermal manipulation of Au-polymer nanocomposites for enhanced plasmonic sensing performance. The samples are subjected to heating-cooling sequences that change the morphology of the aggregates.

The final configuration of the thermally manipulated nanocomposites will not be the same because of the different thermal properties of the polymers and the interactions between the nanoparticles and the polymer chains. It is found that, among the polymers studied in this work, PVA, and SU–82 show the largest shift of the Au localized surface plasmon resonance (LSPR) band upon the incremental heating as well as the highest plasmonic sensitivity. Investigation of the gold structure was done by hyperspectral imaging by using Cyto Viva's hyperspectral microscope (Figure 2.13).

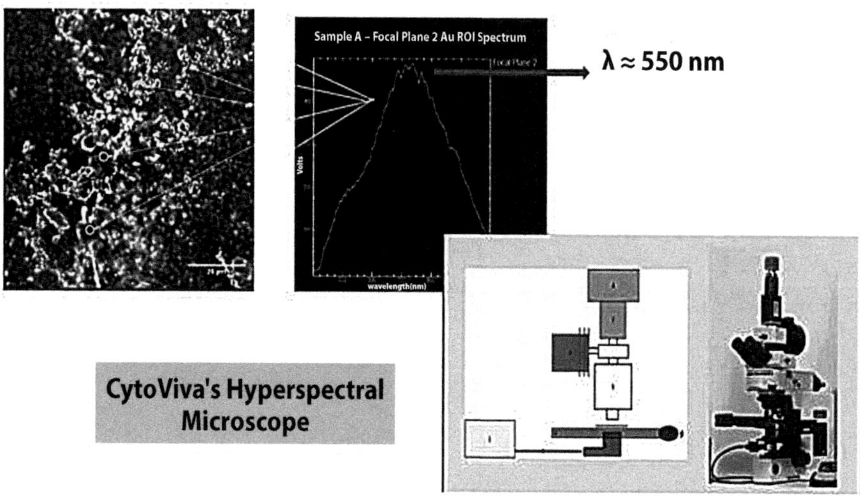

FIGURE 2.13 Hyperspectral image and spectral profiles of AuNPs on the surface of PVA.

Figure 2.14 shows the refractive index sensitivity for the two of the most sensitive platforms, the Au-PVA and Au-PDMS nanocomposites.

It is thought that the thermal manipulation may be a useful method for increasing the plasmonic sensitivity of a platform. The results of this work will be helpful in selecting the best (more sensitive) material for microfluidic sensing experiments.

The results pointed to the following conclusions: the incremental heating, alternating with short cooling periods, leads to the formations of small gold aggregates on the surface, together with a partial or total embedding of a part of the particles into the surface layer of the polymer. This configuration results in a large redshift of the Au LSPR band and, consequently, in an enhanced sensitivity of the nanocomposite platform.

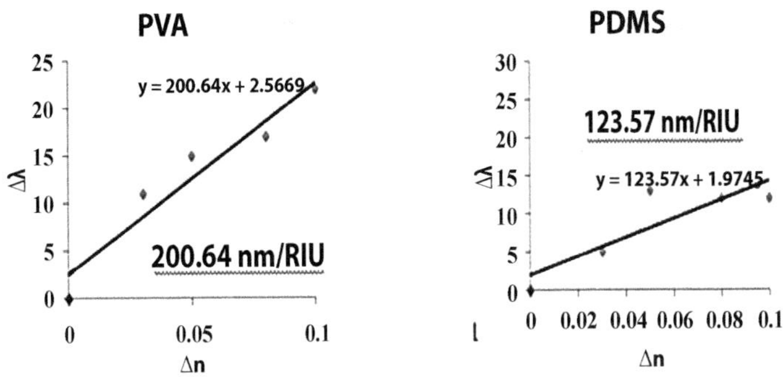

FIGURE 2.14 Refractive index sensitivity of Au-PVA and Au-PDMS surface nanocomposites.

2.4 GOLD RING AND NANOHOLE STRUCTURES

Goldring and nanohole structures have been fabricated by using a novel experimentally simple and convenient nanosphere lithography method. The method is based on the simultaneous self-assembly of polystyrene microspheres and gold colloids in multilayers on glass substrates by a vertical deposition method (thermal convection) (Figure 2.15).

By removing the PS microspheres, instead of the original multilayers, a monolayer, formed by nanoparticles surrounding holes, is formed. This new method can be carried out in any laboratory, without thermal evaporation or sputtering equipment. The sensitivity of the sensor platforms has been found dependent on the size of the holes and their density on a given area. It is thought that, when the holes are small, their number on a given area is high and, therefore, there are more privileged protein adsorption sites. Sensing experiments have demonstrated a high sensitivity of the nanoparticle hole structures towards fibrinogen, ADDLs, and the plant protein AT5G07010.1 from A. thaliana. The antigen-antibody recognition event resulted in significant changes in the position and the shape of the LSPR band, showing a good potential for the ring and hole structures to be further used to study these interactions. The dependency of the sensitivity of the sensor platform on the size of the holes and their density has been demonstrated. Furthermore, sensing experiments have shown a high sensitivity of the whole structure toward fibrinogen, amyloid-derived diffusible ligands, and a plant protein (AT5G07010.1). It was found that the position and shape of the localized surface plasmon resonance band changed significantly as a result of the antigen-antibody recognition event.

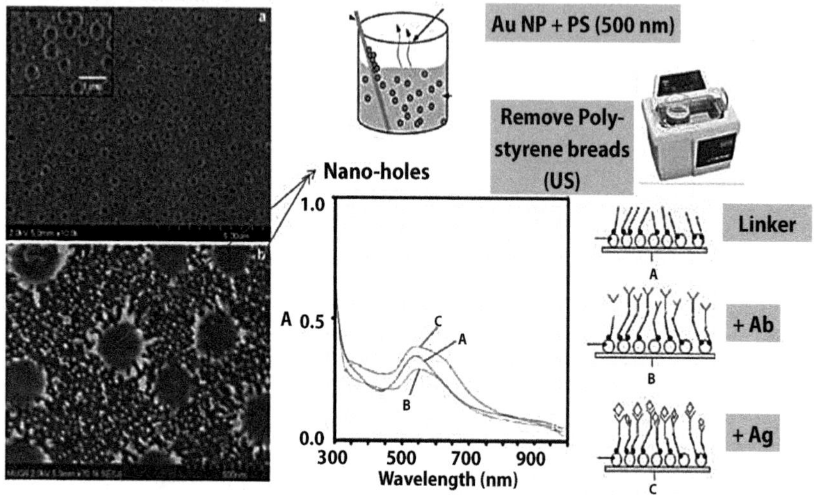

FIGURE 2.15 Gold nanoring and nanohole structures (SEM images) and the sensing performance as shown by the shift of Au plasmon band at different sensing steps.

A novel and simple method of preparation of multilayers of gold nanoparticles on glass substrates are reported. The nanostructure obtained from the angled deposition method shows chain shaped structures with a broad UV-visible absorbance spectrum. The morphology of the non-annealed structures was modified to an island-like structure by annealing at various temperatures. The sensitivity of both the non-annealed and annealed platforms was investigated by using solvents with known refractive indices. The sensing results showed a higher sensitivity for the annealed samples. The annealed platform was used for the sensing of bovine somatotropin (bST) by using an immunoassay format. The proposed sensing platform showed a detection limit as low as 5 ng/ml of bST. Further, the sensing platform was integrated into a microfluidic device and sensing experiments were carried out. The results demonstrated the suitability of nanoisland structures, integrated into a lab-on-a-chip device to detect bovine somatotropin with a good sensitivity. Gold nano island structures were integrated into a lab-on-a-chip for plasmonic detection of bovine growth hormone.

Gold-poly (dimethyl siloxoxane) (Au-PDMS) nanocomposite films with a high elasticity were fabricated through a novel microwave-assisted reduction reaction by using chloroauric acid as a gold precursor and ethanol as a carrier fluid. The high rate of permeation of the ethanol solution in the polymer film, compared to that of an aqueous solution, allows the introduction of the gold precursor into the polymer network with a higher rate and, thus, the reduction reaction is considerably accelerated. The results indicate that when using microwaves, the reduction process is initiated by the formation of sub-nanometric gold seeds on the PDMS surface layers. To improve the distribution of the nanoparticles in the polymer matrix, an annealing process was employed. The mechanical properties of the nanocomposites were also investigated and the module of elasticity of both, as-prepared, and heat treated sample was calculated. The formation of individual nanoparticles with narrow size distribution and with minimum aggregation makes the nanocomposite suitable for label-free protein biosensing. For this purpose, the sensitivity test of the proposed nanocomposite platform was performed and the results demonstrate sensitivity as high as 77 nm/RIU for the annealed sample. The main advantages of the proposed method are the simplicity of the fabrication process as well as the rapid synthesis of gold under these conditions.

Au and Ag complex nanostructures having different sizes and shapes such as shells, rings, disks, etc. have been fabricated and studied for biomedical applications. We have fabricated a novel biosensor based on nanoparticle ring and hole structures by using a nanosphere lithography (NSL) technique developed by our group. In comparison with the NSL methods pioneered and developed by other groups that involves preparing a polystyrene monolayer mask and then evaporating silver onto its interstices, our technique is based on the simultaneous assembly of polystyrene (PS) microspheres and gold nanoparticles in multilayers by a vertical deposition method. The gold nanostructure obtained after the gold-capped spheres are removed consists of nanometer-sized holes surrounded by gold nanoparticles. Sensor platforms based on nanoparticle ring and hole structures were developed and applied for protein sensing The biomolecules studied in this work were: amyloid-derived diffusible ligands (ADDLs; small soluble oligomers prepared from the polypeptide (β-amyloid 1–42), fibrinogen, and a plant protein, AT5G07010.1, from Arabidopsis thaliana). ForAT5G07010.1, the antibody-antigen interaction has been studied as well. It has to be noted that it is for the first time that biosensing of a plant protein and its interaction with a rabbit polyclonal antibody is studied. We have fabricated gold ring and nanohole structures by using a novel experimentally simple and convenient nanosphere lithography method. The method is based on the simultaneous self-assembly of polystyrene microspheres and gold colloids in multilayers on glass substrates by a vertical deposition method.

In conclusion, the topic of the gold nanocomposite is extensive and it comprises a variety of synthesis methods. The properties of the nanocomposites stem from the optical properties of gold nanoparticles, particularly the localized surface plasmon resonance. In this chapter, we described briefly some of the nanocomposites prepared and characterized in our Optical-Bio Microsystems Laboratory. The integration of the synthesis in a microfluidic environment has also been discussed.

KEYWORDS

- **bovine somatotropin**
- **gold-poly (dimethyl siloxoxane)**
- **localized surface plasmon resonance**
- **nanosphere lithography**
- **polystyrene**

REFERENCES

1. Mohammed, A., Simona, B., Abhilash, P., Vo-Van, T., & Muthukumaram, P., (2011). Gold-poly(methyl methacrylate) nanocomposite films for plasmonic biosensing applications, *Polymers*, *3*, 1833–1848. doi: 10.3390/polym3041833.
2. Michael, F., Simona, B., & Muthukumaran, P. (2012). Thermal manipulation of gold nanocomposites for microfluidic platform optimization, *Plasmonics*. doi: 10.1007/s11468-017-0515-3.
3. Simona, B., & Muthukumaran, P., (2012). Microfluidics-nano-integration for synthesis and sensing, *Polymers*, *4*(2), 1278–1310. doi: 10.3390/polym4021278.
4. Jayan, O., Simona, B., & Muthukumaran, P., (2012). Gold nanoisland structures integrated into a lab-on-a-chip for plasmonic detection of bovine growth hormone, *Journal of Biomedical Optics 17*(7), 077001.
5. Hamid, S., Simona, B., Muthukumaran, P., & Rolf, W. (2013). Integration of gold nanoparticles in PDMS microfluidics for lab-on-a-chip plasmonic biosensing of growth hormones, *Biosensors*, and *Bioelectronics*. http://dx.doi.org/10.1016/j.bios.2013.01.016.
6. Fida, F., Varin, L., Badilescu, S., Kahrizi, M., & Vo-Van, T., (2009). Gold nanoparticle ring and hole structures for sensing proteins and antigen-antibody interactions, *Plasmonics, 4*, 201–207, doi: 10.1007/s11468-009-9093-3.

CHAPTER 3

GROWTH OF MESENCHYMAL STEM CELLS ON SURFACE-TREATED 2D POLY(GLYCEROL-SEBACATE) BIO-ELASTOMERS OF VARYING STIFFNESS

RAJU B. MALIGER, PETER J. HALLEY, JUSTIN J. COOPER-WHITE, and DONNA DINNES

AIBN/School of Chemical Engineering, University of Queensland, St. Lucia, Brisbane, Queensland–4072, Australia, Tel.: +61450425510, E-mail: r.maliger@gmail.com

ABSTRACT

Poly(glycerol-sebacate) bioelastomers are regarded as potential materials for tissue engineering applications. This paper deals with the effect of poly(glycerol-sebacate) substrate stiffness and surface treatment methods on the morphology and lineage of mesenchymal stem cells (MSC). Three types of PGS films of varying stiffness (glycerol: sebacic acid = 0.6, 0.8, 1.0) were subjected to aminolysis, hydrolysis, and multilayering (HA-Chitosan). Cell culture studies were performed on treated PGS substrates at four different time points (4 h, 24 h, 7 d, 14 d). The cell-seeded untreated, aminolyzed, hydrolyzed, and multilayered films were analyzed using fluorescence microscopy to investigate cell attachment, proliferation, and morphology. After removing unattached cells, cell lysates were centrifuged and the protein and DNA contents of the supernatants were quantified using Bradford and DNA assays, respectively. At different time-points, the adherent cells of treated PGS substrates (of varying stiffness) exhibited varying morphology. At the highest time-point (14 d) for this study, the PGS surfaces became more confluent with cells (~60–70%) on untreated and hydrolyzed surfaces and less confluent on aminolyzed and multilayered surfaces. Statistical analysis of DNA assay was performed using the Software SAS, and the significant variables were determined from it. Bradford protein assay at different time-points indicated that PGS surfaces did not appear to be cytotoxic towards the cells. For better attachment and proliferation of MSCs on the PGS substrates, the optimal treatment methods and time points required are assessed.

3.1 INTRODUCTION

The clonal nature of marrow cells was first revealed in 1963 by the pioneering work of two scientists, McCullosch and Till [1]. In subsequent years, Friedenstein, and coworkers [2] identified mesenchymal stem cells (MSC), which reside within the stromal compartment of bone marrow. They have also been identified in fetal blood, umbilical cord blood, tibial, and femoral marrow compartments, and thoracic and lumbar spine [3, 4]. These self-renewable, multipotent progenitor cells have the capacity to differentiate into multiple mesenchymal lineages to form bone, cartilage, adipose, tendon, and muscle tissues. They also have the potential to differentiate into other types of tissue-forming cells such as hepatic, renal, cardiac, and neural cells [5, 6]. They also are an adherent cell type and have a fibroblastic morphology. Although MSCs represent a very small fraction (0.001–0.01%) of the total population of nucleated cells in the marrow, they can be isolated and expanded with high efficiency, and induced to differentiate into multiple lineages under defined culture conditions [3]. Thus, these cells are highly attractive candidates for tissue engineering approaches in mesenchymal tissue regeneration. These cells are highly attractive candidates for tissue engineering approaches in mesenchymal tissue regeneration. Further, the low degree of vascularization in cartilage makes MSCs an ideal candidate for cartilage tissue engineering. Marrow-derived MSCs find potential applications in cardiovascular repair, treatment of lung fibrosis, stroke, traumatic injury, and spinal cord injury [7]. To date, MSCs have been used for the treatment of experimental models of diabetes, rheumatoid arthritis (RA), systemic lupus erythematosus (SLE), and multiple sclerosis (MS) [8].

One of the topics of special importance is how MSCs sense matrix elasticity and transduces that information into morphological changes and lineage specification [9]. Physical and chemical properties of the adhesion substrate can profoundly affect cell locomotion, growth, and differentiation [10]. An additional important factor is the mechanical nature of the underlying substrate. The typical dimensions of focal adhesions formed with soft substrates are considerably smaller than those formed following attachment to a rigid surface. This ability to distinguish between soft and rigid substrate enables cells to become oriented whenever they sense a gradient in substrate rigidity and move along the substrate in the direction of higher rigidity [11]. It is reported that the naive MSCs develop into branched, spindle, or polygonal shapes when grown respectively on soft matrices (0.1–1 kPa) that suit neuronal-like cells, moderately elastic surfaces (8–17 kPa) that promote myogenic differentiation, and rigid matrices (25–40 kPa) that stimulate osteogenic differentiation [12]. This ability of MSCs to differentiate themselves into various lineages makes them ideal candidates for various tissue-engineering applications [13].

3.1.1 SURFACE MODIFICATION

Often, the mechanical and degradative properties of substrates are ideal, but may not be conducive to the growth of cells due to poor cytocompatibility of the polymer surface with the cells. Therefore, surface modification of such substrates becomes essential for improved cell attachment and growth. There has been a significant amount of research on the grafting of end-functionalized polymers on inorganic and polymeric substrates in the form of films, fibers, scaffolds, particles, and other geometries [14]. Some of the commonly used surface modification techniques are UV treatment, electron beam-induced grafting, electrostatic self-assembly, surface entrapment modification, polymer-on-polymer stamping, plasma treatment, ion implantation, aminolysis, hydrolysis, and layer-by-layer deposition [15–19]. For this study aminolysis, hydrolysis, and layer-by-layer deposition techniques were chosen for surface modification of PGS substrates.

3.2 EXPERIMENTAL PART

3.2.1 MATERIALS

Glycerol (Reagent Plus > 99% pure) and sebacic acid (99% pure) were obtained from Sigma-Aldrich, Sydney. They were used without further purification. Dimethyl sulfoxide (DMSO) and dimethylformamide (DMF) were obtained from Wako Fine Chemicals. Toluene and acetone were obtained from Sigma-Aldrich. 1-ethyl–3-(3-dimethyl aminopropyl) carbodiimide (EDAC), N-hydroxysuccinimide (NHS), 1,4-diazabicyclo[2,2,2]-octane (DABCO), morpholino (ethane-sulfonic acid) (MES), paraformaldehyde (PFA), sodium bicarbonate, sodium chloride, sodium citrate, and sodium hydroxide were obtained from Sigma-Aldrich and used as received. Chitosan 500 was obtained from Wako Fine Chemicals. Hyaluronic acid (HA), MW = 1.6 MDa, was obtained from Lifecore Biomedical. Branched polyethyleneimine (PEI), MW= 1200 Da, was obtained from Sigma-Aldrich. Collagen I (from human placenta) and bovine serum albumin (BSA) were obtained from Sigma-Aldrich. Dulbecco's Modified Eagle Medium (DMEM), fetal bovine serum (FBS), and phosphate buffered saline were obtained from Gibco. Penicillin, streptomycin, L-glutamine, and ethylenediaminetetraacetic acid (EDTA) were obtained from GIBCO. Alexa Fluor 488 Phalloidin and Hoeschst 33342 were obtained from Invitrogen. Triton X–100, calf thymus DNA and Tris base was obtained from Sigma.

3.2.2 METHODS

Preparation of films: Non-stoichiometric quantities (Glycerol: Sebacic acid = 0.6, 0.8, 1.0-molar ratio) of the monomers were mixed with toluene (40% of combined monomer weight) in the presence of nitrogen gas in separate round-bottomed flasks, which were fitted with a Dean-Stark trap and a reflux condenser to shift the esterification towards high conversion. Esterification was carried out at 130°C for 48 h to synthesize the PGS prepolymers. A BUCHI R–210 Rotavapor, operated at 77 mbar and 40°C, was used for separating toluene from the reaction mixture. Remaining traces of toluene were removed by keeping the prepolymer in a Salvis-LAB vacuum oven (Extech Equipment Pty Ltd, Boronia, Australia) for 12 hours at 10 mbar and 80°C. The purified prepolymers were dissolved in 1,4-dioxane to make up 25% (w/v) solution. The requisite volume of the solution was transferred into a Petrie dish and allowed to cast in an oven (Napco, Model 5851) for 30 h at 130°C under inert atmosphere (nitrogen) to make films of approximate thickness (1.5 mm). The bioelastomer films thus synthesized from three different prepolymers are designated as PGS 0.6, PGS 0.8, and PGS 1.0, respectively. The films were then cut into dimensions of 15 mm diameter and kept in a desiccator until further use.

Preparation of surfaces for cell culture: PGS films were cleaned in 0.1% SDS at 40°C for 20 min, followed by thorough rinsing with ddH_2O, and sterilizing in 70% v/v isopropanol for 10 min. Then 70% v/v ethanol solution was replaced with solutions containing decreasing levels of ethanol with 10 min shaking periods after each addition. The films were moved to a biological safety cabinet to maintain sterility and rinsed with sterile ddH_2O. They were then transferred to 12-well sterile polystyrene tissue culture plates with MES buffer overnight. All further treatment steps were done within this cabinet using filter-sterilized solutions.

Cell culture: Human MSCs, derived from human placenta (Mater Institute, Brisbane), was expanded from 4 to 6 passages. Cryopreserved MSCs (10^6/mL) from liquid N_2 were thawed, and cultured in DMEM containing 10% FBS, 1% Penicillin/ Streptomycin, and 200 mM L-Glutamine in T 75 flask (Greiner) to approximately 70–80% confluence. The cells were then gently lifted by trypsinization and resuspended at a density of 10^5/mL. The cells were incubated at 37°C with 100% humidity and 5% CO_2. The culture medium was changed at 24 h and subsequently every other day. 1 mL of the cell suspension was seeded into each well of a 24-well tissue culture plate (Greiner) with the appropriate untreated or surface modified PGS film. The cells were cultured phenol red-free DMEM to avoid the interference of phenol red with PGS films. Cell samples were either lysed or fixed as outlined below at 4 h, 24 h, 7 d, and 14 d.

Aminolysis and Hydrolysis: The PGS films were aminolyzed using 1 mL per well of a 20 mg/mL solution of PEI in water for 10 min and rinsed thoroughly with 5 mM MES buffer [2-(*N*-morpholino) ethanesulfonic acid], pH 5.5. Another set of films were hydrolyzed using 0.01 M sodium hydroxide for 10 min and rinsed thoroughly with MES buffer 3 times. They were kept in the MES buffer in 12-well plates overnight before using for cell culture. Before cell-seeding, each of the untreated, aminolyzed, and hydrolyzed films were washed twice with PBS (Phosphate buffered saline).

Multilayering: Films to be multilayered were initially subjected to aminolysis as outlined above. Hyaluronic acid (HA) stock solution was prepared by dissolving 5mg/mL sodium hyaluronate (SA) in 5mM MES buffer. MES buffer was added while stirring SA fibers. Then the solution was filtered using a 0.22 μm Sterivex sterile filter unit (Millipore). Chitosan stock solution was prepared by dissolving Chitosan (500) solution in H_2O, followed by adding few drops of acetic acid in chitosan suspension to make a 0.05M final AA-chitosan-H_2O solution concentration. The solution was filtered using a 0.45 mm syringe filter. HA solution (50 μg/mL in MES) was mixed with 1 part in 100 of 50 mg/mL EDAC, 100 mg/mL NHS in DMF and added to the wells (1 mL/well) in which PGS films were placed and incubated for 15 min. Then HA solution was removed and replaced with 1 mL of MES buffer as a rinse, incubated for 5 min, followed by replacement with 1 mL of chitosan solution (50 μg/mL in MES) for 15 min, and another rinse with 1 mL of MES buffer. The process was repeated until 7 layers had been deposited, with HA as the outermost layer. The films were once again incubated in MES buffer for 5 min. The outermost layer (HA) of the surface was functionalized with Collagen, which had EDAC/NHS cross-linker as well. For this Collagen I solution (2.5 mg/mL in 0.1M acetic acid) was diluted to 50 μg/mL in MES buffer, incubated with the surfaces (1 mL/well) overnight in the refrigerator, and rinsed with MES followed by PBS. The surfaces were stored in MES buffer at 4°C until further use. Before cell-seeding, the multilayered films were rinsed 3 times in PBS to ensure removal of unbound collagen–1.

1 mL of the cell suspension was pipetted into each well with gentle agitation to distribute cells, and the plates returned to the incubator. This was marked as time zero for each plate. Four-time points (4 h, 24 h, 7 d, 14 d) were chosen for the cell culture studies.

Fluorescence microscopy: After the desired time, the films were rinsed twice with PBS, fixed using 4% paraformaldehyde in PBS for 10 min, rinsed well in PBS, and stored in PBS in a refrigerator until further analysis. The films were rinsed three times in PBS and incubated with 0.1% Triton X–100 for 5 min to permeabilize cell membranes, followed by rinsing again with PBS. The films were placed, cell side down, onto 30 μL of diluted Alexa 488® Phalloidin in parafilm covered chamber for 30 min to stain F-actin and rinsed with PBS buffer. They were then removed and mounted on to glass microscope slides with mounting medium (1% DABCO, 9% PBS, 90% glycerol) over the films (cell side up) and covered with 18 X 18 mm glass coverslips. The edges of the coverslips were sealed with clear nail polish to prevent drying of the samples.

Microscopic analysis was carried out using BX61 fluorescence microscope fitted with a RETIGA EXi camera (Olympus), and a 20 X objective and 0.50 aperture. The In Vivo software was used for capturing images, version 3.2.0 (MediaCybernetics).

Sample preparation for DNA and protein assay: Each well was rinsed once with PBS to remove unattached cells. The cells from each well for different time-points (4 h, 24 h, 7 d, and 14 d) were lysed using 200 μL of lysis buffer (5 mL 100 mM EDTA, 2.5 mL 1% Triton-X 100, and 42.5 mL PBS) on ice for 1 hour. Cell lysates were centrifuged at 10,000 g for 5 min to palletize cell membranes. The supernatants were collected into microcentrifuge tubes and stored in a deep freezer at –80°C until further analysis. The protein and DNA contents were quantified using Bradford and DNA assays, respectively.

DNA assay: The DNA assay was performed in a 96-well black flat-bottomed plate (Perkin-Elmer). First, a DNA standard curve was prepared using calf thymus DNA (Sigma Aldrich, Sydney). 10 μL of lysis buffer was added to each well-containing DNA standard. This was done to compensate for the lysis buffer contained in the cell lysate samples. 10 μL of sample from each cell lysate was added to each sample well. Working DNA dye solution was prepared by mixing 1 mL 10 X TNE (100mM Tris; 2.0M NaCl; 10mM EDTA; pH 7.4), 9 mL ddH$_2$O, and 10 μL Hoechst dye (1 mg/mL Hoechst stock). Then 100 μL of working DNA dye solution was added to each standard and sample well to be analyzed. The analysis was performed by using SpectraMax M5 (Molecular Devices) Platereader and setting UV excitation and emission wavelengths at 360 and 460 nm, respectively. The DNA assay was performed three times independently for every experimental sample.

Bradford (protein) assay: Bradford assay was performed in a clear 96-well flat-bottomed plate. Initially, the dye reagent was prepared by diluting 1 part Dye Reagent Concentrate (Biorad) with 4 parts DD H$_2$O. The particulates were removed by filtering through a Whatman #1 filter. A protein standard curve was prepared from a 1 mg/mL Bovine Serum Albumin (BSA) stock. 10 μL of each standard and diluted sample solution were pipetted into separate wells. 200 μL of diluted dye reagent was added to each well. The sample and the reagent were gently mixed followed by 5 min incubation at room temperature. The absorbance was measured using the SpectraMax M5 at 595 nm.

3.3 RESULTS AND DISCUSSION

3.3.1 FLUORESCENCE MICROSCOPY

MSCs were cultured upon three types of PGS films (PGS 0.6, PGS 0.8, PGS 1.0) of varying stiffness (718 kPa, 570 kPa, 253 kPa) with a range of surface treatments in 24-well plates for 4 h, 24 h, 7 d, and 14 d time-points to investigate cell attachment, proliferation, and morphology. Fluorescence images of phalloidin-stained F-actin were obtained using a 20X objective.

Figures 3.1–3.4 compare the morphology of cells cultured for 4 h, 24 h, 7 d, and 14 d, respectively, on three types of PGS films (PGS 0.6, PGS 0.8, PGS 1.0) treated by aminolysis (A), hydrolysis (H), and 7-layer HA-chitosan multilayer (M) constructs.

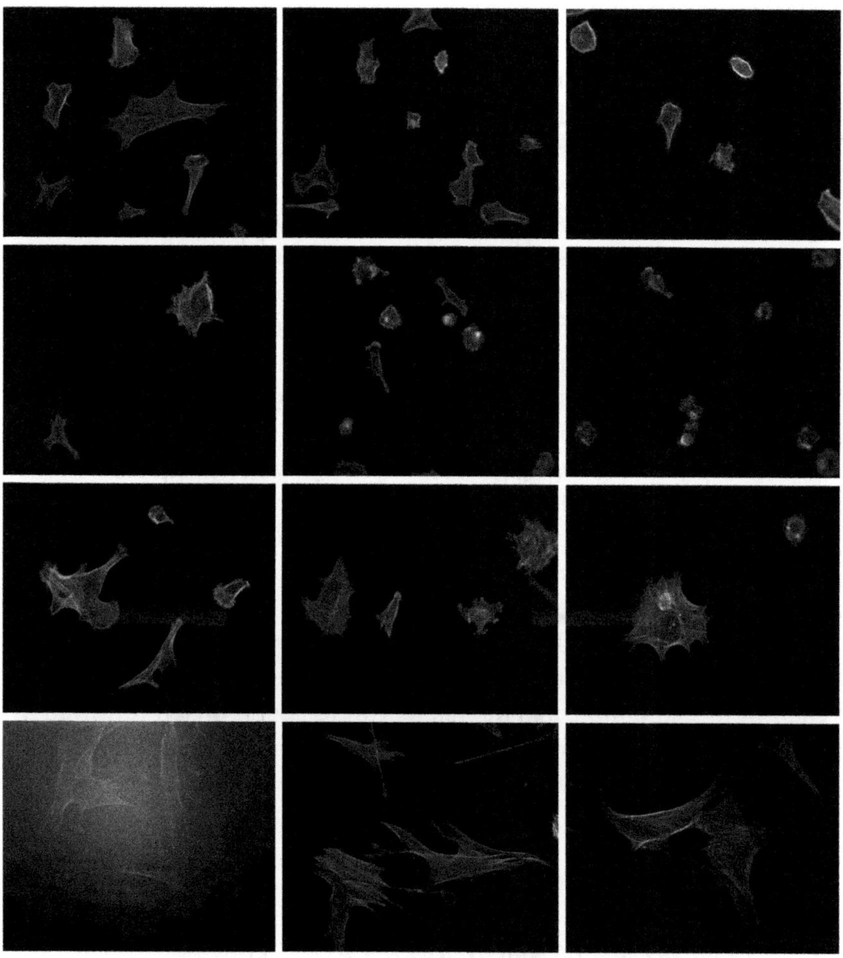

FIGURE 3.1 Fluorescence microscopy images of F-actin stained MSC cultured on untreated (U), aminolyzed (A), hydrolyzed (H), multilayered (M) PGS 0.6, PGS 0.8, and PGS 1.0 surfaces after 4 h. All images were obtained using a 20 X objective.

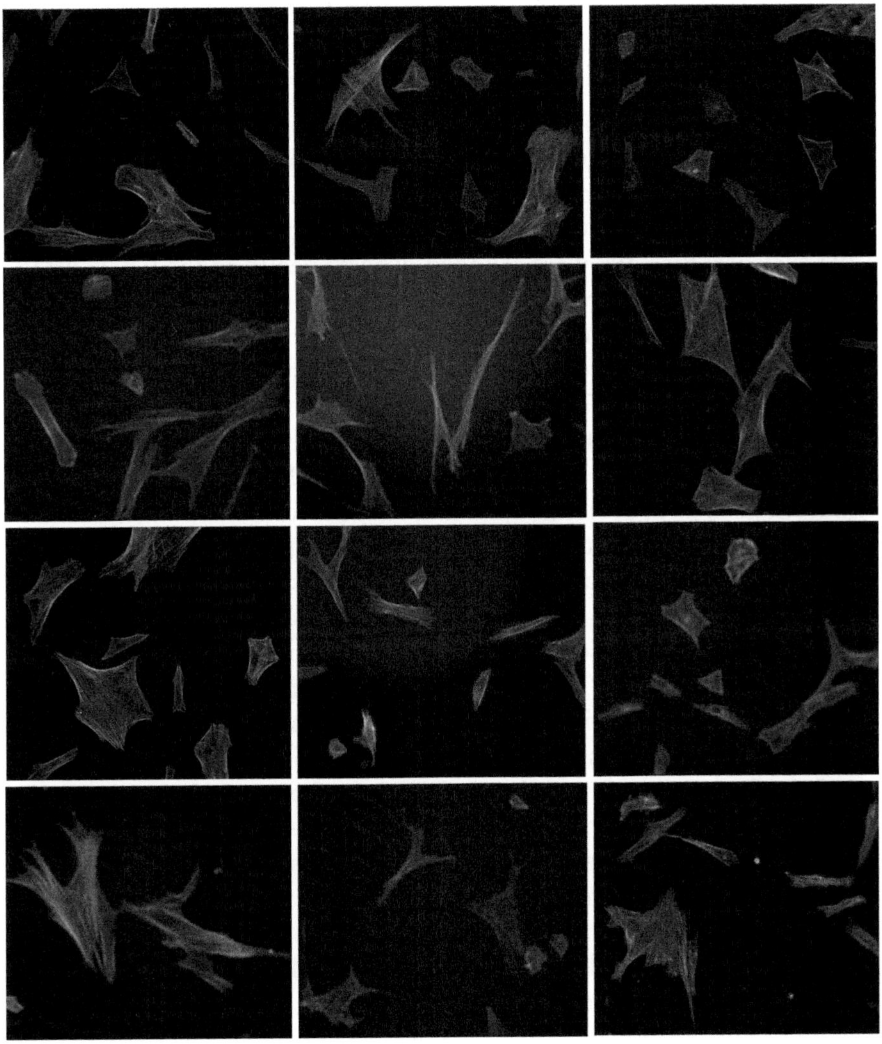

FIGURE 3.2 Fluorescence microscopy images of F-actin stained MSCs cultured on untreated (U), aminolyzed (A), hydrolyzed (H), and HA-chitosan (M) PGS 0.6, PGS 0.8, and PGS 1.0 surfaces after 24 h. All images were obtained using a 20 X objective.

Fluorescence images of cells cultured for 4 h on untreated, aminolyzed, hydrolyzed, and multilayered PGS surfaces are shown in Figure 3.1. It is to be noted that the stiffness of the PGS films decreased in the following order: PGS 0.6 > PGS 0.8 > PGS 1.0. The Young's moduli of these films were 718, 570, and 253 kPa, respectively. These rigid matrices may help stimulate osteogenic differentiation [12]. Majority of the adherent cells on the untreated surface of the films 0.8U and 1.0U displayed a rounded morphology, whereas the 0.6U surface displayed a mixture of rounded and spreading cells (Figure 3.1). In the case of aminolyzed surfaces, the attached cells were rounded, with cells on 0.6A being more spread-out than on the other two PGS surfaces. Similarly, on hydrolyzed 0.8H and 1.0H surfaces, the cells were rounded, whereas cells on the 0.6H surface were characterized by protrusions that may assist in cell motility [20]. The cells on untreated and aminolyzed surfaces exhibited rounded morphology, whereas the cells on hydrolyzed surfaces were polygonal in shape. The cells on hydrolyzed surfaces seem to be more motile than on aminolyzed or untreated surfaces. At 4 h, on all multilayered surfaces, the cells were well spread-out and displayed typical MSC phenotype. Cells require extracellular matrix proteins, such as collagen, for better attachment and growth. Collagen I, which was crosslinked with the outermost layer HA, may have facilitated better cell attachment and spreading of cells on the multilayered constructs. Morphology of the cells remained similar on all multilayered PGS surfaces. Thus, it appears that on multilayered surfaces, the effect of stiffness of the underlying substrate on cell attachment and proliferation is less dominant during the early time-points.

Fluorescence images of cells cultured for 24 h on untreated, aminolyzed, hydrolyzed, and multilayered PGS surfaces are shown in Figure 3.2. It can be observed that on the untreated PGS surfaces, cell attachment was better at 24 h than at 4 h. The cells at 24 h spread well on untreated PGS surfaces. The aminolyzed surfaces at 24 h on 0.6A and 0.8A were characterized by the formation of a mixture of spindle and rod-shaped cells, whereas on 1.0A the cells developed into a slightly more polygonal shape – resembling an osteoblast phenotype. However, further experiments are required to be performed to get any conclusive evidence. On aminolyzed surfaces, at 24 h the cell distribution was better than at 4 h time-point. On hydrolyzed PGS surfaces, good cell attachment was observed, with cells on 0.6H showing greater spreading. The cell attachment on multilayered surfaces at 24 h remained identical to that at 4 h, but very few adherent cells remained rounded or filament-shaped.

The fluorescence microscopy images of MSCs grown on treated PGS surfaces at 7 d time-point are shown in Figure 3.3. Due to issues associated with autofluorescence of the film, the image 1.0A was unattainable. The fluorescence micrographs of untreated (U) surfaces after 7 days were characterized by the presence of rounded and bone-shaped MSCs, whereas the cells on aminolyzed surfaces were polygonal and spread-out in all directions. The hydrolyzed surface, 0.6H, was noted to have long filament-shaped MSC phenotype, whereas the 0.8H surface had smaller rod-shaped cells with minimal cells remaining. On 1.0H, the cells were spindle and polygonal in shape. The micrographs of the multilayered surfaces at 7 days showed filament-shaped adherent cells with little evidence of spreading within any of the viewing fields analyzed.

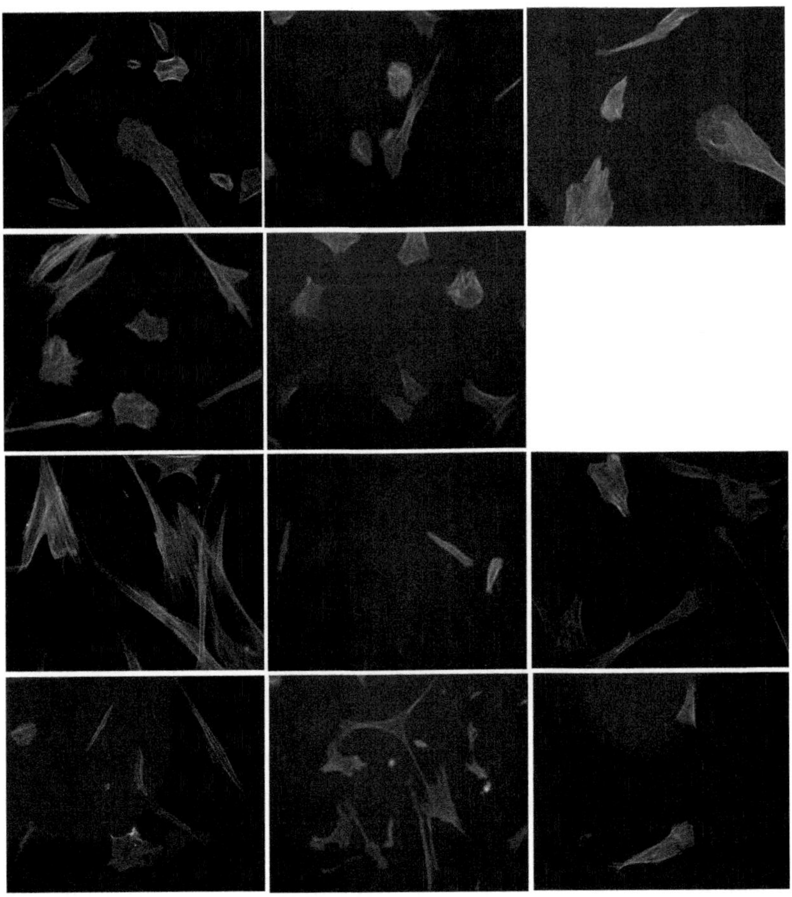

FIGURE 3.3 Fluorescence microscopy images of F-actin stained MSCs cultured on untreated (U), aminolyzed (A), hydrolyzed (H), and HA-chitosan (M) PGS 0.6, PGS 0.8, and PGS 1.0 surfaces after 7 days. All images were obtained using a 20 X objective.

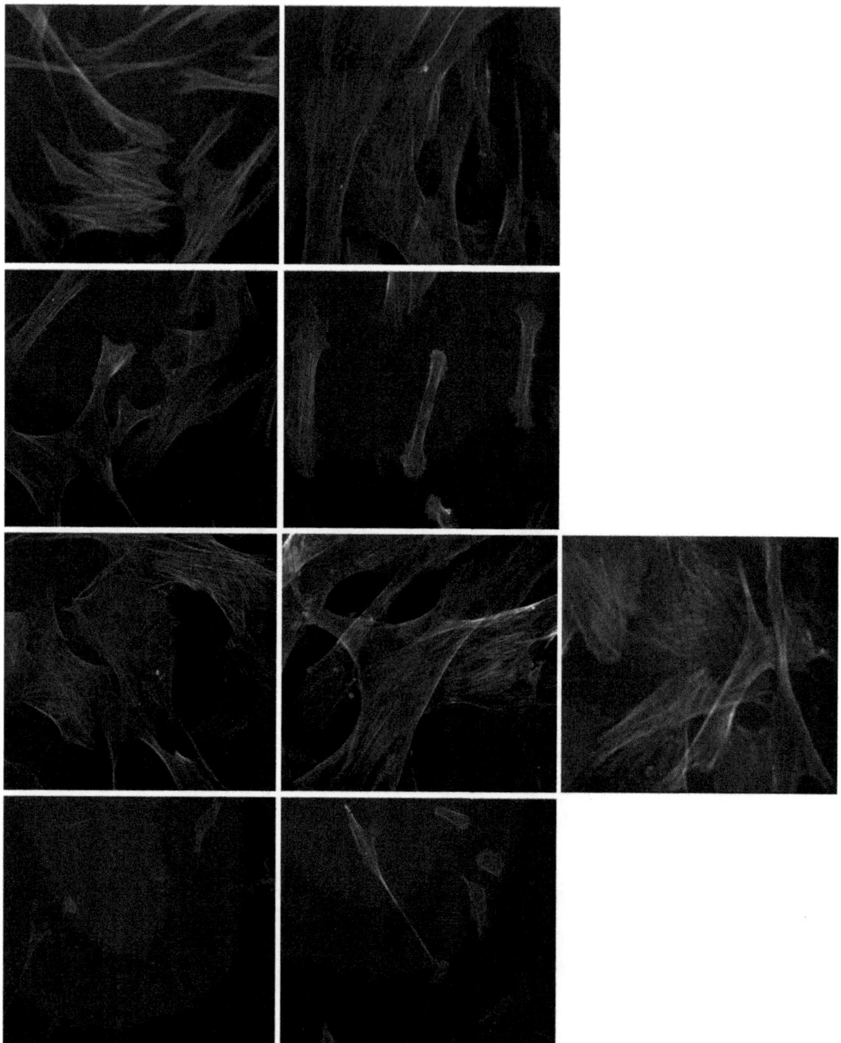

FIGURE 3.4 Fluorescence microscopy images of F-actin stained MSCs cultured on untreated (U), aminolyzed (A), hydrolyzed (H), and HA-chitosan (M) PGS 0.6, PGS 0.8, and PGS 1.0 surfaces after 14 days. All images were obtained using a 20 X objective.

The fluorescence microscopy images of MSCs cultured on various treated PGS surfaces at 14 d are shown in Figure 3.4. Due to problems associated with autofluorescence several images were unattainable.

It can be observed from the fluorescence images of day 14 (see Figure 3.4) that on the untreated surfaces (0.6U & 0.8U), the cells are interconnected and well spread out. On the aminolyzed surface, 0.8A, the cells were rod-shaped and appeared motile. On 0.6A, the cells were polygonal in shape and interconnected. It can be noticed that the attachment and spreading of cells on hydrolyzed surfaces was faster between 7 and 14 d than during other time points, confirming that there was a noticeable increase in cell number during that period. At 14 d, MSCs proliferated quite well on all hydrolyzed surfaces and were well interconnected. The cells were polygonal in shape, and may potentially be directed along an osteoblast phenotype. However, experiments at further time-points need to be considered to confirm this. PGS surfaces, when hydrolyzed, become more hydrophilic, an environment much similar to that of the extracellular matrix. This is one of the reasons why MSCs may attach and proliferate faster on hydrolyzed surfaces. It can be seen from Figure 3.4 that the growth of MSCs on multilayered surfaces diminished, with little evidence of spreading within any of the viewing fields analyzed. This suggests that although collagen-I supported the growth of cells in the early phase of culture by masking the role of the stiffness of the underlying substrate, at later stages the stiffness of the films may have become a vital parameter for any substantial growth of the cells. At 14 d, the surfaces became more confluent with cells (~ 60–70%) on untreated and hydrolyzed surfaces and less confluent on aminolyzed and multilayered surfaces. Therefore, for the time-points considered for this study, untreated, and multilayered surfaces seem to be more supportive of MSC cell attachment and proliferation.

3.3.2 DNA ASSAY

A DNA assay was used for relative quantification of cell number at successive stages of cell culture. The results of DNA assay are shown in Figure 3.5.

The amount of DNA on poly (styrene) surface (control cell culture surface) was higher than on any other PGS surface after 7 and 14 days due to cells becoming confluent on poly (styrene) surfaces. The growth of the cells on untreated, aminolyzed, hydrolyzed, and multilayered PGS surfaces remained unchanged for up to 7 days. After 14 d, there was an increase in the amount of DNA on 0.6H, 0.6M, and 0.8A surfaces. Hydrolysis of PGS films turns the matrix more hydrophilic, thereby making the environment similar to that of the extracellular matrix. This may encourage cell attachment and differentiation. It can be observed from Figure 3.4 that MSCs on 0.6H are polygonal in shape and likely to be differentiating into an osteoblast phenotype. Similarly, an increase in DNA amount on 0.8A at 14 d (Refer to Figure 3.5) is marked by the formation of elongated rod-shaped cells (see Figure 3.4). Statistical analysis of the data was performed using the software SAS, version 9.1 (Cary, North Carolina, USA). Time (4 h, 24 h, 7 d, 14 d), treatment method (U, A, H, M), surface (PGS 0.6, PGS 0.8, PGS 1.0), and the number of replicates were used as variables for the analysis. The significance level of these parameters with respect to DNA amount is given in Table 3.1.

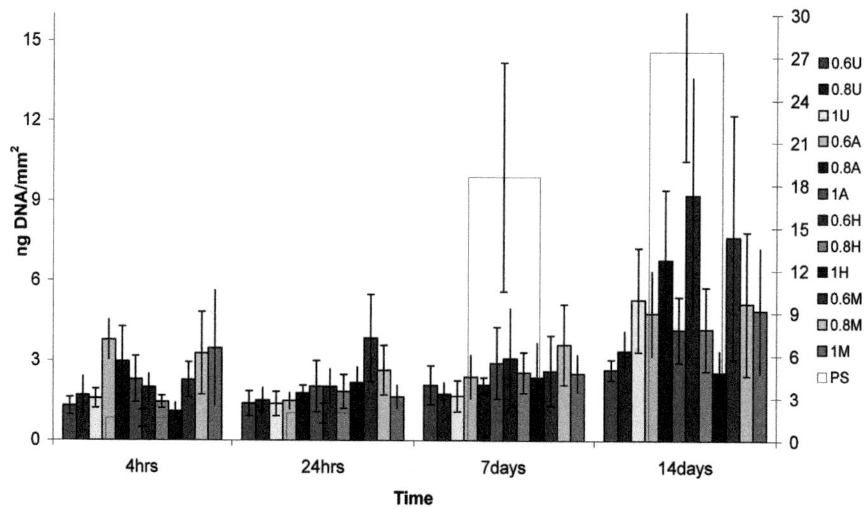

FIGURE 3.5 (See color insert.) DNA quantification in cell lysates at 4 h, 24 h, 7 days, and 14 days. A – aminolyzed; H – hydrolyzed; M – multilayered; U – untreated film; PS – polystyrene.

It can be observed from Table 3.1 that treatment and time are the significant variables ($p < 0.05$). The calculated least square mean values of DNA at different time-points (excluding controls) are graphically shown in Figure 3.6.

TABLE 3.1 Statistical Analysis of DNA Assay

Source	DF	Mean Square (ng DNA/mm2)	F Value	Pr > F
Replicate	2	107.858	24.86	<.0001
Conc	2	5.223	1.20	0.3047
Treatment	3	13.911	3.21	0.0267
Conc*Treatment	6	4.464	1.03	0.4115
Time	3	72.971	16.82	<.0001
Conc*Time	6	2.031	0.47	0.8304
Treatment*Time	9	1.589	0.37	0.9483
Conc*Treatment*Time	18	4.364	1.01	0.4607

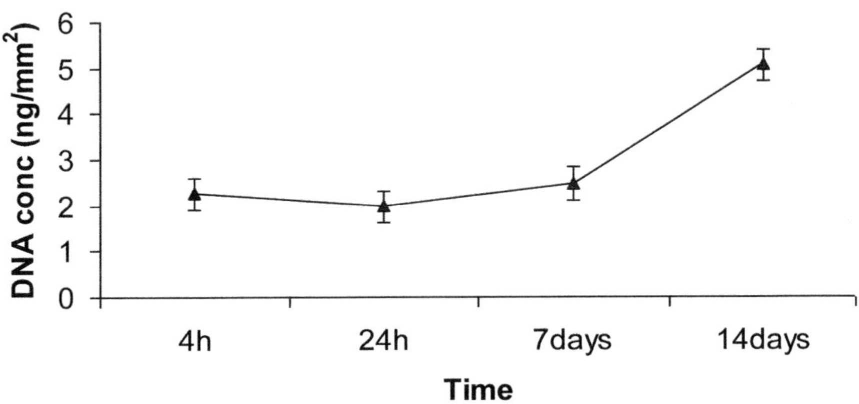

FIGURE 3.6 Average DNA concentration (ng/mm^2) at various time-points.

MSCs proliferated, although with a very long lag phase in their growth, for up to 7 days. However, the DNA concentration increased between 7 and 14 d, suggesting an increase in cell number. Thus, PGS surfaces do not appear to be cytotoxic towards the cells. The dependence of DNA concentration on various surface treatments is shown in Figure 3.7.

Figure 3.7 indicates that the average DNA concentration on aminolyzed, hydrolyzed, and multilayered surfaces were higher than on the untreated surfaces. The minimum surface stiffness required for osteogenic differentiation of MSCs to form collagenous bone is 100 kPa [12]. The stiffness range of the PGS surfaces used for this study is 253–718 kPa. The surface treatment on PGS surfaces did not lead to any conclusive phenotype differentiation at the end of the 14-day time-point, however, according to morphological studies, the cell attachment was better on untreated and hydrolyzed PGS surfaces. Statistical analysis of the DNA results confirmed that concentration (glycerol: sebacic acid) or substrate stiffness was not a significant variable (Refer to Table 3.1) for the time-points chosen for this study. Further experiments are required to be performed at time-points greater than 14 days to determine the effect of PGS substrate stiffness on the differentiation of cells into various phenotypes.

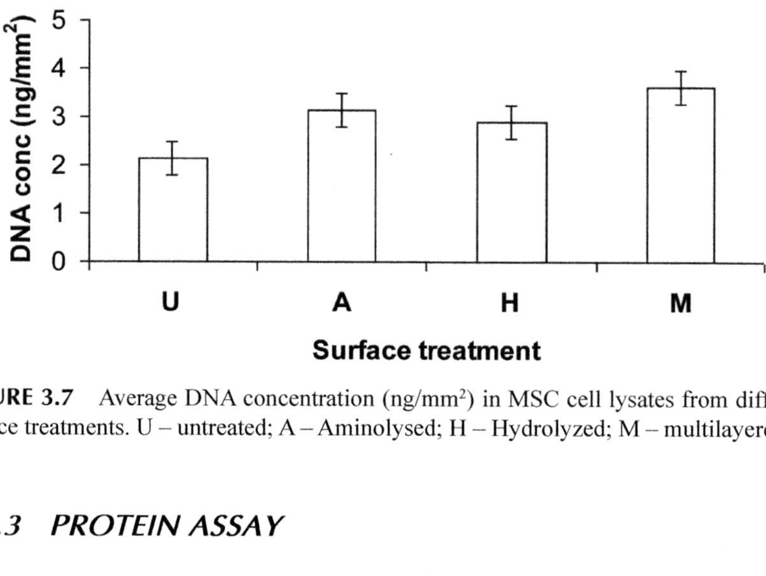

FIGURE 3.7 Average DNA concentration (ng/mm²) in MSC cell lysates from different surface treatments. U – untreated; A – Aminolysed; H – Hydrolyzed; M – multilayered.

3.3.3 PROTEIN ASSAY

Protein quantification was done to assess changes in cellular processes at different time-points. The results of the Bradford protein assay at 4 h, 24 h, 7 d, and 14 d time-points are shown in Figure 3. 8 and Table 3.2.

Similar to the DNA results, the protein concentration remained almost constant on all surfaces at 4 h and 24 h. It increased on all surfaces by the 7 d time-point and underwent marginal changes until 14 d (Figure 3.8). This indicates that although various PGS surfaces did not trigger the growth of MSCs, the protein concentration did not decrease with time. From protein results, it is evident that PGS surfaces do not appear to be cytotoxic towards the cells. This is in agreement with the results published earlier [16]. The cell lysates of PGS surfaces, 1.0H and 0.6M at 24 h, and 0.8A and 0.8M at 7d, were found to contain higher protein concentration than on the same surfaces at earlier time-points. It suggests an increase in the cellular processes or differentiation on these films between 7 and 14 days. However, further experiments on live-dead cell assay need to be performed to get any conclusive evidence on cell proliferation. Statistical analysis on Bradford assay showed that 'time' was the only significant variable (see Table 3.2). The average protein concentrations combining all surfaces (except controls) together over a period of 14 days are graphically shown in Figure 3.9.

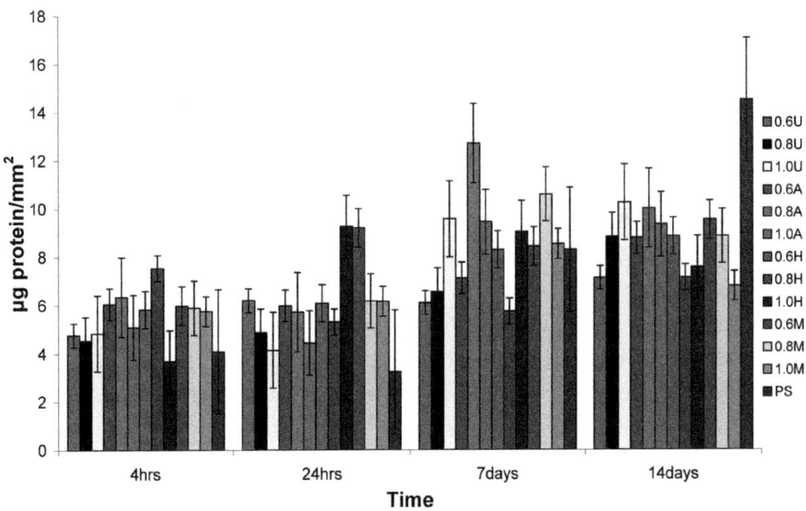

FIGURE 3.8 **(See color insert.)** Quantification of protein in MSC cell lysates at 4 h, 24 h, 7 days, and 14 days time- points. A – aminolyzed, H – hydrolyzed, M – multilayered, U – untreated film, PS – polystyrene.

TABLE 3.2 Statistical Analysis of Protein Assay

Source	DF	Mean Square	F Value	Pr > F
Replicate	1	73.746	9.68	0.0032
Conc	2	0.311	0.04	0.9600
Treatment	3	7.194	0.94	0.4269
Conc*Treatment	6	5.674	0.74	0.6166
Time	3	60.658	7.96	0.0002
Conc*Time	6	4.3707245	0.57	0.7493
Treatment*Time	9	3.8325759	0.50	0.8649
Conc*Treatment*Time	18	4.3903119	0.58	0.8989

The average protein concentration increased for the first three time-points (4 h, 24 h, 7 d), which were likely contributed by the adsorption of different FBS proteins onto the PGS film. These proteins in the cell lysates can be separated and quantified using sodium dodecyl sulfate-polyacrylamide gel electrophoresis (SDS-PAGE) and Western blot. From day 7 to 14, there was a marginal increase in the protein concentration, indicating quiescence in further cellular process.

3.4 CONCLUSION

Morphological studies using fluorescence microscopy suggested that cell attachment and spreading was better on hydrolyzed surfaces. The protrusions of cells and subsequent formation of polygonal shapes indicated they were likely directed along osteoblast differentiation. However, further experiments at higher time-points, in addition to studying differentiation markers, will provide conclusive evidence. On multilayered surfaces, collagen on the HA-chitosan layers supported the growth of cells in the early stages of cell culture, but at later stages, the cell growth diminished with little evidence of spreading within any of the viewing fields analyzed. The cells on untreated surfaces displayed rounded morphology in the initial stages of culture, and at 14 days the cells were interconnected and spread out. However, except at 14 d time-point, there were no striking differences in phenotypes of MSCs on various PGS films. The attachment of cells on aminolyzed and multilayered surfaces was characterized by the formation of thin rod-shaped and filament-shaped cells. However, further experiments need to be performed to confirm such phenotype differentiation.

DNA and protein assay results indicated that the cells proliferated well on polystyrene surfaces by 7 d. Statistical analysis of the DNA results showed that aminolyzed, hydrolyzed, and multilayered surfaces, to a certain extent, enhanced cell growth and attachment. For further studies, bromodeoxyuridine can be used in the detection of the proliferation of cells. Surface treatment and time emerged as the significant parameters during the cell-growth when statistical analysis was performed on the DNA results. Statistical analysis of protein results showed that time was the most significant variable and that the protein content on PGS surfaces increased with time. Also, differentiation of cells on PGS surfaces did not lead to any specific phenotypes at the end of the 14-day time-point, indicating that higher time-points need to be considered to study the effect of matrix stiffness and surface treatment on cell attachment and phenotype differentiation. To understand the definitive changes in cellular processes, techniques such as SDS-PAGE, western blotting, real-time PCR, immunofluorescent staining, and flow cytometry (surface markers) can be used. The stiffness of the surfaces decreased in the following order: 0.6U (718 kPa) > 0.8U (570 kPa) > 1.0U (253 kPa). The cells were spread more on stiffer untreated surfaces (0.6U and 0.8U). On aminolyzed surfaces, the cells developed into the rod, spindle, and polygonal shapes. The spreading and interconnectivity were higher on 0.6U than on other surfaces. On hydrolyzed surfaces, the cells were rounded in shape, and at later stages became confluent and spread quite well on surfaces of varying stiffness. On multilayered surfaces, collagen I supported the growth of cells in the initial stages on all surfaces, thereby making the role of stiffness. However, at later stages, the cell growth diminished indicating that matrix stiffness became an important parameter for cell differentiation. MSCs on PGS surfaces did not differentiate into definite phenotypes, however the cell attachment and spreading varied on different treated surfaces. PGS surfaces did not trigger rapid proliferation of MSCs, and further, the protein and DNA concentrations in cell lysates did not increase or decrease rapidly with time. The growth and differentiation of MSCs on hydrolyzed and aminolyzed 3D PGS scaffolds are to be studied over a range of time-points to appreciate the effect of substrate stiffness on cell growth and phenotype differentiation. Further, to analyze various phenotype differentiation, cells can be stained with alkaline phosphatase (for bone), Oil Red O (for adipose), and Alcian Blue (for cartilage).

KEYWORDS

- **aminolysis**
- **bradford assay**
- **DNA assay**
- **hydrolysis**
- **mesenchymal stem cells**
- **multilayering**
- **poly(glycerol-sebacate)**
- **tissue engineering**

REFERENCES

1. Becker, A. J., McCulloch, E. A., & Till, J. E., (1963). *Nature, 197*, 452–454.
2. Friedenstein, A. J., Piatetzk, S. I., & Petrakova, K. V., (1966). *Journal of Embryology and Experimental Morphology, 16*, 381–390.
3. Barry, F. P., & Murphy, J. M., (2004). *International Journal of Biochemistry & Cell Biology, 36*, 568–584.
4. Fibbe, W. E., (2002). *Annals of the Rheumatic Diseases, 61*, 29–31.
5. Alhadlaq, A., & Mao, J. J., (2004). *Stem Cells and Development, 13*, 436–448.
6. Bielby, R., Jones, E., & McGonagle, D., (2007). *Injury-International Journal of the Care of the Injured, 38*, 26–32.
7. Tae, S. K., Lee, S. H., Park, J. S., & Im, G. I., (2006). *Biomedical Materials, 1*, 63–71.
8. Uccelli, A., Pistoia, V., & Moretta, L., (2007). *Trends in Immunology, 28*, 219–226.
9. Abercrombie, M., & Dunn, G. A., (1975). *Experimental Cell Research, 92*, 57–62.
10. Pelham, R. J., & Wang, Y. L., (1997). *Proceedings of the National Academy of Sciences of the United States of America, 94*, 13661–13665.
11. Bershadsky, A. D., Balaban, N. Q., & Geiger, B., (2003). *Annual Review of Cell and Developmental Biology, 19*, 677–695.
12. Engler, A. J., Sen, S., Sweeney, H. L., & Discher, D. E., (2006). *Cell, 126*, 677–689.
13. Caplan, A. I., (2006). In: Bronzino, J. D., (ed.), *Tissue Engineering and Artificial Organs* (Vol. 30, pp. 1–7). Taylor & Francis Group: Boca Raton.
14. Walters, K. B., & Hirt, D. E., (2007). *Macromolecules, 40*, 4829–4838.
15. Pompe, T., Keller, K., Mothes, G., Nitschke, M., Teese, M., Zimmermann, R., & Werner, C., (2007). *Biomaterials, 28*, 28–37.
16. Croll, T. I., O'Connor, A. J., Stevens, G. W., & Cooper-White, J. J., (2006). *Biomacromolecules, 7*, 1610–1622.
17. Sun, H., & Onneby, S., (2006). *Polymer International, 55*, 1336–1340.
18. Berg, M. C., Yang, S. Y., Hammond, P. T., & Rubner, M. F., (2004). *Langmuir, 20*, 1362–1368.
19. Zhu, H. G., Ji, J., & Shen, J. C., (2004). *Biomaterials, 25*, 109–117.
20. Beningo, K. A., Dembo, M., Kaverina, I., Small, J. V., & Wang, Y. L., (2001). *Journal of Cell Biology, 153*, 881–887.

CHAPTER 4

ENHANCEMENT OF IMPACT STRENGTH IN JUTE FIBER-REINFORCED POLYPROPYLENE-SPENT COFFEE GROUND COMPOSITES

N. JAYA CHITRA

Professor and Head, Department of Chemical Engineering, Dr. M.G.R Educational and Research Institute (University), Maduravoyal, Chennai–95, India,
E-mails: jayakarun7696@gmail.com; hod-chem@drmgrdu.ac.in

ABSTRACT

The most abundant biomass material in the world is lignocellulosic biomass. This project aims in the utilization of agro waste/biowaste as fillers in thermoplastics due to the increased needs in overcoming the environmental problems caused by their by-product. Most recent developments in using a biomass material are with jute fibers, palm coir, banana fibers and sisal, etc., as reinforcement. All these natural fibers have excellent physical and mechanical properties and can be utilized more effectively in the development of composite materials for various applications. Various processing techniques and conditions; Compression molding process, Injection molding and extrusion methods are used in composite production.

The aim of the present work is to develop Jute sandwich composite through compression molding for which already compounded spent coffee ground composite of previous work was used as matrix and braided jute fiber were used as reinforcing the material. Mechanical properties like Tensile strength and Impact strength properties were tested and the results show there is a decrease in the former properties compared to later. Further chemical testing was also done to determine its resistance and to develop composites for various application.

4.1 INTRODUCTION

Composite materials were known to mankind from the Old Stone Age. Product made from natural fiber-reinforced biodegradable polymer composites are yet to be seen in high magnitude. In recent years, polymeric based composite materials are being used in many applications, such as automotive, sporting goods, marine, electrical, industrial, construction, household appliances, etc. Polymeric composites have high strength and stiffness, lightweight, and high corrosion resistance. Natural fibers are available in abundance in nature and can be used to reinforce polymers to obtain light and strong materials. Natural fibers from plants are beginning to find their way into commercial applications such as automotive industries, household application [1]. Government regulations and a growing environmental awareness throughout the world have triggered a paradigm shift towards designing materials compatible with the environment. The use of bio fibers, derived from annually renewable resources, as reinforcing fibers in both thermoplastic and thermoset matrix composites provides positive environmental benefits with respect to ultimate disposability and raw material utilization [2].

The depletion of petroleum resources coupled with an awareness of global environmental problem provides the alternatives for new green materials that are compatible with the environment and their development is independent of petroleum-based resources. The development of natural fiber-reinforced biodegradable polymer composites promotes the use of environmentally friendly materials. The use of green materials provides an alternative way to solve the problems associated with agriculture residues. Agricultural crop residues such as oil palm, pineapple leaf, banana, and sugar palm produced in billions of tons around the world. The vital alternative to solve this problem is to use the agriculture residues as reinforcement in the development of polymer composites [3, 4].

Thermoplastics widely used for biofibers are polyethylene [5], polypropylene (PP) [6], and polyvinyl chloride (PVC); here as phenolic, polyester, and epoxy resins are mostly utilized thermosetting matrices [7]. Different factors can affect the characteristics and performance of NFPCs.

The hydrophilic nature of the natural fiber and the fiber loading also has impacts on the composite properties. Usually, high fiber loading is needed to attain good properties of NFPCs. Generally, notice that the rise in fiber content causes improving in the tensile properties of the composites [8]. Another vital factor that considerably impacts the properties and surface characteristics of the composites is the process parameters utilized.

For that reason, appropriate process techniques and parameters should be rigorously chosen in order to get the best characteristics of producing composite [9].

4.2 POLYPROPYLENE

Polypropylene (PP) is a thermoplastic "addition polymer" made from the combination of propylene monomers. Globally, most propylene monomer comes from the steam-cracking process using naphtha which is a valuable fraction of crude oil. The target product used form naphtha cracking is ethylene monomer in which propylene is a by-product of the cracking process which is produced along with various other byproducts. Recently, a new process by which propane is dehydrogenated to propylene monomer is being used. Propylene is used in a variety of applications like the packaging for consumer products, plastic parts for various industries including the automotive industry and textiles. It has exceptional resistance at room temperature to organic solvents like fats.

4.2.1 NEED FOR POLYPROPYLENE

Polypropylene is used both in the household and industrial applications because of its invaluable characteristic ability to function as both a plastic material and as a fiber.

Polypropylene had been used in a number of applications across a range of industries. Perhaps the most interesting example to include in the living hinge development. Polypropylene is a very flexible, soft material with a relatively low melting point.

As a matrix material, PP is widely used because it has some excellent characters for composite fabrication because of its suitability for filling, reinforcing, and blending. PP with natural fibrous polymers is one of the most promising routes to create natural–synthetic polymer composites.

4.2.2 ADVANTAGES OF POLYPROPYLENE

- Polypropylene is readily available and inexpensive.
- Polypropylene has high flexural strength due to its semi-crystalline nature.
- Polypropylene has a relatively slippery surface.
- Polypropylene is resistant to absorbing moisture.
- Polypropylene has good chemical resistance over a wide range of bases and acids.
- Polypropylene possesses good fatigue resistance.
- Polypropylene has good impact strength.
- Polypropylene act as good insulators.

4.2.3 DISADVANTAGES OF POLYPROPYLENE

- Polypropylene has a high thermal expansion coefficient which limits its high-temperature applications.
- Polypropylene is susceptible to UV degradation.
- Polypropylene has poor resistance to chlorinated solvents and aromatics.
- Polypropylene is highly flammable and susceptible to oxidation.

Despite its shortcomings, polypropylene is a great material overall. It has a unique blend of qualities that aren't found in any other material which makes it an ideal choice for this project.

4.3 JUTE FIBER

Jute is a bast fiber used for sacking, burlap, and twine as a backing material for tufted carpets. It is a long, soft, shiny fiber that can be spun into coarse, strong threads. It is one of the cheapest natural fibers and is second only to cotton in amount produced and variety of uses. Jute fiber has some unique physical properties like high tenacity, bulkiness, sound & heat insulation property, low thermal conductivity, antistatic property etc. Due to these qualities, jute fiber is more suited for the manufacture of technical textiles in certain specific areas. Moreover, the image of jute as a hard and unattractive fiber does not affect its usage in technical textiles.

Jute is 100% biodegradable and thus environment- friendly. It is available in India at competitive prices. Now it is not just a major textile fiber, but also a raw material for non-textile products, which helps to protect the environment, which is an integral part of any development planning.

Jute fibers are always known as strong, coarse, environment-friendly, and organic. The use of jute was primarily confined to marginal and small manufacturers and growers, but now it is used as important raw materials for several industries. It is unfortunate that jute still lags behind other fibers like silk, wool, and cotton. However, at present time, jute is termed as a favorite fabric for packaging materials and furnishings and as golden fibers for the national and international fashion world. Jute fibers are used for making mats, gunny cloth, cordage, hangings, paper, and decorative articles. Prevalent uses of jute in handicraft stuff, in order to give an aesthetic appeal, have made it popular across the globe.

Jute fibers are composed primarily of the plant materials cellulose, lignin, and pectin. Both the fiber and the plant from which it comes are commonly called jute. It belongs to the genus Corchorus in the basswood family, Tiliaceae.

4.3.1 PROPERTIES OF JUTE

Jute has little elasticity and is a very breathable fabric. Contrary to most textile fibers, which consist mainly of cellulose, jute fibers also include lignin. Usually found in wood fibers, lignin brings extra strength and durability, making it ideal for making hard-working rugs and doormats.

The chemical composition of jute fiber: Cellulose 58–63%, Hemicellulose 21–24%, Lignin 12–14%, Wax 0.4–0.8%, Pectin, 0.2–0.5%, Protin 0.8–2.5%, Mineral matter 0.6–1.2%.

4.3.2 APPLICATION OF JUTE FIBER

Jute is the second most important vegetable fiber after cotton; not only for cultivation but also for various uses.

Jute is used chiefly to make cloth for wrapping bales of raw cotton and to make sacks and coarse cloth.

- The fibers are also woven into curtains, chair coverings, carpets, area rugs, hessian cloth, and backing for linoleum.
- While jute is being replaced by synthetic materials in many of these uses, some uses take advantage of jute's biodegradable nature, where synthetics would be unsuitable.
- Jute butts, the course ends of the plants, are used to make inexpensive cloth.
- Traditionally jute was used in traditional textile machinery as textile fibers having cellulose (vegetable fiber content) and lignin (wood fiber content). But, the major breakthrough came when the automobile, pulp, and paper, and the furniture and bedding industries started to use jute and its allied fibers with their non-woven and composite technology to manufacture nonwovens, technical textiles, and composites.
- Jute can be used to create a number of fabrics such as Hessian cloth, sacking, scrim, carpet backing cloth (CBC), and canvas.
- Hessian, lighter than sacking, is used for bags, wrappers, wall-coverings, upholstery, and home furnishings.
- Diversified jute products are becoming more and more valuable to the consumer today. Among these are espadrilles, floor coverings, home textiles, high-performance technical textiles, Geotextiles, composites, and more.
- Jute is also used in the making of ghillie suits which are used as camouflage and resemble grasses or brush.

Thus, jute is the most environment-friendly fiber starting from the seed to expired fiber, as the expired fibers can be recycled more than once. Considering all the above properties braided jute mat have been used as reinforcement in this project, for better anisotropic property to the composite

4.4 SPENT COFFEE GROUNDS

4.4.1 INTRODUCTION

Coffee production in India is found to be dominated in the hill tracts of southern state, Karnataka accounting 53% followed by Kerela 28% and Tamil Nadu 11% of the production of 8,200 tonnes. Amongst the coffee production in different countries Indian coffee is said to be the finest coffee grown in the shade rather than direct sunlight anywhere in the world.

Spent coffee ground (SCG) contains large amounts of organic compounds (i.e., fatty acids, amino acids, polyphenols, minerals, and polysaccharides) that justify its valorization. Earlier innovation explored the extraction of specific components such as oil, flavor, terpenes, and alcohols as value-added products. However, by-products of coffee fruit and bean processing can also be considered as potential functional ingredients for the food industry. There is an urgent need for practical and innovative ideas to use this low-cost SCG and exploit its full potential increasing the overall sustainability of the coffee agro-industry. Spent coffee ground polypropylene composite had been prepared by injection molding for my earlier investigation and its mechanical, thermal, and morphological properties were studied.

4.4.2 SCOPE AND OBJECTIVE

1. Utilization spent coffee ground polypropylene biocomposites for the preparation of the jute sand witched composites through compression molding.
2. To evaluate and report the efficiency of jute as reinforcement in the spent coffee ground biocomposite by testing its mechanical properties.
3. To study the resistance to various chemicals, this is essential for their acceptance as marketable products.

4.4.3 MATERIALS AND METHOD

The commercially available and industrially important polypropylene was used in the present study as a matrix material. Homo-polymer grade polypropylene (Repol HIIO) supplied by Reliance Industries and spent coffee grounds were collected from coffee day shops was used for the study. Renewable reinforcing filler materials like netted jute fibers collected from the local market. Chemicals like concentrated hydrochloric acid, 20% Sodium Hydroxide, and 20% Glacial acetic acid of analytical grade were supplied by Sigma Aldrich.

4.5 COMPRESSION MOLDING

Specifically designed to facilitate the replacement of metal components with polymers (and other composites), the compression molding process is a method of molding in which a preheated polymer is placed into an open, heated mold cavity. The mold is closed with a top plug and pressure is applied to force the material to contact all areas of the mold. Throughout the process heat and pressure are maintained until the polymer has cured.

While the compression molding process can be employed with both thermosets and thermoplastics, today most applications use thermoset polymers. Advanced composite thermoplastics can also be compression molded with unidirectional tapes, woven fabrics, randomly orientated fiber mat or chopped strand. Compression molding is a high-volume, high-pressure plastic molding method that is suitable for molding complex, high-strength objects. The compounded and injection molded coffee composite along with braided jute fibers were set for compression molding at a temperature ranging from 180–200 degree Celsius. Hydraulic presses are usually employed to provide the pressure which may range from 20–30 *MPa* or even higher up to 80 *MPa* time required to harden the mold piece ranges from 1–15 minutes. The cooling time for the mold is about 2 hours.

4.6 CHEMICAL TESTING

The molded samples were dipped in various chemicals like 20% Concentrated Hydrochloric acid, 20% Sodium Hydroxide and 20% Glacial acetic acid and the effect of chemicals over the composites were analyzed (Figures 4.1 and 4.2)

FIGURE 4.1 Molded SCGPP composites.

FIGURE 4.2 Chemical testing of molded SCGPP.

4.7 RESULTS AND DISCUSSION

4.7.1 EFFECT OF BRAIDED JUTE FIBER ON TENSILE AND IMPACT PROPERTIES OF SCG/PP COMPOSITES

Table 4.1 shows the effect of braided jute SCGPP composites. It was found that there is a drastic decrease in Tensile strength of the braided jute and SCG/PP Composite and controversially there is a drastic increase in the impact strength compared to the neat polypropylene

TABLE 4.1 Tensile and Impact Strength of Jute Reinforced Spent Coffee Grounds Composites

Serial no.	Sample	Tensile strength (MPa) ASTMD 638	Impact strength (KJ/m$^{2)}$ ASTMD 256
1	Neat Polypropylene (PP)	24.95	4.12
2	Jute/ SCGPP	8.78	15.2

In view of the above results, it can be noted that sandwiching chemically treated jute fiber between two molded spent coffee ground sheets had decreased the tensile strength drastically. It may be inferred due to the reason that the polar groups present in the composite had an interfacial reaction with the coupling agent and lack of polar groups from the matrix remained insufficient for the jute cellulose to be binded strongly in spite of chemical treatment (Figure 4.3).

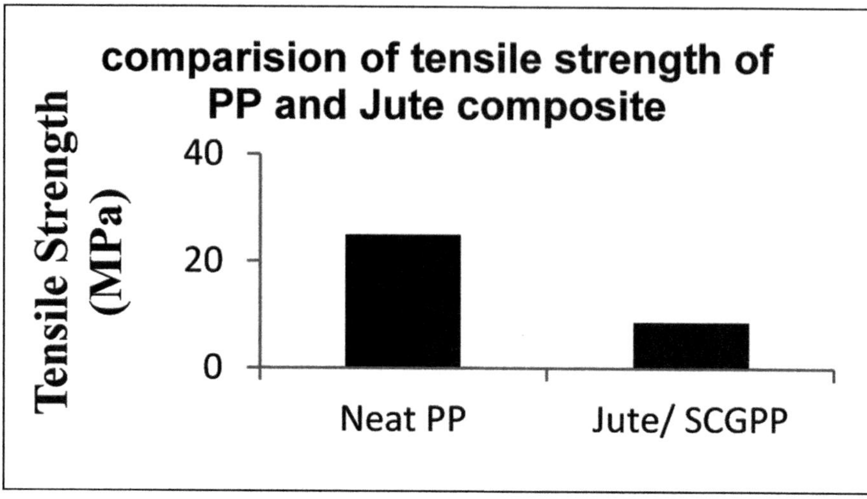

FIGURE 4.3 Tensile strength of Neat PP and Jute/ SCGPP composites.

From the above results, it is observed that there is a tremendous increase in impact strength due to the increase in stiffness of the braided Jute fibers sandwiched in between two composite layers, the anisotropic property of jute fibers may be the reason for its high impact strength thereby decreases in the relative deformation of the Jute/SCGPP composite material (Figure 4.4).

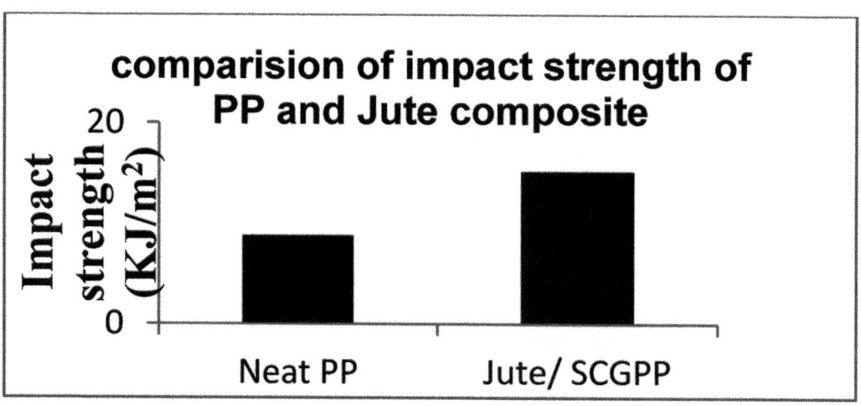

FIGURE 4.4 Impact strength of Neat PP and Jute/SCGPP composites.

4.7.2 EFFECT OF CHEMICALS ON BRAIDED JUTE FIBER SCG/PP COMPOSITES

The effect of chemicals on jute sandwiched SCG/PP composites like concentrated Hydrochloric acid (HCl), Sodium hydroxide (NaOH) and Glacial acetic acid (CH_3COOH) were noted. On careful investigation, it can be seen that Glacial acetic acid had more leaching effect on the composite as compared to HCl and NaOH. This shows that the Jute/SCGPP composites are highly resistant to strong acid and bases which makes it applicable for Laboratory and Marine applications (Figure 4.5).

FIGURE 4.5 **(See color insert.)** Samples after chemical treatment.

4.8 CONCLUSIONS

Jute sandwiched SCG/PP composites has:

- decreases the tensile strength;
- increases the impact strength;
- high impact strength of the composites makes it applicable in the automobile and machinery parts. Few molds had been tried as shown in Figure 4.6;
- jute/SCGPP composites is highly resistant to strong acid and bases which makes it applicable for laboratory and marine applications.

FIGURE 4.6 Compression molded sheets of Jute/ SCGPP samples.

KEYWORDS

- **chemical properties**
- **coffee grounds**
- **jute**
- **mechanical properties**
- **polypropylene**

REFERENCES

1. Allenberger, F. T. W., & Weston, N., (2004). *Natural Fibers, Plastics, and Composites Natural.* Materials sourcebook from C.H.I.P.S. Texas.
2. Narayan, R., (1992). *Biomass (Renewable) Resources for Production of Materials.* Chemicals, and fuels-A paradigm Shqt. ACS Symp. Ser. 476.1.
3. Abdul, K. H. P. S., Siti, A. M., Rizuan, R., Kamarudin, H., & Khairul, A., (2008). *Polymer Plastic Tech. Engg., 47,* 237.
4. Mwaikambo, L. Y., & Ansell, M. P., (2002). Chemical modification of hemp, sisal, jute, and kapok fiber by alkalization. *J. Applied Polymer Science, 84*(12), 2222–2234.
5. Arrakhiz, F. Z., El Achaby, M., Malha, M., et al., (2013). Mechanical and thermal properties of natural fibers reinforced polymer composites: Doum/low-density polyethylene, *Materials & Design, 43,* 200–205.
6. Di Bella, G., Fiore, V., Galtieri, G., Borsellino, C., & Valenza, A., (2014). Effects of natural fibers reinforcement in lime plasters (kenaf and sisal vs. polypropylene). *Construction and Building Materials, 58,* 159–165.

7. Ku, H., Wang, H., Pattarachaiyakoop, N., & Trada, M., (2011). A review on the tensile properties of natural fiber-reinforced polymer composites, *Composites Part B: Engineering, 42*(4), 856–873.
8. Norul, I. M. A., Paridah, M. T., Anwar, U. M. K., Mohd, N. M. Y., & H'Ng, P. S., (2013). Effects of fiber treatment on morphology, tensile, and thermogravimetric analysis of oil palm empty fruit bunches fibers, *Composites Part B: Engineering, 45*(1), 1251–1257.
9. Tawakkal, I. S. M. A., Cran, M. J., & Bigger, S. W., (2014). Effect of kenaf fiber loading and thymol concentration on the mechanical and thermal properties of PLA/kenaf/thymol composites. *Industrial Crops and Products, 61*, 74–83.
10. Ruys, D., Crosky, A., & Evans, W. J., (2002). Natural fiber structure. *Int. J. Mater Product Technol., 17*(1–2), 2–10.
11. Kandachar, P., & Brouwer, R., (2002). Applications of bio-composites in industrial products. *Mater Res. Soc. Symptoms Process, 702*, 101–112.

CHAPTER 5

PROCESSING AND MECHANICAL CHARACTERIZATION OF SHORT BANANA-JUTE HYBRID FIBER-REINFORCED POLYESTER COMPOSITES

SIVA BHASKARA RAO DEVIREDDY[1] and
SANDHYARANI BISWAS[2]

[1]*Department of Mechanical Engineering, St. Ann's College of Engineering and Technology, Chirala, Andhra Pradesh, 523187, India, E-mail: sivabhaskararao@gmail.com*

[2]*Department of Mechanical Engineering, National Institute of Technology, Rourkela, Odisha, 769008, India*

ABSTRACT

Natural fibers such as jute, cotton, kenaf, flax, banana, oil palm, sisal, and pineapple leaf fibers have been in considerable demand in recent years due to its eco-friendly and renewable nature. In the present work, the mechanical characteristics of short randomly oriented banana-jute hybrid fiber-reinforced polyester composites were studied as a function of the overall fiber loading (0, 10, 20, 30, and 40 wt.%) and different weight ratios of banana and jute fibers (1:1, 1:3, and 3:1). The hybrid composites are fabricated using a hand lay-up technique and experimental tests are carried out as per ASTM standards to find out the tensile modulus and strength, flexural modulus and strength, impact energy and micro-hardness properties. The incorporation of both natural fibers into polyester matrix resulted in an increase in mechanical properties up to 30 wt.% of fiber loading. The tensile strength of neat polyester composites increases by 125%, 171%, and 106% with the addition of 30 wt.% of fiber loading with a weight ratio of banana and jute as 1:1, 1:3, and 3:1 respectively. Similarly, the increase in flexural strength about 83%, 98%, and 67% was observed as the fiber loading in polyester increased from 0 to 30 wt.% with a weight ratio of banana and jute as 1:1, 1:3, and 3:1, respectively. The energy absorbed by the hybrid composite due to impact load is 3.17, 3.03 and 3.39 times of pure polyester resin for hybrid composites with the weight ratio of banana and jute 1:1, 1:3, and 3:1, respectively at 40 wt.% of fiber loading. With the addition of 40 wt.% of fiber loading, the micro-hardness polyester composites improve to 113%, 123% and 97% with corresponding weight ratios of banana and jute as 1:1, 1:3, and 3:1, respectively. The interfacial analysis was also carried out with the help of a scanning electron microscope to study the microstructural behavior of the tested specimen. The experimental investigation reveals that the mechanical properties show their maximum values with weight ratio banana and jute as 1:3.

5.1 INTRODUCTION

Composites materials consist of one or more discontinuous phases embedded in a continuous phase. The discontinuous phases are the principal load carrying members while the surrounding continuous phase helps to keep them in their desired locations, prevent them from environmental damages, and also acts as stress transfer medium [1]. The demand for the composite materials is increasing day-by-day in buildings, automotive parts, aerospace elements, marine structures and structural members in order to improve the energy efficiency. Due to increasing interest in environmental concerns, the engineers and scientists have turned over on utilizing natural fibers as reinforcement material in the polymer matrix to produce fiber-reinforced polymer composites for structural, building, and other needs. These fibers are obtained from the natural resources having many advantages such as low cost, low density, environmentally friendly, lightweight, good thermal insulation, and high specific mechanical performance [2]. The natural fibers like kenaf, coir, banana, hemp, sisal, jute, vakka, flax, and pineapple leaf have shown better mechanical properties with a thermoset matrix, thus attracting the attention of researchers and material scientists for the application in civil structures, furniture, consumer goods, food packaging, and automotive components [3, 4]. Tensile modulus and strength, flexural modulus and strength, inter-laminar shear strength, impact strength, and hardness are the important mechanical properties of the fiber-reinforced polymer composites. Most of the studies reveal that the mechanical properties of natural fiber-reinforced composites are affected by a number of parameters such as fiber loading, fiber orientation, fiber aspect ratio, fiber dispersion, fiber-matrix adhesion, fiber geometry and stress transfer at the interface [5]. A great deal of work has already been done on the effect of various factors on the mechanical behavior of natural fiber-reinforced polymer composites [6–12].

The conventional composites generally possess only one type of reinforcement and termed as nanocomposites. Hybridization with more than one fiber type in the same matrix provides another dimension to the potential versatility of fiber-reinforced composite materials [13]. Many researchers in the past have studied the mechanical performance of hybrid composites using synthetic-natural fibers such as banana/glass [14], jute/glass [15], sisal/glass [16], oil palm/glass [17], pineapple/glass [18], kenaf/glass [19], and sugar palm/glass [20]. From these studies, they concluded that a hybrid composite offers better resistance to water absorption, cost reduction, weight saving, and increased modulus. The tensile properties of composites mainly depend on the modulus and strength of fibers, the chemical stability and strength of the resin and the bonding between the fibers and matrix in transferring stress across the interface [21]. Khanam et al. [22] evaluated the effect of fiber length on the tensile strength of short sisal/silk fiber-reinforced hybrid composites and observed that the tensile strength is slightly higher for the 20 mm fiber length based composites as compared to the 10 mm and 30 mm fiber length based composites. Kiran et al. [23] evaluated the tensile strength of natural fibers like sun hemp, sisal, and banana fiber-reinforced polyester composites and found that the tensile strength of 30 mm fiber length based composites increased gradually from 0 wt.% to 55 wt.% of fiber and then there is a drop in tensile strength. Athijayamani et al. [24] investigated the tensile strength of Roselle/sisal fiber-reinforced polyester hybrid composites at different fiber lengths and weight ratio and concluded that the tensile strength increased with the fiber loading and fiber length. Kumar et al. [25] prepared tri-layer hybrid fiber-reinforced polyester composites based on sisal and coconut sheath fibers with six different stacking sequences and observed that the tensile strength of hybrid composite having coconut sheath as skin and sisal as a core material is slightly higher than the other stacking sequences.

Flexural properties are also another important mechanical property for any material. In flexural testing, different mechanisms such as compression, tension, and shearing take place instantaneously [26]. Khanam et al. [22] studied the effect of the fiber length on the flexural strength of short sisal/silk fiber-reinforced hybrid composites and found that 20 mm fiber length based composites exhibit higher flexural strength than 10 mm and 30 mm based composites. Sana et al. [27] investigated the effect fiber weight ratio on the flexural behavior of Typha fiber-reinforced polyester composites and found the maximum flexural modulus and strength of 6.16 GPa and 69.8 MPa, respectively at 12.6% fiber weight ratio. Tao et al. [28] studied the flexural strength of short jute/PLA composites and reported an increase in flexural strength at fiber content up to 30%. Mohanty et al. [29] investigated the flexural strength of jute-polyester amide composites and reported an increase in flexural strength for fiber loading from 20 to 32 wt.%. Impact strength shows the ability of the material to absorb impact energy. The fiber reinforcement plays an important role in the impact resistance of the composite materials as they interact with the crack formation in the matrix and act as stress transferring medium [30]. Romanzini et al. [31] evaluated the impact strength of glass-ramie fiber-reinforced polyester composites at 10, 21 and 31 vol.% of fiber content and found that the hybrid composites showed higher impact strength than the pure polyester resin. Öztürk et al. [32] studied the impact behavior of jute/Rockwool reinforced phenol formaldehyde (PF) composites by varying fiber loadings (16, 25, 34, 42, 50, and 60 vol.%) and found that the impact strength of jute/PF composite increased with jute fiber loading up to 50 vol.%. An attempt has been made by Kumar et al. [25] to examine the impact strength of coconut sheath (CS) and sisal (S) fiber-reinforced hybrid composites and found that the impact strength of hybrid composites is higher than the pure coconut sheath and sisal fiber based composites. Also, reported that the hybrid composite with CS/S/CS sequence shows better impact strength than composites with other sequences.

Hardness is defined as the resistance of a material to permanent deformation such as scratch or indentation. Kumar et al. [16] evaluated the effect of fiber length on the hardness of sisal-glass fiber-reinforced epoxy based hybrid composites using Rockwell hardness tester and found that the 20 mm fiber length based composites had a higher hardness than 10 and 30 mm fiber length based composites. Rezaei et al. [33] evaluated the effect of fiber length and fiber content on the hardness of short carbon fiber-reinforced polypropylene composite and reported that the hardness of composite increases with carbon fiber and it is relative to fiber loading and modulus of the composite. Srinivasa and Bharath [34] studied the hardness of areca fiber-reinforced epoxy composites and found that the incorporation of areca fibers inside epoxy increases the hardness of composites. Reddy et al. [35] determined the hardness of the short uniaxially oriented kapok/glass polyester hybrid composites by keeping the volume ratios of kapok to glass at 3:1, 1:1, and 1:3 at constant fiber volume percentage of 9 vol% and found that the maximum hardness of kapok/glass composites is observed at 1:3 volume ratio of kapok and glass fiber. The micro-hardness of the bagasse/sugarcane fiber-reinforced unsaturated polyester composites is studied by Oladele [36]. The investigation revealed that the micro-hardness of composite increases due to adequate wetting and bonding between the sugarcane fiber and the polyester. Among various natural fibers, jute, and banana fibers have the potential to be used as reinforcement in polymer composites which is abundantly available in countries like India, Sri Lanka, and some of the African countries but are not optimally utilized [37]. Jute is an important agro-fiber which has gained worldwide attention as a potential material for polymer reinforcement due to its natural properties such as high tensile modulus, low density, and low elongation at break and its specific stiffness and strength. Banana fiber is a waste product of banana cultivation. Hence, without any additional cost, these fibers can be obtained in bulk quantity and used for industrial purposes. The literature survey on hybrid composites indicated that no work has been reported on polyester-based hybrid composites of short banana and jute fibers. This research work attempts to explore the potential utilization of short jute fibers as composite reinforcement with a combination of locally available banana fibers as reinforcement in the polyester matrix. The present work deals with the mechanical characterization of short randomly oriented banana-jute fiber-reinforced polyester composites at different fiber loadings as well as by varying the weight ratio of banana and jute as 1:1, 1:3, and 3:1. Fiber and matrix interaction was studied by observing the fracture surfaces of the hybrid composites using a scanning electron microscope.

5.2 MATERIALS AND METHODS

5.2.1 MATERIALS DESCRIPTION

The unsaturated polyester resin of grade ECMALON 4411 is used as matrix material for the present investigation. For a proper chemical reaction, methyl ethyl ketone peroxide and cobalt naphthenate are used as catalyst and accelerator respectively. The resin and the corresponding catalyst and accelerator are supplied by Ecmass resin (Pvt) Ltd., Hyderabad, India. Banana (Musa sapientum) and jute (Corchoruscapsularis) fibers are used as reinforcement materials for fabricating the composite specimen. Banana fiber has been obtained from V. K Enterprise, Gujarat, and jute fiber has been obtained from the local supplier. Figure 5.1(a) and (b) shows the short banana fiber and jute fiber, respectively. The intrinsic properties of constituent materials are given in Table 5.1.

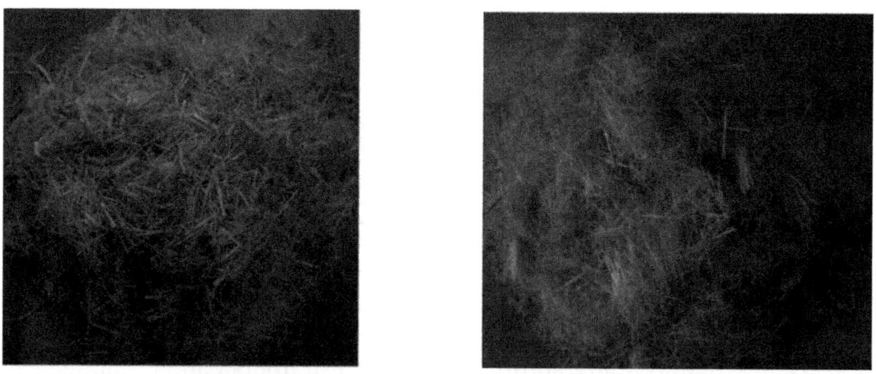

(a) (b)

FIGURE 5.1 (See color insert.) Short (a) banana fiber and (b) jute fiber.

5.2.2 FABRICATION OF HYBRID COMPOSITE

The fabrication of the composite slabs is done by conventional hand lay-up technique followed by light compression molding technique. Banana and jute fibers were cut into 15 mm length. Keeping the weight ratio of banana and jute as 1:1, 1:3, and 3:1 short randomly oriented intimately mixed hybrid composites were fabricated at different fiber weight percentages (0, 10, 20, 30, and 40 wt.%). The polyester resin is mixed with 1.5 wt.%

of the catalyst and 1.5 wt.% of the accelerator for cure initiation. Banana and jute fibers were pre-impregnated with the matrix material. For quick and easy removal of the composite material, a mold release sheet and mold release spray are used on the top and bottom of the wooden mold. Care was taken to avoid the formation of air bubbles during preparation. The pressure was applied from the top and then the mold was allowed to cure at room temperature for 48 hr. After 48 hr, the samples were taken out of the mold and cut into required size by a diamond cutter for calculating mechanical properties.

TABLE 5.1 Physical, Mechanical, and Chemical Composition of Constituent Materials [38]

Property	Banana fiber	Jute fiber	Polyester
Density (g/cm^3)	1.35	1.4	1.13
Diameter (μm)	80–250	25–120	-
Tensile strength (MPa)	529–759	393–773	21
Tensile modulus (GPa)	6.2 ± 0.87	8.5 ± 0.59	1.48
Elongation (%)	1–3.5	1.5–1.8	2.5
Cellulous content (%)	62–64	61–71	–
Hemicelluloses (%)	10–24	14–20	–
Lignin content (%)	5	12–13	–
Moisture content (%)	10–12	–	–
Lumen size (μm)	5	3.40	–
Microfibrillar angle (°)	12	8	–

5.2.3 MECHANICAL CHARACTERIZATION

5.2.3.1 TENSILE TEST

The tensile properties of the composites are an important indicator of the material's behavior under loading in tension. The static uniaxial tensile test is most widely used to determine the modulus of elasticity and tensile strength of the material. The testing process involves placing the test specimen in the testing machine and applying tension to it until it fractures. In the case of composite material, the tensile test is generally performed on flat specimens. The ASTM standard test for tensile properties of fiber-reinforced composites has the designation D3039–76. The dimension of the specimen is 153 mm × 12.7 mm × 4 mm and a span length of 70 mm was employed. In the present work, the tensile test is performed by universal testing machine Instron 1195 at a crosshead speed of 10 mm/min. The tensile modulus and tensile strength of the hybrid composites were determined from the stress-strain curves. For each composite type, three identical specimens were tested and the mean value is reported.

5.2.3.2 FLEXURAL TEST

The three-point bend test is carried out to obtain the flexural properties of all the hybrid composites. A flexural test imposes compressive stress on the concave side and tensile stress on the convex side of the composite specimen which causes a shear stress along the center line. The specimen bends and fractures when the load is applied at the middle of the beam. The flexural test was carried out on rectangular specimens of composite samples using universal testing machine Instron 1195 according to the procedure described in ASTM D790. The dimension of the specimen is 100 mm × 12.7 mm × 4 mm and a loading span of 40 mm was employed. The test was conducted using the load cell of 10kN at 10 mm/min rate of loading. For each composite type, three identical specimens were tested and the mean value is reported as the property of that composite. The flexural strength of the composite specimen is determined using the following Eq. (5.1).

$$\frac{3PL}{2bt^2} \tag{5.1}$$

where P is the maximum load at failure (N), L is the span length of the sample (mm), and b and t are the width and thickness of the specimen (mm), respectively. Flexural modulus was calculated from Eq. (5.2).

$$\frac{mL^3}{4bt^3} \tag{5.2}$$

where m is the slope of the tangent to the initial straight-line portion of the load-deflection curve (N/mm).

5.2.3.3 IMPACT TEST

The energy required to fracture a standard test piece under an impact load is called impact energy. Low velocity instrumented impact tests are carried out on the hybrid composite specimens. The test is conducted as per ASTM standard has the designation D 256 by using Charpy impact tester supplied by VEEKAY test lab, India. The dimension of the specimen is 64 × mm 12.7 × mm 4 × mm and the depth of the notch is 2 mm at 45° angle on one side at the center. The impact-testing machine determines the notch impact strength of the material by shattering the V-notched specimen with a pendulum hammer, measuring the spent energy and relating it to the cross-section of the specimen. The impact energy absorbed in breaking the hybrid composite specimen is recorded directly on the dial indicator.

5.2.3.4 MICRO-HARDNESS

Hardness is a measure of material's resistance to permanent deformation or damage. In the present study, a Vickers hardness tester is used to measure the micro-hardness of the hybrid composite samples as per ASTM D785 test standards. A diamond indenter, in the form of a right pyramid with a square base and an angle of 136° between opposite faces, is forced into the material under a load. The two diagonals X and Y of the indentation left on the surface of the material after removal of the load are measured and their arithmetic means L is calculated. In the present study, the load is considered F = 24.54 N and Vickers hardness number is calculated using the following Eq. (5.3).

$$H_v = 0.1889\frac{F}{L^2} \tag{5.3}$$

and

$$L = \frac{X+Y}{2}$$

where F is the applied load in N, L is the diagonal of a square impression in mm, X is the horizontal length in mm and Y is the vertical length in mm. For each composite type, three identical specimens were tested and the mean value is reported as the property of that composite.

5.2.3.5 SCANNING ELECTRON MICROSCOPY

The surface morphology of the hybrid fiber-reinforced composites has been studied using a scanning electron microscope JEOL JSM–6480LV. The composite samples are mounted on stubs with silver paste. To improve the penetration of light and for better surface micrographs, the thermal conductivity of the samples is enhanced by vacuum-evaporated a thin film of platinum onto them before the photomicrographs are taken.

5.3 RESULTS AND DISCUSSION

5.3.1 TENSILE PROPERTIES OF HYBRID COMPOSITES

The tensile strength of the short banana-jute fiber-reinforced polyester based composites with different fiber loading and weight ratio is shown in Figure 5.2. The result revealed that with the addition of fibers up to 30 wt.%, a gradual increase in tensile strength is observed. This may be due to the proper adhesion between the fiber and the matrix and the reinforcement imparted by the fibers which allow stress transfer from the matrix to the fibers. However, a further increase in fiber loading, i.e., 40 wt.%, there is a decrease in the tensile strength. The reason may be due to the fact that the matrix content is highly reduced by the accumulation of excess fibers in the hybrid composites. Moreover, the possibility of fiber entanglements and agglomeration results in the hybrid composite that leads to a decrease in stress transfer between the matrix and the fiber. The tensile strength of neat polyester is 20.7 MPa which is increased to 46.6 MPa, 56.2 MPa, and 42.8 MPa with the addition of banana and jute fibers of 30 wt.% with weight ratio as 1:1, 1:3, and 3:1, respectively. Hybrid composites with a weight ratio of banana and jute fiber as 1:3 show higher tensile strength irrespective of fiber loading and matrix types. The surface area of the fiber in a unit area of the composite is higher in a jute fiber than that of a banana fiber because the diameter of jute fiber is less than that of banana fiber. Hence, stress-transfer in the unit area, as well as physical interaction, is higher in the case of weight ratio as 1:3 of banana-jute fiber-based composites [39].

The tensile modulus of natural fiber-reinforced hybrid composites depend on many parameters such as the type of reinforcement, type of matrix, fiber length, fiber orientation, fiber strength and modulus, aspect ratio and the interfacial bonding between fiber and matrix [40]. The influence of fiber loading and weight ratio on the tensile modulus of the banana-jute fiber-reinforced polyester composites are shown in Figure 5.3. It is clearly visible that with the increase in fiber loading, the tensile modulus of the composites also increases. Similar observations have also been observed in case of randomly oriented short banana and sisal hybrid composites by Idicula et al. [41]. The brittle nature of polymer matrix decreases with the addition of banana and jute fibers and therefore elastic modulus increases with fiber loading. The maximum tensile modulus of the composite is found to be 2.506 GPa, 2.648 GPa, and 2.342 GPa at 40 wt.% of fiber loading with a weight ratio of banana and jute fiber as 1:1, 1:3, and 3:1, respectively. The percentage increment in elastic modulus of the composites with a weight ratio of banana and jute fiber as 1:1, 1:3, and 3:1 is found to be 69.32%, 78.91%, and 58.24%, respectively, over the neat polyester.

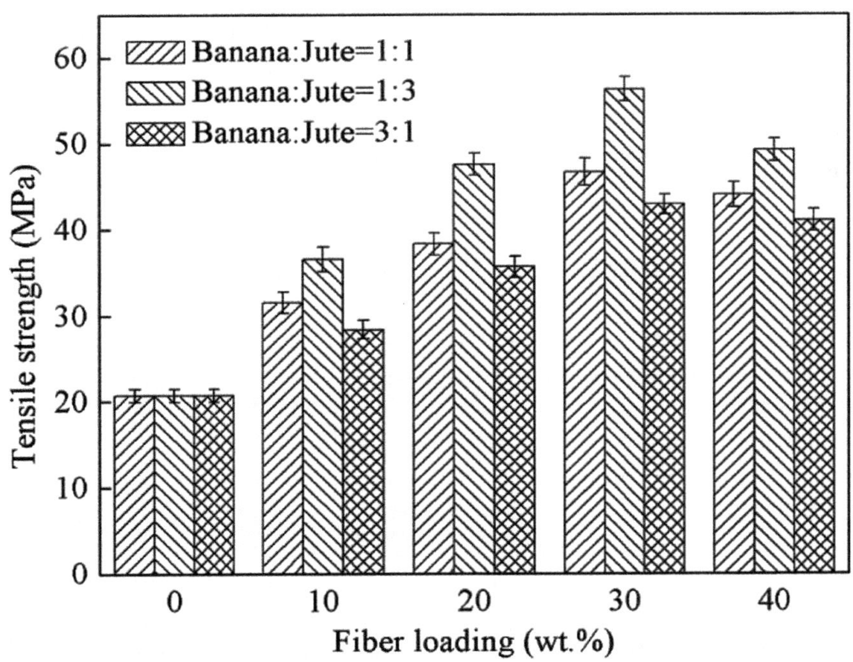

FIGURE 5.2 Effect of fiber loading and weight ratio on tensile strength of hybrid composites.

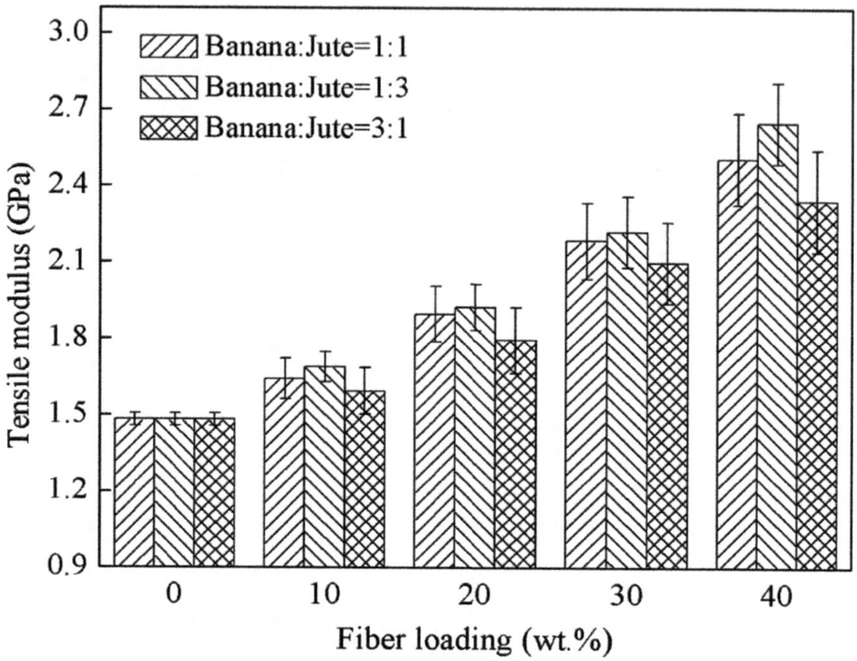

FIGURE 5.3 Effect of fiber loading and weight ratio on tensile modulus of hybrid composites.

5.3.2 FLEXURAL PROPERTIES OF HYBRID COMPOSITES

The effect of fiber loading on the flexural strength and flexural modulus of short banana-jute fiber-reinforced polyester composites are shown in Figures 5.4 and 5.5, respectively. It is observed that the flexural properties of all hybrid composites considered in the present study increase with the increase in fiber loading up to 30 wt.%. However, further increase in fiber loading the properties decreases. The decrease in flexural properties at higher weight percentages of fiber may be due to weak interfacial bonding, fiber agglomeration, increased fiber-to-fiber interactions and also dispersion problem [11]. According to Mohanty et al. [29], the optimum fiber loading varies with the fiber aspect ratio, nature of fiber and matrix, fiber-matrix adhesion, etc. The increase in flexural strength of 83%, 98%, and 67% is observed as the fiber content in polyester increased from 0 to 30 wt.% fiber loading with a weight ratio of banana and jute fiber as 1:1, 1:3, and 3:1, respectively. Similarly, the flexural modulus of the neat polyester matrix is found to be 1.61 GPa and is increased by 78%, 91%, and 70% for polyester based composites at 30 wt.% of fiber loading with weight ratio of banana and jute fiber as 1:1, 1:3, and 3:1, respectively. The high strength jute fiber layers are able to bear the applied tensile and compressive stresses subjected on the hybrid composites. This leads to an increase in the flexural properties of the hybrid composites with a weight ratio of banana and jute fiber as 1:3. However, the flexural properties of the composites also influenced by various other parameters such as the degree of crystallinity, the porosity content, and the size of lumen Bledzki and Gassan [42].

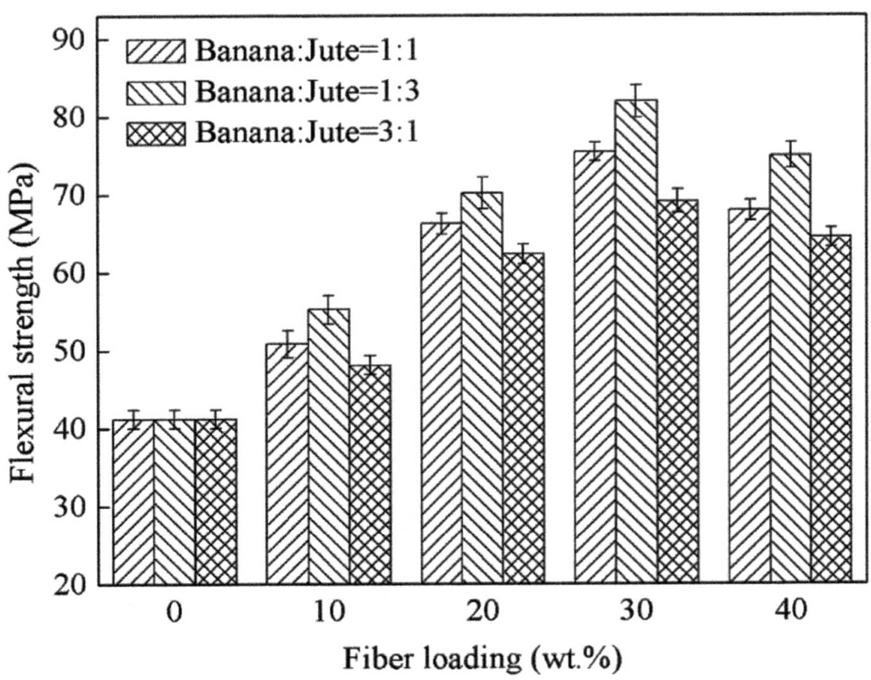

FIGURE 5.4 Effect of fiber loading and weight ratio on flexural strength of hybrid composites.

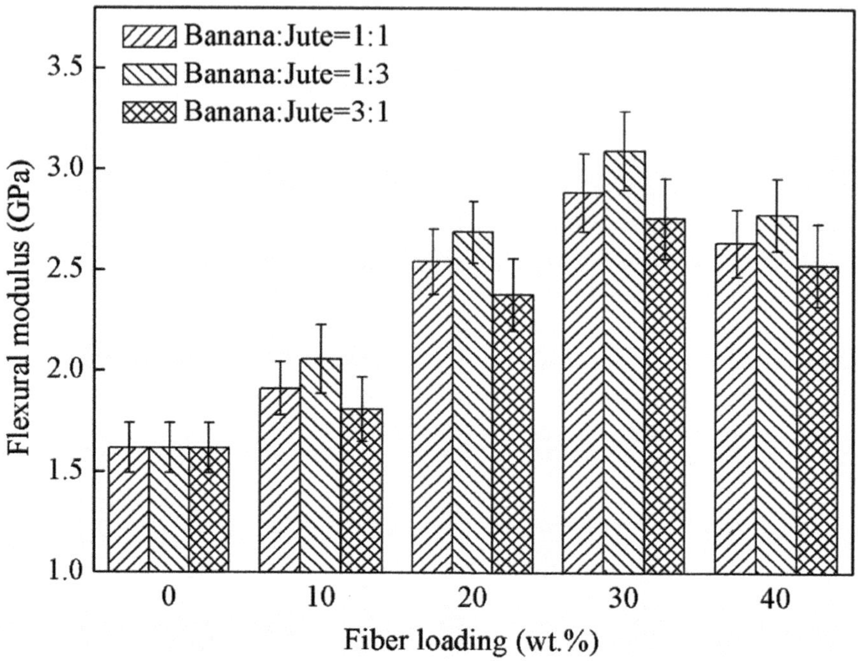

FIGURE 5.5 Effect of fiber loading and weight ratio on flexural modulus of hybrid composites.

5.3.3 *IMPACT ENERGY OF HYBRID COMPOSITES*

The effect of fiber loading and weight ratio on the impact energy of the short banana-jute fiber-reinforced polyester based composites is shown in Figure 5.6. It is evident from the Figure 5.6 that the impact energy increases with the increase in fiber loading, and the maximum impact energy value is observed at the fiber loading of 40 wt.%. The high impact energy of the hybrid composites attributed to the extra energy dissipation mechanism. At lower fiber loading, short banana and jute fibers are embedded in the resin and, hence, fibers pull out and fiber breakage occurs in the application of a sudden load. At higher fiber loading, the contact area between the reinforcement and matrix will increase [43]. The fiber content in polyester increased from 0 to 40 wt.%, the energy absorbed by the hybrid composite is found to be 202%, 181%, and 209% higher than that of pure polyester resin for hybrid composites with the weight ratio of banana and jute fiber as 1:1, 1:3, and 3:1, respectively. It is observed that the impact energy of the composites with a weight ratio of banana and jute fiber as 3:1 is more compared to that of the composites with a weight ratio of 1:1 and 1:3. Generally, the natural fibers having a high microfibrillar angle shows a higher composite fracture-toughness than those with a small microfibrillar angle. The microfibrillar angle of banana fiber is 11° which is more than the jute fiber microfibrillar angle i.e., 8°. Therefore, the increase in impact energy of the composites with more percentage of banana fiber is obtained may be due to the above reason.

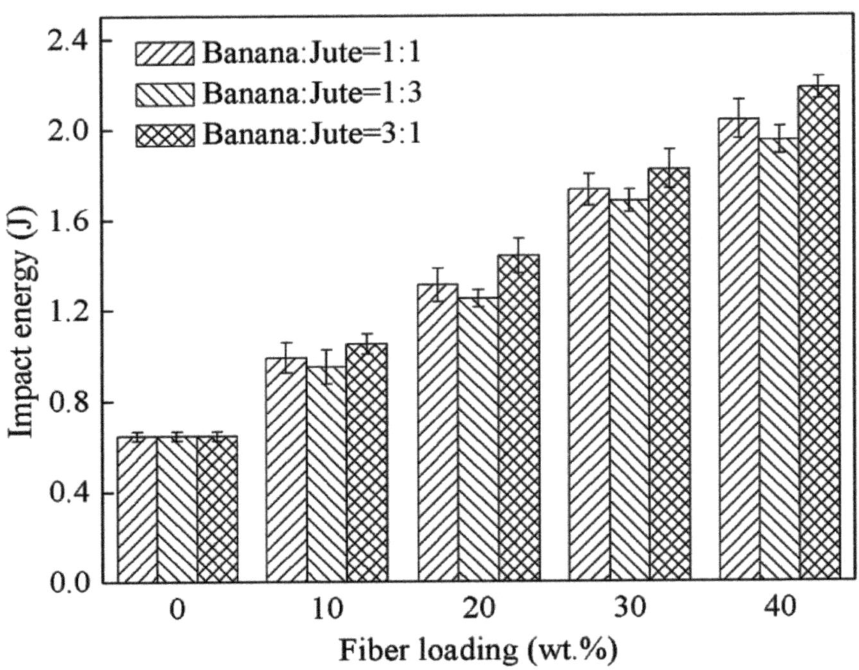

FIGURE 5.6 Effect of fiber loading and weight ratio on impact energy of hybrid composites.

5.3.4 *MICRO-HARDNESS OF HYBRID COMPOSITES*

The effect of fiber loading and weight ratio on micro-hardness values of the polyester-based composites is shown in Figure 5.7. It is observed from the Figure 5.7 that with the increase in fiber loading, micro-hardness of the composites increases irrespective of the reinforcement and matrix type. This may be attributed to a compressive force applied during hardness testing which presses the reinforcement and matrix together which result in a decrease in indentation and effective transfer of load. In general, the fibers that increase the tensile modulus of composites also increase the hardness of composites. This is because hardness is a function of relative fiber volume and modulus [34]. The micro-hardness of neat polyester is found 16.88 Hv. As the fiber loading increases, micro-hardness of the composite enhances. Among all the composites under this investigation, composites with a weight ratio of banana and jute fiber as 1:3 shows the highest micro-hardness, compared with a weight ratio of banana and jute fiber as 1:1 and 3:1. This may be due to the brittle nature of the lignocellulose fiber. A similar trend of increase in hardness value with the increase in fiber loading is also observed by few researchers [35].

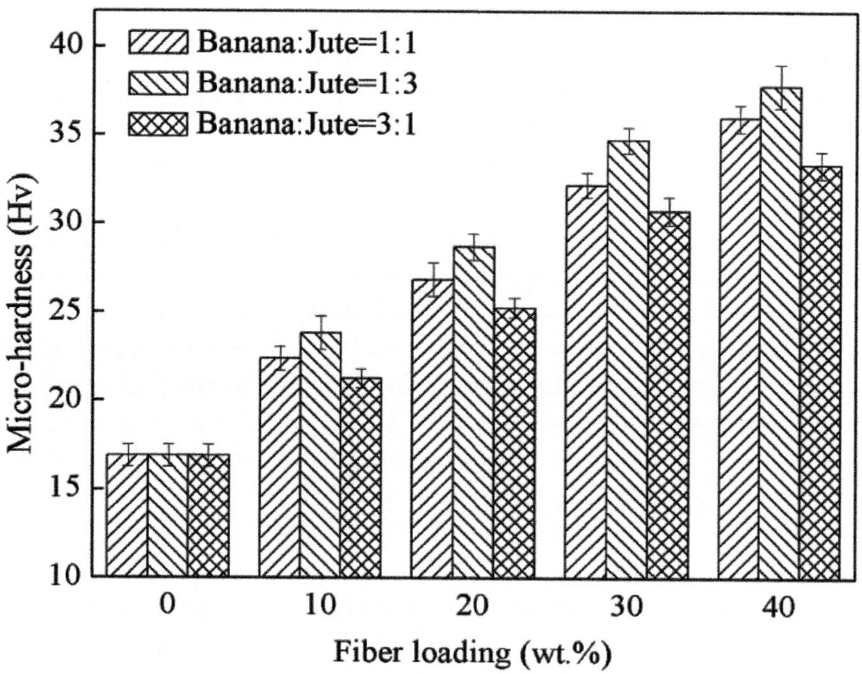

FIGURE 5.7 Effect of fiber loading and weight ratio on micro-hardness of hybrid composites.

5.3.5 SURFACE MORPHOLOGY

It is well known that the properties of the composite material mainly depend on the bonding between the reinforcement and the matrix. In order to evaluate the fiber-matrix bonding, the microstructure of the fibers and the dispersion of the banana and jute fibers in the matrix material are observed under scanning electron microscope. The cross-sectional view of banana and jute fiber is shown in Figure 5.8. These micrographs confirm the cylindrical shape of the fibers used as reinforcement in the matrix body. It is evident from Figure 5.8(a) that the surface of the banana fiber is porous and is covered with a layer of waxy and pithy substances. It can be observed from the Figure 5.8(b) that jute fiber has a compact structure and the filaments are packed together due to the intercellular binding materials such as waxes and hemicelluloses. Generally, the jute fiber strength and its mechanical property are more due to the lowest porosity and close packing of the cellulose chain. That is why composites with a weight ratio of banana and jute fiber as 1:3 shows better properties as compared to others.

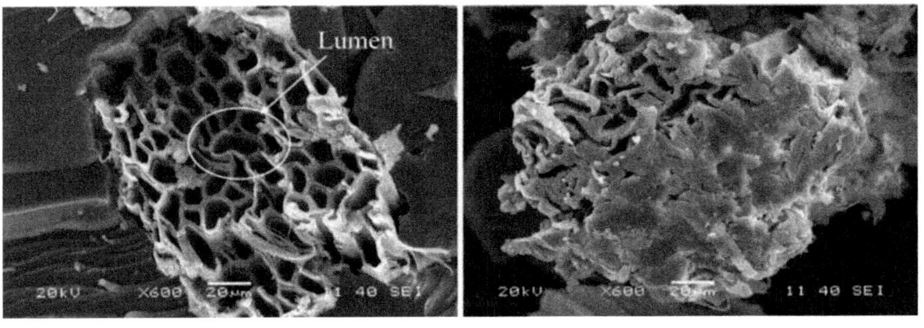

(a) (b)

FIGURE 5.8 Scanning electron micrographs for cross-sectional views of (a) banana fiber, and (b) jute fiber.

The scanning electron micrographs of longitudinal direction view of banana and jute fibers are shown in Figure 5.9. It is clear from the image that the average diameter of jute fiber is around 80–90 micron (Figure 5.9(a)) and banana fiber is around 170–190 micron (Figure 5.9(b)). Figure 5.10 shows the morphological cross-section of neat polyester and composite reinforced with a different weight ratio of banana and jute fiber. All the micrographs are taken for composites with fiber loading of 30 wt.%. The morphological structure of the neat polyester is presented in Figure 5.10(a). From the figure, no pulling and stretching of polymer on the surface of the polyester composite are observed, but there is a sharp cut surface due to the brittle nature of polyester composite. The brittle nature of the neat polymer composite is studied and confirmed by many researchers. Figure 5.10 (b), (c), and (d) shows the composites with a weight ratio of banana and jute fiber as 1:1, 1:3, and 3:1, respectively. From the figures it is observed that the distributions of banana and jute fibers in the matrix material are more or less uniform and further increase of fiber loading beyond 30 wt.%, reduces the inter-fiber distance up to the limit that fibers start to interfere with each other, which may decrease the properties of the fiber as well as the composites.

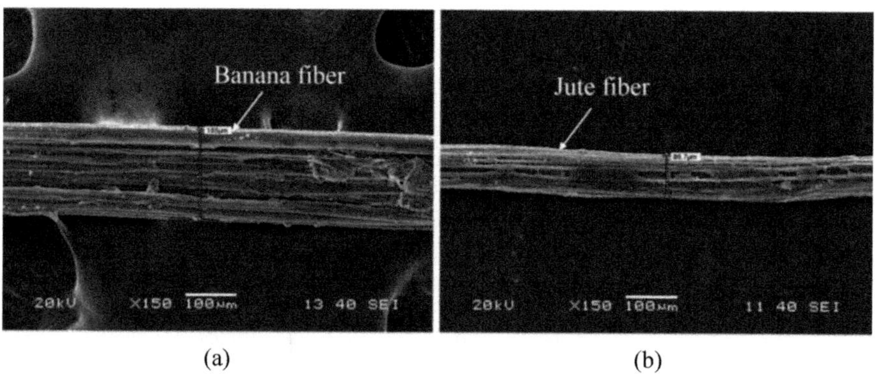

FIGURE 5.9 Scanning electron micrographs in the longitudinal direction of (a) banana fiber, and (b) jute fiber.

Scanning electron microscopy image for the tensile fracture of the short banana-jute hybrid fiber composite is compared with the 30 wt.% and 40 wt.% fiber loading in Figure 5.11. It can be clearly observed from the Figure 5.11(a) there is good interfacial bonding between the fiber and the matrix at 30 wt.% of fiber loading that leads to better of strength properties of hybrid composites. Figure 5.11(b) shows the scanning electron microscopy image of 40 wt.% fiber loading. The presence of fiber pull out, fiber breakage and voids between fiber and matrix, indicate weak interfacial bonding between the fiber and matrix which in turn results in poor mechanical properties.

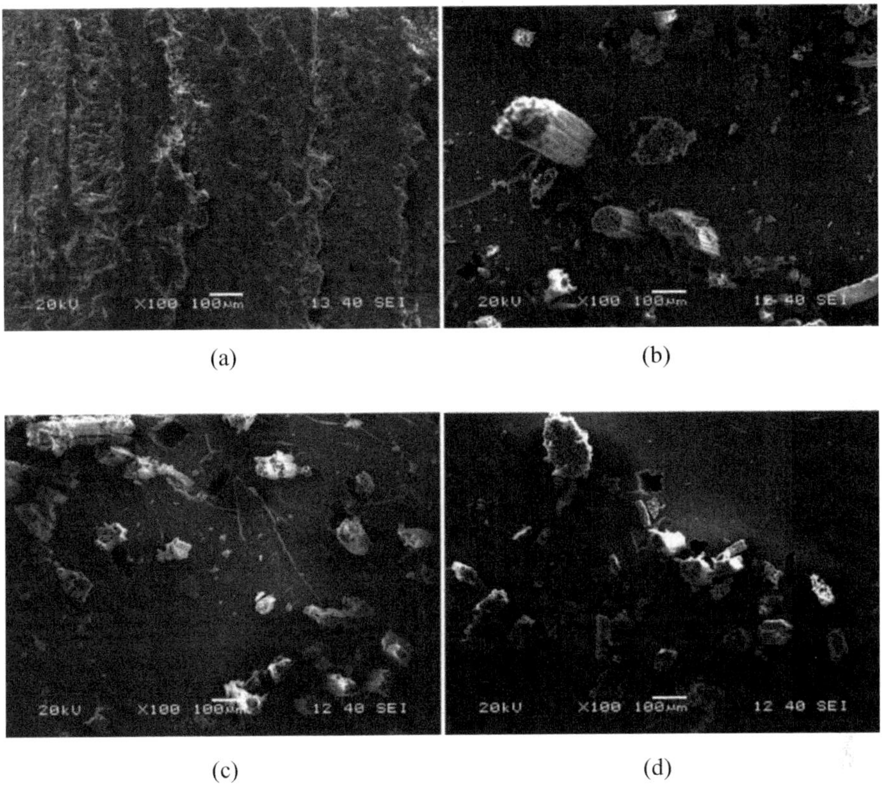

FIGURE 5.10 Scanning electron micrographs of the morphological cross-section of (a) neat polyester and the composites with a weight ratio of banana and jute fiber as (b) 1:1, (c) 1:3 and (d) 3:1.

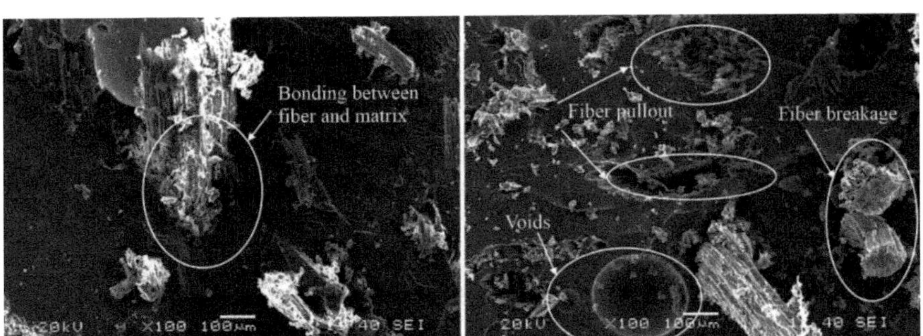

FIGURE 5.11 Tensile fracture surfaces of hybrid composites at (a) 30 wt.%, and (b) 40 wt.% having a weight ratio of banana and jute 1:3.

5.4 CONCLUSION

The experimental investigation of the mechanical characteristics of short banana-jute fiber-reinforced polyester based hybrid composites lead us to the following conclusions obtained from this study:

- Successful fabrication of short banana-jute fiber-reinforced polyester based hybrid composites is possible by varying the fiber loading and weight ratio using simple hand lay-up technique.
- The tensile strength of neat polyester composites increases 20.7 MPa to 56.2 MPa with addition of 30 wt.% of fiber loading with weight ratio of banana and jute as 1:3 and the maximum tensile modulus of the composite is found to be 2.648 GPa at 40 wt.% of fiber loading with weight ratio of banana and jute fiber as 1:3.
- The flexural strength and modulus of the banana-jute composites are increased up to 30 wt.% of fiber loading and then decreased with the increase in fiber loading. The flexural strength and modulus of neat polyester composites increase by 98% and 91% with the addition of 30 wt.% of fiber loading with a weight ratio of banana and jute as 1:3.
- The impact energy and micro-hardness of the banana-jute composites is increased up to 40 wt.% of fiber loading. The maximum impact energy absorbed by the hybrid composite due to impact load is 3.39 times of pure polyester resin for hybrid composites with the weight ratio of banana and jute as 3:1. The micro-hardness polyester composites improve to 113%, 123% and 97% with corresponding weight ratios of banana and jute 1:1, 1:3, and 3:1, respectively.
- The Scanning electron microscope images for the tensile fractured specimens are analyzed for the voids, fiber breakage and fiber pull outs during the loading at 30 wt.% and 40 wt.% fiber loading. Higher mechanical strength has the lesser fiber pullouts and better adhesion between fiber and matrix.
- Finally, banana and jute fibers as reinforcement in polyester composites results in a positive hybrid effect for mechanical properties. Therefore, cost-effective, and value-added hybrid composites having high mechanical properties could be well developed by the judicious selection of banana and jute fiber.

KEYWORDS

- **banana fiber**
- **composite**
- **fiber loading**
- **hybrid composite**
- **jute fiber**
- **mechanical properties**
- **natural fiber**
- **scanning electron microscope**
- **short fiber**
- **weight ratio**

REFERENCES

1. Singh, S., Kumar, P., & Jain, S., (2013). An experimental and numerical investigation of the mechanical properties of glass fiber-reinforced epoxy composites. *Adv. Mat. Lett., 4,* 567–572.
2. Saheb, D. N., & Jog, J. P., (1999). Natural fiber polymer composites: A review. *Adv. Polym. Technol., 18*(4), 351–363.
3. John, M. J., & Thomas, S., (2008). Biofibres and biocomposites. *Carbohydr. Polym., 71*(3), 343–364.
4. Murali, M. R. K., Mohana, R. K., & Ratna, P. A. V., (2010). *Fabrication and Testing of Natural Fibre Composites: Vakka, Sisal, Bamboo, and Banana* (Vol. 31),28-64.
5. Kahraman, R., Abbasi, S., & Abu-Sharkh, B., (2005). Influence of epolene G–3003 as a coupling agent on the mechanical behavior of palm fiber-polypropylene composites. *Int. J. Polym. Mater., 54*(6), 483–503.
6. Jindal, U. C., (1986). Development and testing of bamboo-fibers reinforced plastic Composites. *J. Compos. Mater., 20*(1), 19–29.
7. Rout, J., Misra, M., Mohanty, A. K., Nayak, S. K., & Tripathy, S. S., (2003). SEM observations of the fractured surfaces of coir composites. *J. Reinf. Plast. Compos., 22*(12), 1083–1100.
8. Mishra, V., & Biswas, S., (2013). Physical and mechanical properties of bi-directional jute fiber-epoxy composites. *Procedia. Eng., 51*, 561–566.
9. Bai, S. L., Li, R. K. Y., Wu, L. C. M., Zeng, H. M., & Mai, Y. W., (1998). Tensile failure mechanisms of sisal fibers in composites. *J. Mater. Sci. Lett., 17*(21), 1805–1807.
10. Pothan, L. A., Thomas, S., & Neelakantan, N. R., (1997). Short banana fiber-reinforced polyester composites: Mechanical, failure, and aging characteristics. *J. Reinf. Plast. Compos., 16*(8), 744–765.

11. Devi, L. U., Bhagawan, S. S., & Thomas, S., (1997). Mechanical properties of pineapple leaf fiber-reinforced polyester composites. *J. Appl. Polym. Sci., 64*,(9), 1739–1748.
12. Shibata, M., Takachiyo, K. I., Ozawa, K., Yosomiya, R., & Takeishi, H., (2002). Biodegradable polyester composites reinforced with short abaca fiber. *J. Appl. Polym. Sci., 85*(1), 129–138.
13. Sreekala, M. S., George, J., Kumaran, M. G., & Thomas, S., (2002). The mechanical performance of hybrid phenol-formaldehyde-based composites reinforced with glass and oil palm fibers. *Compos. Sci. Technol., 62*(3), 339–353.
14. Haneefa, A., Bindu, P., Aravind, I., & Thomas, S., (2008). Studies on the tensile and flexural properties of short banana/glass hybrid fiber-reinforced polystyrene composites. *J. Compos. Mater., 42*(15), 1471–1489.
15. Gujjala, R., Ojha, S., Acharya, S., & Pal, S., (2014). Mechanical properties of the woven jute-glass hybrid-reinforced epoxy composite. *J. Compos. Mater., 48*(28), 3445–3455.
16. Ashok, K. M., Ramachandra, R. G., Siva, B. Y., Venkata, N. S., & Naga, P. N. V., (2010). Frictional coefficient, hardness, impact strength, and chemical resistance of reinforced sisal-glass fiber epoxy hybrid composites. *J. Compos. Mater., 44*(26), 3195–3202.
17. Khalil, H. P. S. A., Hanida, S., Kang, C. W., & Fuaad, N. A. N., (2007). Agro-hybrid composite: The effects on mechanical and physical properties of oil palm fiber (EFB)/ glass hybrid reinforced polyester composites. *J. Reinf. Plast. Compos., 26*(2), 203–218.
18. Devi, L. U., Bhagawan, S. S., & Thomas, S., (2009). Dynamic mechanical analysis of pineapple leaf/glass hybrid fiber-reinforced polyester composites. *Polym. Compos., 31*(6), 956–965.
19. Davoodi, M. M., Sapuan, S. M., Ahmad, D., Ali, A., Khalina, A., & Jonoobi, M., (2010). Mechanical properties of hybrid kenaf/glass reinforced epoxy composite for passenger car bumper beam. *Mater. Des., 31*(10), 4927–4932.
20. Sapuan, S. M., Lok, H. Y., Ishak, M. R., & Misri, S., (2013). Mechanical properties of hybrid glass/sugar palm fiber-reinforced unsaturated polyester composites. *Chinese J. Polym. Sci., 31*(10), 1394–1403.
21. Munikenche, G. T., Naidu, A. C. B., & Chhaya, R., (1999). Some mechanical properties of untreated jute fabric-reinforced polyester composites. *Compos. Part A Appl. Sci. Manuf., 30*(3), 277–284.
22. Noorunnisa, K. P., Mohan, R. M., Raghu, K., John, K., & Venkata, N. S., (2007). Tensile, flexural, and compressive properties of sisal/silk hybrid composites. *J. Reinf. Plast. Compos., 26*(10), 1065–1070.
23. Udaya, K. C., Ramachandra, R. G., Dabade, B. M., & Rajesham, S., (2007). Tensile properties of sun hemp, banana, and sisal fiber-reinforced polyester composites. *J. Reinf. Plast. Compos., 26*(10), 1043–1050.
24. Athijayamani, A., Thiruchitrambalam, M., Manikandan, V., & Pazhanivel, B., (2010). Mechanical properties of natural fibers reinforced polyester hybrid composite. *Int. J. Plast. Technol., 14*(1), 104–116.
25. Kumar, K. S., Siva, I., Rajini, N., Jeyaraj, P., & Jappes, J. W., (2014). Tensile, impact, and vibration properties of coconut sheath/sisal hybrid composites: Effect of stacking sequence. *J. Reinf. Plast. Compos., 33*(19), 1802–1812.
26. John, M. J., & Anandjiwala, R. D., (2009). Chemical modification of flax reinforced polypropylene composites. *Compos. Part A Appl. Sci. Manuf., 40*(4), 442–448.

27. Sana, R., Foued, K., Yosser, B. M., Mounir, J., Slah, M., & Bernard, D., (2015). Flexural properties of Typha natural fiber-reinforced polyester composites. *Fibers Polym., 16*(11), 2451–2457.
28. Yu, T., Li, Y., & Ren, J., (2009). Preparation and properties of short natural fiber-reinforced poly(lactic Acid) composites. *Trans. Nonferrous Met. Soc. China, 19*, 651–655.
29. Mohanty, A. K., Khan, M. A., & Hinrichsen, G., (2000). Influence of chemical surface modification on the properties of biodegradable jute fabrics-polyester amide composites. *Compos. Part A Appl. Sci. Manuf., 31*(2), 143–150.
30. Thomas, S., & Pothan, L. A., (2008). *Natural Fiber Reinforced Polymer Composites: From Macro to Nanoscale*, Archives contemporariness Ltd: Paris.
31. Romanzini, D., Lavoratti, A., Ornaghi, H. L., Amico, S. C., & Zattera, A. J., (2013). Influence of fiber content on the mechanical and dynamic mechanical properties of glass/ramie polymer composites. *Mater. Des., 47*, 9–15.
32. Öztürk, B., (2010). The hybrid effect in the mechanical properties of Jute/Rockwool hybrid fibers reinforced phenol formaldehyde composites. *Fibers Polym., 11*(3), 464–473.
33. Rezaei, F., Yunus, R., Ibrahim, N. A., & Mahdi, E. S., (2007). Effect of fiber loading and fiber length on mechanical and thermal properties of short carbon fiber-reinforced polypropylene composite. *Malaysian J. Anal. Sci., 11*(1), 181–188.
34. Srinivasa, C. V., & Bharath, K. N., (2011). Impact and hardness properties of areca fiber-epoxy reinforced composites. *J. Mater. Environ. Sci., 2*(4), 351–356.
35. Venkata, R. G., Venkata, N. S., & Shobha, R. T., (2008). Kapok/glass polyester hybrid composites: Tensile and hardness properties. *J. Reinf. Plast. Compos., 27*(16–17), 1775–1787.
36. Oladele, I. O., (2014). Effect of bagasse fiber reinforcement on the mechanical properties of polyester composites. *J. Assoc. Prof. Eng. Trinidad Tobago, 42*(1), 12–15.
37. Venkateshwaran, N., & Elayaperumal, A., (2010). Banana fiber-reinforced polymer composites – A review. *J. Reinf. Plast. Compos., 29*(15), 2387–2396.
38. Devireddy, S. B. R., & Biswas, S., (2015). The physical and mechanical behavior of unidirectional banana/jute fiber-reinforced epoxy based hybrid composites. *Polymer Composites.* John Wiley and Sons Inc.
39. Idicula, M., Neelakantan, N. R., Oommen, Z., Joseph, K., & Thomas, S., (2005). A study of the mechanical properties of randomly oriented short banana and sisal hybrid fiber-reinforced polyester composites. *J. Appl. Polym. Sci., 96*(5), 1699–1709.
40. Arib, R. M. N., Sapuan, S. M., Ahmad, M. M. H. M., Paridah, M. T., & Zaman, H. M. D. K., (2006). Mechanical properties of pineapple leaf fiber-reinforced polypropylene composites. *Mater. Des., 27*(5), 391–396.
41. Idicula, M., Joseph, K., & Thomas, S., (2010). Mechanical performance of short banana/sisal hybrid fiber-reinforced polyester composites. *J. Reinf. Plast. Compos., 29*(1), 12–29.
42. Bledzki, A. K., & Gassan, J., (1999). Composites reinforced with cellulose-based fibers. *Prog. Polym. Sci., 24*(2), 221–274.
43. Åkesson, D., Skrifvars, M., Seppälä, J., & Turunen, M., (2011). Thermoset lactic acid-based resin as a matrix for flax fibers. *J. Appl. Polym. Sci., 119*(5), 3004–3009.

CHAPTER 6

GRAPHENE AND GRAPHENE-BASED POLYMER NANOCOMPOSITES: THE NEW WONDER MATERIALS OF THE NANOWORLD

UJJAL K. SUR

Department of Chemistry, Behala College, University of Calcutta, Kolkata–60, India, E-mail: uksur99@yahoo.co.in

ABSTRACT

Graphene, the one-atom-thick planar sheet of sp^2 bonded carbon atoms packed in a honeycomb lattice, is considered to be the mother of all graphitic materials like fullerenes, carbon nanotubes, and graphite. Graphene, the two-dimensional form of graphite was first isolated in 2004 by Geim and Novoselov of Manchester University, United Kingdom. Today, graphene is the most attractive nanomaterial not only because it is the thinnest known material and the strongest material ever measured in the universe, but also due to its excellent electrical, thermal, mechanical, electronic, and optical properties. It has potential applications ranging from sensors, field-effect transistors, displays, energy storage and photovoltaic devices. In this paper, a brief overview on various aspects of graphene such as synthesis, functionalization, self-assembly, some of its amazing properties along with its various applications ranging from sensors to energy storage devices had been illustrated. This paper also reviews recent advances in the graphene-based polymer nanocomposites. The structure, synthesis, and various interesting properties of polymer/graphene nanocomposites are illustrated in general along with detailed examples drawn from the scientific literature. The challenges and outlook of these graphene-based polymer nanocomposites are also discussed.

6.1 INTRODUCTION

The introduction of nanoscience and nanotechnology in the mid–90's and subsequent development of this versatile area has become prominent over recent few years, as miniaturization of both materials and devices becomes more significant in practical areas, such as computation, sensors, biomedical applications, energy storage, and many other varied applications. Nanomaterials provide an extensive collection of applications due to their structural features and interesting physicochemical properties. On the other hand, researchers are probing materials with superior physicochemical properties which are dimensionally more appropriate in the area of nanoscience and nanotechnology. By the way, the discovery of graphene and graphene-based nanocomposites particularly polymer is a significant extension to novel nanomaterials.

Ever since fullerene (C_{60}) was isolated in 1985 by Smalley and his co-workers [1], quite a few novel carbon nanomaterials have been isolated. In 1991, carbon nanotube (CNT) was discovered by Japanese scientist Iijima [2]. On the other hand, graphene, the two-dimensional form of graphite was isolated in 2004 by Geim and Novoselov of Manchester University, United Kingdom [3]. Graphene has become the latest super-material in recent years due to its unique physicochemical properties such as high specific surface area, high chemical stability, high optical transmittance, high elasticity, high porosity, biocompatibility, tunable bandgap and tunable properties due to easiness in chemical functionalization [4, 5].

Graphene is a flat single sheet from graphite, has the ideal two-dimensional (2D) structure with a monolayer of carbon atoms packed into a honeycomb crystal plane. Graphene is considered as the fundamental building block for graphitic materials of all other dimensions. It can be wrapped up into zero-dimensional (0D) fullerenes, rolled into one-dimensional (1D) nanotubes and stacked into three dimensional (3D) graphite. Therefore, graphene is called the mother of all graphitic carbon-based nanomaterials.

Graphene was accidentally and unexpectedly discovered by Geim and Novoselov of Manchester University in the United Kingdom in 2004 after peeling off highly oriented pyrolytic graphite [3]. This technique is known as micromechanical cleavage in the literature. In this technique, a cellophane tape is employed to peel off graphene layers from a graphite flake, followed by pressing the tape against a substrate such as silicon. A single sheet of graphene is obtained by removal of the tape. This is also referred to as "Scotch tape" or "Peel-off" method.

The existence of 2D crystal is highly controversial. Scientists Peirls and Landau had argued that 2D crystals were thermodynamically unstable and could not exist in nature [6]. Thermal fluctuations will destroy the long-range order, causing the melting of the 2D crystal lattice. It was also presumed that graphene can not exist in free state and was believed to be thermodynamically unstable with respect to the formation of curved structures such as soot, fullerenes, and nanotubes. However, later it was found that the stability of 2D crystals of graphene can be attributed to the gentle crumpling in the third dimension as revealed by transmission electron microscopic (TEM) studies.

6.2 SYNTHESIS OF GRAPHENE

Though in 2004, Novoselov, and Geim synthesized graphene by micromechanical cleavage of highly oriented pyrolytic graphite with a poor yield, several methods have been employed to produce graphene. The mechanical exfoliation method is the most primitive technique to deposit graphene monolayers and bilayers onto a hydrogen passivated Si(100) surface. On the other hand, mechanical exfoliation method has a limitation in terms of its poor yield. Therefore, substitute methods are essential to produce graphene with a high yield.

The different methods which have been cited in the literature hitherto to produce graphene sheets include micromechanical cleavage of graphite, unzipping of carbon nanotubes, chemical exfoliation of graphite, solvothermal synthesis, epitaxial growth on silicon carbide (SiC) surfaces and metal surfaces, chemical vapor deposition (CVD) of hydrocarbons on metal surfaces, bottom-up organic synthesis and the reduction of graphene oxide obtained from graphite oxide by a variety of reducing agents [7–10]. The last mentioned synthetic protocol gives up only chemically modified graphene.

The chemical method is well known as a scalable method to attain graphene and graphene derivatives at a large scale due to its simplicity, reliability, and extremely low cost. This method has been widely employed to produce chemically derived graphene. In this method, graphite is first oxidized to graphite oxide using either the Hummers method [11] or the modified Hummers method in the presence of strong acids and oxidants like conc. H_2SO_4 and $KMnO_4$. Graphite oxide can be readily exfoliated as individual graphene oxide (GO) sheets by sonicating in an aqueous medium. GO, which is an oxidized form of graphene is adorned by hydroxyl and epoxy functional groups on the hexagonal network of carbon atoms with the presence of carboxyl groups at the edges. GO is highly hydrophilic in nature and can form stable aqueous dispersions as a result of the presence of a large number of oxygen-containing functional groups on its surface. Chemically derived graphene can be obtained by reducing the GO using hydrazine solution or any other reducing agents.

There have been reports in the literature on the reduction of GO in solution phases using different reducing agents such as hydrazine, dimethylhydrazine, hydroquinone, ethylene glycol, sodium borohydride, lithium borohydride and in the vapor phase using hydrazine/hydrogen or just by thermal annealing [12,13]. Therefore, GO is an excellent precursor to synthesize graphene nanosheets. Few-layer graphene can also be synthesized using sugars such as glucose, fructose, and sucrose as the reducing agents [14]. Microwave, laser, plasmas, sonochemical, electrochemical as well as hydrothermal techniques were also employed to synthesize graphene from GO [15, 16].

6.3 SOME OF THE FASCINATING PROPERTIES OF GRAPHENE

The outstanding physicochemical properties of graphene have been utilized for several versatile applications ranging from electronic devices to electrode materials. It shows outstanding electronic properties, permitting electricity to flow rapidly through the materials. In fact, it has been demonstrated that electrons in graphene and its derivatives act as massless particles analogous to photons, zipping across a graphene layer without being scattered.

Graphene also illustrates exceptional optical property with high optical transmittance. The optical transmission through the graphene surface in the visible range is more than 95% as observed from experimental studies. Therefore, graphene has the ideal optical property of high optical transmission and can be used as transparent electrodes in liquid-crystal displays (LCDs) to replace indium-tin-oxide (ITO) as an electrode material.

The unique properties of graphene happen due to the combined actions of electrons. The interaction between electrons and the honeycomb lattice in graphene causes the electrons to behave as massless relativistic particles.

Due to this reason, the electrons in graphene are governed by the famous Dirac equation, the equation to describe the relativistic motion of electrons. Therefore, these electrons in graphene are known as Dirac fermions. The relativistic behavior of electrons in graphene was first predicted in 1947 by Canadian physicist Philip Wallace, long before the isolation of graphene [17]. We are already accustomed with massless Dirac fermions in neutrinos in high-energy particle physics. However, neutrinos have no electric charge and do not interact strongly with any kind of matter. On the contrary, the Dirac fermions in graphene carry one unit of electric charge and can interact strongly with the electromagnetic field. The manipulation of electrons in graphene using electromagnetic fields may allow us to go beyond the limits of silicon semiconductor technology.

Electrons in graphene move with an effective speed of light 300 times less than the speed of light in vacuum, allowing relativistic effects to be observed. The electrons in graphene can travel long distances with scattering, making it the ideal material for the fabrication of fast electronic components.

Graphene also exhibits unusual "half-integer" Quantum Hall Effect (QHE), which distinguishes itself from an ordinary metal and semiconductor. In the original Hall effect, a current flowing along the surface of a metal in the presence of a transverse magnetic field causes a potential drop. The ratio of a potential drop to the current flowing is called the Hall resistivity, which is directly proportional to the applied magnetic field. In a 2D electron gas, the Hall resistivity becomes quantized at a temperature close to absolute zero. This is called QHE. The Hall resistivity is represented by h/ne^2, where h is Planck's constant, e is the electric charge and n is a positive integer. In case of graphene, QHE arises due to a quantum-mechanical effect called Berry's phase and the Hall resistivity should be quantized in terms of odd integers only. QHE is observed at room temperature in case of graphene, whereas it is only observed in ordinary metal at very low temperature.

Graphene can also exhibit the property of fluorescence quenching. It has the ability to quench the fluorescence emitted by various organic and biological molecules. which is associated with photoinduced electron transfer [18]. The major features of graphene, in a nutshell, is shown below.

- Graphene is a one-atom-thick monolayer of carbon, which was isolated in 2004 for the first time.

- Graphene is 2D and has a honeycomb structure.
- Graphene is optically transparent.
- Electrons in graphene behave as massless particles similar to photons, zipping across a graphene layer without scattering.
- QHE is observed in graphene at room temperature. It is called "half-integer" QHE.
- Electrons in graphene can travel large distance without scattering.

6.4 APPLICATIONS OF GRAPHENE

The most significant application for graphene and graphene derivatives is most probably its use in composite materials. In fact, it has been demonstrated that by dispersing a small amount of graphene throughout polymers, tough lightweight materials can be designed and fabricated. These fabricated composites have good electrical and thermal conductivity and can endure much higher temperatures compared to normal polymers. In recent times, a research group at Northwestern University, USA synthesized graphene-based polymer composite materials. These graphene-based polymer composite materials could be ideal to develop lightweight gasoline tanks and plastic containers and can be potentially utilized to build lighter, more fuel-efficient aircraft and car parts, stronger wind turbines, medical implants and sports equipment.

Graphene powders are used in electric batteries due to the high surface-to-volume ratio and high conductivity provided by graphene powder leading to the enhancement of the overall efficiency of modern batteries.

High conductivity and high optical transparency make graphene suitable for fabricating transparent conducting coating in LCDs and solar cells. Recently, researchers from Korea and the USA developed ultraviolet (UV) nitride light-emitting diode which uses a few layers of graphene as a transparent conducting layer.

Graphene has an extremely high specific surface area and high porosity, facilitating the ideal environment for adsorption of different gases such as hydrogen (H_2), methane (CH_4) and carbon dioxide (CO_2). Graphene can be an ideal material for hydrogen storage due to the lightweight, high surface area and chemical stability. Hydrogen gas can be chemically stored in graphene by physisorption or chemisorption and can be alternative to other hydrogen storage materials.

Recent research and development on electrochemical power sources are mostly focused on fuel cells, batteries, and electrochemical capacitors and are projected for attaining high specific energy, high specific power, long cycle life at somewhat low cost. The high rate of environmental pollution created by electric vehicles and recent growth of portable electronic devices has facilitated the development of high-performance supercapacitor as flexible energy storage devices. Electrochemical capacitors, which are also known as supercapacitors or ultracapacitors, store energy using either ion adsorption (electrochemical double layer capacitors) or fast surface redox reactions or faradaic reactions (pseudo-capacitors/redox-capacitors). They can replace batteries in electrical energy storage and producing applications when high power delivery or uptake is essential. As a result of their high specific power, supercapacitors have applications including automobiles, electric vehicles, and various hybrid electric vehicles. Recently, graphene-based electrode material has been used for supercapacitor applications [19, 20]. In contrast to the conventional high surface materials, the effective surface area of graphene-based materials as capacitor electrode materials are not dependent on the distribution of pores at the solid state, which is dissimilar from the other carbon-based supercapacitors produced with activated carbons and carbon nanotubes. Obviously, the effective surface area of graphene materials should depend vastly on the layers. Consequently, the single or few layered graphenes should be anticipated to show a higher effective surface area and superior supercapacitive performance.

6.5 GRAPHENE-POLYMER NANOCOMPOSITES

A polymer nanocomposite can be defined as a multiphase solid material, where one of the phases has one, two, or three dimensions less than 100 nm and can be dispersed in different polymeric matrices. Nanocomposites are suitable for applications as high-performance composites, where the good dispersion of the filler can be accomplished and the properties of the nanoscale filler are significantly different or superior to those of the polymer matrix. The development of graphene-based polymer nanocomposites can provide new kind of composite materials with various interesting properties [21]. Polymer nanocomposites were discovered by the Toyota research group, which has brought a new dimension to the field of materials science [22]. Particularly, the use of inorganic nanomaterials as fillers to prepare polymer/inorganic composites has generated a lot of interest due to their exceptional properties and several potential applications. Until now, the most of the research has focused on polymer nanocomposites based on natural layered materials, such as layered silicate compounds or synthetic clay (layered double hydroxide (LDH)) [23]. However, the electrical and thermal conductivities of clay minerals are very poor and to solve these restrictions, carbon-based nanofillers, such as CNTs and carbon nanofibers (CNFs), have been employed to produce the polymer nanocomposites. Among these, CNTs are well known to be very effective as conductive fillers.

Graphene sheets can offer a substitute choice to produce functional nanocomposites due to their outstanding properties and the natural abundance of their precursor graphite. Graphene and graphene derivative as a nanofiller may be favored over other conventional nanofillers due to the high surface area, thermal, and electrical conductivity, flexibility, and transparency. Polymer/graphene nanocomposites demonstrate better-quality mechanical, thermal, and electrical properties compared to the pure polymeric materials. It was also reported that the improvement in mechanical and electrical properties of graphene-based polymer nanocomposites is much better in comparison to that of clay or other carbon filler based polymer nanocomposites [24]. On the other hand, the enhancement in the physicochemical properties of the nanocomposites is dependent on the distribution of graphene layers in the polymer matrix as well as interfacial bonding between the graphene layers and the polymer matrix. Interfacial bonding between graphene and the host polymer matrix controls the overall properties of the graphene-based polymer nanocomposites.

6.6 SYNTHESIS AND PROPERTIES OF GRAPHENE-POLYMER NANOCOMPOSITES

Graphene-polymer nanocomposites can be synthesized by employing three different methods such as (i) *in situ* intercalative polymerization, (ii) solution intercalation, and (iii) melt intercalation.

In the *in situ* intercalative polymerization method, graphene or graphene derivative is first swollen within the liquid monomer in the presence of a solvent. Next, an appropriate initiator is diffused and polymerization is initiated by either heating or radiation. A range of polymer nanocomposites has been prepared using this method, such as PS/graphene, PMMA/expanded graphite (EG).

Solution intercalation method can be employed to synthesize graphene-polymer nanocomposites. In this method, the polymer or pre-polymer is solubilized on a solvent system and graphene or modified graphene layers are allowed to swell. Polymer nanocomposites like polyethylene grafted maleic anhydride (PE-g-MA)/graphite, PP, poly(vinyl alcohol) (PVA)/graphene, poly(vinyl chloride) (PVC)/CNT, ethylene vinyl acetate (EVA)/LDH have been prepared using this method. In the melt intercalation technique, the solvent is not required, and graphite, graphene, or graphene derivative can be mixed with the polymer matrix in the molten state. A thermoplastic polymer can be mixed mechanically with graphite, graphene, or graphene derivative at high temperatures using conventional methods, such as extrusion and injection molding. The polymer chains are then intercalated or exfoliated to form nanocomposites. This is a popular method for preparing thermoplastic nanocomposites. Melt intercalation method is much more commercially striking compared to the other two methods, as both solvent processing and *in situ* polymerization are less versatile and less environmentally friendly. A broad range of polymer nanocomposites, such as PP/EG, high-density polyethylene (HDPE)/EG, PPS/EG, and PA6/EG, have been prepared using this method.

Most of the properties of polymer/graphene nanocomposites were better compared to the polymer matrix and other carbon filler (CNT, carbon nano fiber (CNF), and graphite)-based composites. These superior properties of the nanocomposites can be achieved with very low graphene content ($\leq$2wt%). Graphene-based polymer nanocomposites display better mechanical properties compared to the neat polymer or conventional graphite-based composites. There was a substantial improvement in the mechanical properties of the nanocomposites prepared by the solution mixing method (solution intercalation) compared to thermal mixing method (melt intercalation). Graphene can also exhibit extremely high thermal stability and can behave as an ideal nanofiller for the production of thermally stable nanocomposites. Graphene-filled polymer nanocomposites display better thermal stability compared to the neat polymer. Graphene-based polymer nanocomposites exhibit a several-fold increase in electrical conductivity. The significant improvement in electrical conductivity can be explained by the formation of a conducting network by graphene sheets in the polymer matrix.

6.7 APPLICATIONS OF GRAPHENE-POLYMER NANOCOMPOSITES

The probable applications of graphene-based polymer nanocomposites are solely dependent on the mechanical, electrical, and thermal properties as discussed above and there is the further scope of enhancement of their properties which can provide a further opening for future research activities and potential applications. The direct use of graphene as a nanofiller has opened a new dimension for the production of lightweight, low cost, and high-performance composite materials for a range of applications. It is expected that these conducting polymer/graphene nanocomposites will exhibit their potentials applications in the future for biomedical application, such as ultra-miniaturized low-cost sensors for the analysis of blood and urine. The inherent high conductivity and high aspect ratios of graphene nanosheets as well as the appealing physicochemical properties of graphene-based polymer nanocomposites can be employed in the development of various sensors such as gas, pH, pressure, and temperature sensors. Conducting polymer/graphene composites can also be widely used as electrode materials in a range of electrochromic and energy storage devices such as lithium-ion batteries, supercapacitors, bipolar plates in fuel cells, organic solar cells, and field emission devices. The polymer/graphene flexible electrode has some commercial applications in light-emitting diodes (LEDs), transparent conducting coatings for solar cells, and displays. Grapheme can be performed as a suitable candidate for conducting and transparent coating material in solar cells. The probable market size for conductive nanofillers and nanocomposites will reach $ 5–10 billion by the end of 2015 as from expectation when graphene-based composite will play an important role.

6.8 SUMMARY

The isolation of graphene in 2004 has brought a flow of interest in research activities as well as practical applications globally due to its unique material properties [25]. The uniqueness in electronic, mechanical, and thermal properties of graphene and its derivatives in conjunction with its cost-effective mass production build it a potential wonder material for composites with different polymers, metals, metal oxides, and other carbon-based nanomaterials. With a huge number of papers being cited in the literature in recent times, this extraordinary development forecasts an explosion in graphene research and development across the whole world. Numerous efforts have been made to further boost the properties of nanocomposites by varying the chemical structure of graphene and its derivatives through chemical functionalization. Production of solar energy directly from sunlight by using the graphene-based nanocomposites has not been achieved so far in laboratories. Therefore, generation of renewable energies like solar and wind energy and the storage of such energies in a well-planned manner provide challenges in the development of various graphene-based nanocomposites.

On the other hand, hydrogen can be stored in graphene and graphene-based composites [26]. It is expected to enhance the storage capacity by employing suitable graphene composites. Carbon dioxide can also be stored in graphene/ZIF–8 nanocomposites [27]. Graphene-ceramics composites are also at the periphery of biomedical engineering research, due to its recyclability, chemically inertness, a high degree of stability and biocompatibility. Graphene-based materials along with their composites have obtained the rank of being the new supermaterial due to their versatile applications in various fields. Graphene is also a material of wonder even in terms of fundamental physics. Due to its unusual electronic property, graphene has led to the appearance of a new pattern of 'relativistic' condensed-matter physics, where quantum relativistic phenomena, some of which are unobservable in high-energy physics, can now be mimicked and tested experimentally even in laboratories.

ACKNOWLEDGMENTS

UKS would like to acknowledge financial support from the projects funded by the UGC, New Delhi (grant no. PSW–045/13–14-ERO) and UGC-DAE CSR, Kolkata center, Collaborative Research Schemes (UGC-DAE-CSR-KC/CRS/13/RC11/0984/0988). UKS would also like to thank INSA, New Delhi (SP/VF–9/2014–15/273 1st April 2014) for INSA visiting Scientist Fellowship for 2014–2015.

KEYWORDS

- **Dirac Fermion**
- **exfoliation**
- **grapheme**
- **graphite**
- **micromechanical cleavage**
- **nanomaterials**
- **polymer nanocomposites**
- **quantum Hall effect**

REFERENCES

1. Kroto, H. W., Heath, J. R., O'Brien, S. C., Curl, R. F., & Smalley, R. E., (1985). C60: Buckminsterfullerene. *Nature, 318*, 162–163.
2. Iijima, S., (1991). Helical microtubules of graphitic carbon. *Nature, 354*, 56–58.
3. Novoselov, K. S., Geim, A. K., Morozov, S. V., Jiang, D., Zhang, Y., Dubonos, S., Grigorieva, I. V., & Firsov, A. A. Electric field effect in atomically thin carbon films. *Science, 306*, 666–669.
4. Rao, C. N. R., Sood, A. K., Subrahmanyam, K. S., & Govindaraj, A., (2009). Graphene: The new two-dimensional nanomaterial. *Angew. Chem., Int. Ed., 48*, 7752–7777.
5. Sur, U. K., (2012). Graphene: A rising star on the horizon of Materials Science. *Int. J. Electrochem.*, 1–12. doi: 10.1155/2012/237689.
6. Peierls, R. E., (1935). Quelques proprietes typiques des corpses solides. *Ann. I. H. Poincare, 5*, 177–222.
7. Park, S., & Ruoff, R. S., (2009). Chemical methods for the production of graphenes. *Nat. Nanotechnol., 4*, 217–224.
8. Ruoff, R. S., (2008). Graphene: Calling all chemists. *Nat. Nanotechnol., 3*, 10–11.
9. Sutter, P. W., Flege, J.-I., & Sutter, E. A., (2008). Epitaxial graphene on ruthenium. *Nat. Mater., 7*, 406–411.
10. Zhang, Y. B., Small, J. P., Pontius, W. V., & Kim, P., (2005). Fabrication and electric-field-dependent transport measurements of mesoscopic graphite devices. *Appl. Phys. Lett., 86*, 073104.
11. Hummers, W. S., & Offeman, R. E., (1958). Preparation of graphite oxide. *J. Am. Chem. Soc., 80*, 1339–1339.
12. Si, Y., & Samulski, E. T., (2008). Synthesis of water soluble graphene. *Nano Lett., 8*, 1679–1682.
13. Rao, C. N. R., Biswas, K., Subrahamanyam, K. S., & Govindaraj, A., (2009). Graphene, the new nanocarbon. *J. Mater. Chem., 19*, 2457–2469.

14. Zhang, J., Yang, H., Shen, G., Cheng, P., Zhang, J., & Guo, S., (2010). Reduction of graphene oxide via ascorbic acid. *Chem. Commun.*, *46*, 1112–1114.
15. Chen, W., Yan, L., & Bangal, P., (2010). Preparation of graphene by the rapid and mild thermal reduction of graphene oxide induced by microwaves. *Carbon*, *48*, 1146–1152.
16. Vinodgopal, K., Neppolian, B., Lightcap, I., Grieser, F., Ashokkumar, M., & Kamat, P., (2010). Sonolytic design of graphene-Au nanocomposites: Simultaneous and sequential reduction of graphene oxide and Au (III). *J. Phys. Chem. Lett.*, *1*, 1987–1993.
17. Wallace, P. R. The band theory of graphite. *Phys. Rev.* 1947, 71, 622–634.
18. Ramakrishna, M. H. S. S., Subrahmanyam, K. S., Venkata, R. K., George, S. J., & Rao, C. N. R., (2011). Quenching of fluorescence of aromatic molecules by graphene due to electron transfer. *Chem. Phys. Lett.*, *506*, 260–264.
19. Stoller, M. D., Park, S., Zhu, Y., An, J., & Ruoff, R. S., (2008). Graphene-based ultracapacitors. *Nano Lett.*, *8*, 3498–3502.
20. Vivekchand, S. R. C., Rout, C. S., Subrahmanyam, K. S., Govindaraj, A., & Rao, C. N. R., (2008). Graphene-based electrochemical supercapacitors. *J. Chem. Sci.*, *120*, 9–13.
21. Verdejo, R., Bernal, M. M., Romasanta, L. J., & Lopez-Manchado, M. A., (2011). Graphene filled polymer nanocomposites. *J. Mater. Chem.*, *21*, 3301–3310.
22. Kuilla, T., Bhadra, S., Yao, D., Kim, N. H., Bose, S., & Lee, J. H., (2010). Recent advances in graphene based polymer composites. *Prog. Polym. Sci.*, *35*, 1350–1375.
23. Ray, S. S., & Okamoto, M., (2003). Polymer/layered silicate nanocomposites: A review from preparation to processing. *Prog. Polym. Sci.*, *28*, 1539–1641.
24. Garcia, N. J., & Bazan, J. C., (2009). The electrical conductivity of montmorillonite as a function of · relative humidity: La-montmorillonite. *Clay. Miner.*, *44*, 81–88.
25. Sur, U. K., (2014).Graphene. In: Scott, R. A., (ed.), *Encyclopedia of Inorganic and Bioinorganic Chemistry in 2014*, John Wiley & Sons, 1–32.
26. Subrahmanyam, K. S., Kumar, P., Maitra, U., Govindaraj, A., Hembram, K. P. S. S., Waghmare, U. V., & Rao, C. N. R., (2011). Chemical storage of hydrogen in few-layer Graphene. *Proc. Natl. Acad. Sci. USA*, *108*, 2674–2677.
27. Kumar, R., Jayaramulu, K., Maji, T. K., & Rao, C. N. R., (2013). Hybrid nanocomposites of ZIF–8 with graphene oxide exhibiting tunable morphology, significant CO_2 uptake, and other novel properties. *Chem. Commun.*, *49*, 4947–4949.
28. Landau, L. D., (1937). II. *Phys. Z. Sowjetunion*, *11*, 26–35.

CHAPTER 7

PREPARATION AND CHARACTERIZATION OF ZnO DOPED CTS/PEG-AG NANO-COMPOSITE FOR IMPACTING PHOTOCATALYTIC RESPONSE

T. KOKILA[1], P. S. RAMESH[2], D. GEETHA[3], and S. KAVITHA[3]

[1]*Department of Physics, Vivekanandha College of Arts and Sciences for Women (Autonomous), Elayampalayam, Namakkal–637205, Tamilnadu, India*

[2]*Department of Physics (DDE Wing), Annamalai University, Annamalai Nagar 608002, Tamilnadu, India*

[3]*Department of Physics, Annamalai University, Annamalai Nagar 608002, Tamilnadu, India, E-mail: psrddephyau@gmail.com*

ABSTRACT

An investigation on ZnO-doped ($x = 0$, 0.03, 0.06, and 0.09) Cts/PEG-Ag nanomaterials have been synthesized by reduction method at 80°C was reported. To study the structural, morphological, particles size, stability, and optical properties of the samples by X-ray diffraction (XRD), scanning electron microscopy (SEM) with energy dispersive spectroscopic (EDS) analysis, Fourier transform infrared (FTIR) spectroscopy and UV-visible spectroscopy have been studied. The XRD spectrum shows that all the samples are facing center cubic (fcc) structure. The presence of functional groups and chemical bonding is confirmed by FT-IR. The ultraviolet-visible measurement showed an increasing ZnO content, probably due to an increase in the lattice parameters. ZnO doped Cts/PEG-Ag nanocomposite showed excellent photocatalytic behavior and the underlying mechanical properties. Hence ZnO doped Cts/PEG-Ag nanocomposite has potential application as photocatalyst and can also be used as a water purifier. The DLS results show that the nanocomposites have a favorable compatibility and stability. In application study, it is proved that this has photoreduction (exposure to sunlight) leaded to the potential aggregation of silver nanoparticles: metal ions that were released tended to aggregate at the surface of the material. Photocatalytic activity of Cts/PEG/Ag and 0.09M ZnO doped Cts/PEG-Ag nanocomposite was examined by studying the photodegradation of the product under natural sunlight irradiation and the results clearly showed that ZnO doped Cts/PEG/Ag sample exhibited higher activity than Cts/PEG-Ag.

7.1 INTRODUCTION

The rapid growth of industries leads to the expansion of environmental problems. One of the main environmental problems is water pollution from industries, agriculture, and so on, which have contaminants such as organic and inorganic compounds. Recently, the new approach for wastewater treatment is an application of photocatalyst through an oxidation process. The photocatalytic process requires the energy of light equal to or greater than the band gap energy of ZnO for production of electron and hole. The generated electron and hole can then react with water molecules and hydroxyl ions, which subsequently produce hydroxyl radical known as a strong oxidant for organic pollutant degradation in water. Apart from the oxidation process, the hole and electron can recombine easily, thus resulting in very low photocatalytic efficiency [1]. Therefore, many attempts have been made to related this electron-hole recombination such as modification of ZnO surface by doping various elements [2]. Surface modification using appropriate element dopant could reduce the recombination rate of the electron-hole pair and increase the photocatalytic activity of ZnO. Zinc oxide is no stranger to scientific study. In the past 100 years, it has produced as a subject of thousands of research papers, dating back as early as 1935. For its potential of ultraviolet absorbance, wide chemistry, piezoelectricity, and luminescence at high temperatures, ZnO has entered into the industry, and now is one of the critical building blocks in today's modern society. It is found in paints, cosmetics, plastic, and rubber manufacturing, electronics, and pharmaceuticals. More recently, however, it has again gained large interest for its semiconducting properties

ZnO nanostructure material has gained much interest owing to its wide applications for various devices such as solar cells, resistors, transducers, transparent conducting electrodes, sensors, and catalysts. However, these properties of the pure bulk ZnO are not stable and cannot meet the increasing needs for the present applications. In order to modify the properties of the ZnO, this semiconductor material was usually doped with some dopants such as Al, Si, and Ga. For example, Al-doped ZnO increases its conductivity without bacteria. They even have antibacterial activity against spores which are resistant to high temperature and high pressure. From the literature, it is evident that the antibacterial activity of ZnO nanoparticles depends on the surface area and concentration, while the crystalline structure and particle shape have little effect. Further, it is also mentioned in the literature that smaller the size of ZnO particles better is its antibacterial activity. Thus higher the concentration and the larger the surface area of the nanoparticles, the better is its antibacterial activity. Recently, many ZnO with Ag modification procedures have successfully shifted the photocatalytic activity of ZnO from UV region to visible light region, therefore enhanced photocatalytic activity could be obtained as a consequence of more complete utilization of solar energy [3]. Among studies on the development of nanocomposite materials with unique optical, electronic, magnetic, and catalytic properties a promising direction is the synthesis of nanomaterials by means of dispersing a metal, in particular, Ag, throughout a polymer blend.

Noble metal nanoparticles exhibit two physical-chemical properties which are not observed either in individual molecules or in bulk metals. It exhibits strong adsorption of electromagnetic waves in the visible range due to SPR. Silver nanoparticles stabilized inappropriately undergo fast oxidation and easily aggregate in a solution. In spite of the novel properties exhibited by the metal nanoparticles due to quantum size effects, their synthesis protocol posses a major environmental problem. Silver nanoparticles are one of the promising products in the nanotechnology industry. The development of consistent processes for the synthesis of silver nanomaterials is an important aspect of current nanotechnology research. One such promising process is green synthesis. Silver nanoparticles can be synthesized by several physical, chemical, and biological methods. However, for the past few years, various rapid chemical methods have been replaced by green synthesis because of avoiding toxicity of the process and increased quality.

Most of the synthetic methods reported to date rely heavily on the use of organic solvents and toxic reducing agents. All these chemicals are highly reactive and pose potential environmental and biological risks. Hence, the development of novel approaches is desirable. So, chitosan and PEG were used as a reducing and stabilizing agent for the preparation of silver nanoparticles. ZnO has attracted wide interest because of its good photocatalytic activity high stability and non-toxicity [1, 2]. ZnO nanoparticle exerts biocidal effects on bacterial, fungal, and viral species [4]. For economical and efficient use of ZnO, ZnO nanocomposites have been developed and tested for antimicrobial and photocatalytic purposes. Additionally, doped (Ag) reduced the ionization energy of acceptors in ZnO and thus enhanced the emission [5]. Due to its intriguing biological properties; chitosan has long been known and used in pharmaceutical and biomedical application [3]. Because of its unusual bioactivity, the formulation of chitosan with drugs has dual therapeutic effects, which make chitosan a novel candidate for drug carriers [6]. Poly (ethylene glycol) is frequently used in the production of polymer blends. PEG could improve the flexibility and ductility of the polymer by blending with the rigid polymer. The incorporation of PEG was expected to improve the biocompatibility of blend [7]. Chitosan/PEG blend may provide additional functionality as well as the flexibility. Chitosan improves the mechanical properties and decreases water solubility of the PEG, while PEG contributes to the formation of more flexible materials.

Silver Nanoparticles are used for purification and quality management of air, biosensing, imaging, drug delivery system. Biologically synthesized silver nanoparticles have many applications like coatings for solar energy absorption and intercalation material for electrical batteries, as optical receptors, as catalysts in chemical reactions, for biolabeling, and as antimicrobials. Though silver nanoparticles are cytotoxic, they have tremendous applications in the field of high sensitivity bimolecular detection and diagnostics, antimicrobials, and therapeutics, catalysis, and microelectronics. It has some potential application like diagnostic biomedical optical imaging, biological implants (like heart valves) and medical application like wound dressings, contraceptive devices, surgical instruments and bone prostheses. Many major consumer goods manufacturers already are producing household items that utilize the antibacterial properties of silver nanoparticles. These products include nanosilver lined refrigerators, air conditioners, and washing machines.

This study aims to investigate the feasibility of a novel nanocomposite of ZnO doped Cts/PEG-Ag nanocomposites for photocatalytic applications. Nanocomposites have been synthesized with doping of low ZnO concentration (0.03, 0.06, and 0.09M). In this study ZnO doped Cts/PEG-Ag nanocomposites with enhanced catalytic activities were successfully prepared via reduction method without any chemical reducing agents. It can be reported that the effect of ZnO doping on structural and morphological properties by X-ray diffractometer (XRD), Fourier transform infrared spectroscopy (FT-IR), scanning electron microscope (SEM)/EDS, and atomic force microscopy (AFM), and the optical absorption studies was carried out by UV-Visible and particles size distribution, stability was discussed by DLS/Zeta potential. The photodegradation of ZnO doped Cts/PEG-Ag was investigated. The photodegradation efficiency of nanocomposite against MB was analyzed for judging the feasibility of their use as catalysts.

7.2 EXPERIMENTAL

7.2.1 MATERIALS

Silver nitrate ($AgNO_3$), Zinc nitrate [Zn $(NO_3)_2$.$6H_2O$], polymers viz., Cts, and PEG, Ethanol, and de-ionized water are analytical grade and purchased from Sd-fine and were used without further purification. All the chemicals were used as received since they were of analytical reagent grade with 99% purity. Ultrapure water used for all dilution and sample preparation.

7.2.2 SYNTHESIS OF SILVER NANOPARTICLES

A solution of 1 wt% Cts and PEG was prepared by dissolving the Cts/PEG in double distilled water. Silver nitrate solution (10^{-2} M) was prepared by adding $AgNO_3$ in double distilled water. Silver nitrate (0.04 wt%) solution was mixed with Cts/PEG solution followed by the addition of 1.2 ml. Then the various concentrations of ZnO such as 0.03, 0.06 and 0.09 mM were added with the above solution separately at 80°C. The transparent colorless solution was converted to the characteristics yellowish-brown color, indicating the formation of silver nanoparticles.

7.2.3 CHARACTERIZATION

The prepared samples were primarily characterized by UV-visible spectroscopy, which has proved to be a very useful technique for the analysis of nanoparticles. Ultraviolet-visible spectra were obtained using a Shimadzu UV–1650pc Spectrophotometer. XRD analysis was carried out on an X-ray diffractometer (X'Pert-PRO). The high resolution on XRD patterns was measured at 3kW with Cu target using a scintillation counter (λ = 1.5406Å) at 40 kVand 40 mA were recorded in the range of 2θ=10–80°. The changes in the surface chemical bondings and surface composition were characterized by using Fourier Transform Infrared (FT-IR) spectroscopy (Nicolet Avatar series 330) ranging from 400 to 4000 cm^{-1}. Scanning Electron Microscopy (SEM) was used to identify the shape by FEI Quanta 200. Atomic Force Microscopy (Agilent-N9410A series 5500) was used to determine the size and topography of AgNPs. The sample for TEM analysis was dispersed in methanol, ultrasonicated for1handde- posited in C-coated Cu grids (JEOL 1010). The average size of AgNPs in aqueous medium was used to determine hydrodynamic diameter by DLS. The zeta potential was measured with a Zeta sizer Nano ZS90 (Melvern International Ltd.) instrument.

7.2.4 PHOTOCATALYTIC ACTIVITY MEASUREMENT

The photocatalytic activity was evaluated by observing the degradation of the solutions of MB. The process was carried out in a 100 ml beaker which was equipped with the tip of a light guide (5 mm in diameter) of the mercury-vapor lamp (OmniCure, EXFO). The UV light of very low intensity (2.2 W/cm^2, 20%) was used in this experiment. The distance between solution and tip of light-guide was 5cm. In each case, 30 ml of dye solution (10 ppm) and 20 mg Ag-ZnO were mixed to make suspension by stirring for 30 min in dark before UV irradiation. Prior to irradiation, different photocatalysts and dye solutions were magnetically stirred for 30 min under dark conditions to establish an adsorption/desorption equilibrium between dyes and photocatalyst surface. At specific time intervals, 1ml of the sample was withdrawn from the system and centrifuged to separate the residual catalyst, and then the absorbance intensity was measured at the corresponding wavelength. The composite photocatalyst without UV irradiation (in dark) was taken as control. For repeated use, Ag–ZnO NCs were separated from the suspended solution by repeated centrifuging and washing process.

7.3 RESULTS AND DISCUSSION

In the present study silver nanoparticles were obtained using the ability of chitosan (Cts) and Ploy (ethylene glycol) (PEG) to form chelate compounds with metal ions. Cts and PEG are characterized by the presence in its structure two polar groups (OH, NH_2) donor of electrons can form a link with inorganic materials enhancing electrical, optical, and biological properties qualify for various applications.

7.3.1 THE MECHANISM OF ZNO AND SILVER GENERATION

Chitosan, PEG, zinc oxide and silver nitrate were dissolved in water. Through adjusting water of the solution, the sol containing Zn^{2+} and Ag^+ ions was obtained. The –OH and $–NH_2$ groups of Cts and PEG chain reduced Ag+ ions to Ag [8]. The possible reactions of the reduction of Ag^+ to Ag by the chitosan, PEG is taken place as follows

$$AgNO_3 + aq.Cts/PEG \rightarrow Ag^+ + Cts/PEG^{OH-} \ \Delta \longrightarrow Ag^0/Cts/PEG \quad (7.1)$$

In order to make Zn^{2+} ions transform into ZnO, the sample was kept at 80°C. The possible reactions of ZnO formation take place as follows [9–11].

$$Zn^{2+} + OH^- \rightarrow Zn\text{-}O\text{-}H^+ \quad (7.2)$$

$$Zn\text{-}O\text{-}H^+ + OH^- \ \Delta \longrightarrow ZnO + H_2O \quad (7.3)$$

Since Zn^{2+} disperses homogeneously in chitosan/PEG sol, ZnO was generated homogeneously within chitosan/PEG polymer blend.

7.3.2 OPTICAL STUDIES (UV-VIS)

Figure 7.1 (a–e) presents UV–Vis absorbance spectra of ZnO Cts/PEG-Ag and ZnO doped Cts/PEG/Ag nanocomposite. Figure 7.1 (a) showed the excitonic absorption peak is observed due to the ZnO nanoparticles at 246 nm. It can be noticed that Cts/PEG/Ag spectrum (Figure 7.1 (b)) exhibited an absorption band at around (400–420) nm, which is a typical plasmon band, suggesting the formation of silver nanoparticles.

Composites arise due to the excitation of surface plasmon vibrations of Ag atoms. Compared with ZnO, Cts/PEG/Ag and Cts/PEG/Ag-ZnO nano-composites show three absorption bands. One absorption band at 246 nm in Figure 7.1 (a) was attributed to the presence of ZnO. Another absorption band at 406 nm was attributed to the presence of Ag nanoparticle. The band ranged from 420–450 nm due to ZnO absorption, blue-shifted by 20 nm. The width of the absorption band for the nanocomposites containing ZnO is broader than those containing ZnO and Cts/PEG/Ag. Furthermore, the center of absorption band also varied and shifted to higher wavelength (400–500) nm. This indicated that ZnO within Cts/PEG/Ag was in the size of a nanometer, which was consistent with the SEM analysis. The other absorption band at about 425nm was due to the presence of silver NPs [12]. Therefore, it further identified Ag and ZnO generated within Cts/PEG, which agrees well with the above analysis.

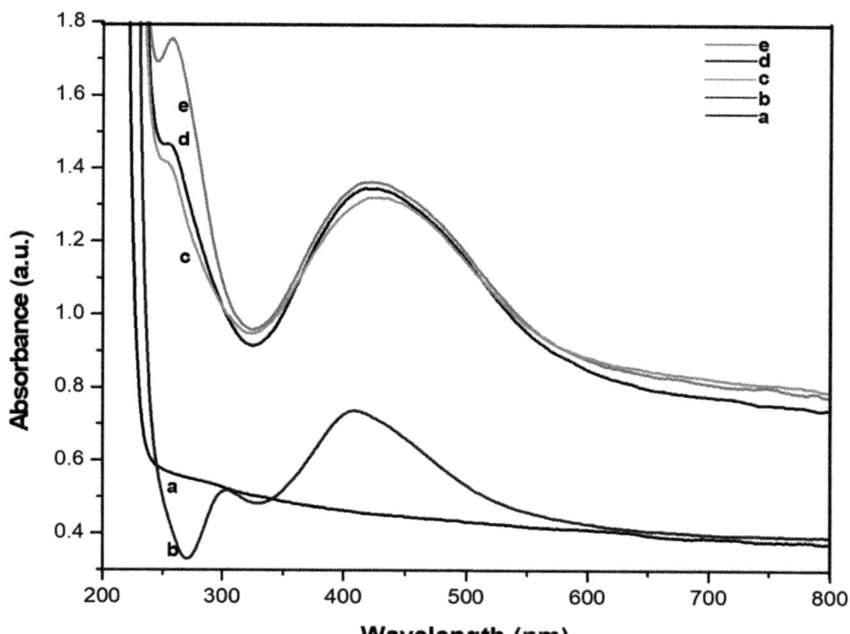

FIGURE 7.1 UV-Vis spectra of (a) pure ZnO and (0.04% w/v) Cts/PEG assisted Ag doped with ZnO: (b) 0.0 (c) 0.03 (d) 0.06 (e) 0.09 mM.

Figure 7.1 (c–e) shows the slight shift was occurred at (0.03, 0.06 and 0.09 mM) ZnO doped Ag nanoparticles, and its absorption edge was around 426–417 nm. However, when the concentration of ZnO increases, there is an increase in the intensity of the SPR band. The increase in the intensity of the SPR band from the Figure 7.1 (c–e) is due to the formation of more AgNPs. Above the 0.09mM $AgNO_3$ shows no considerable variation in the SPR band.

7.3.3 STRUCTURAL ANALYSIS (XRD)

Figure 7.2 (a–e) depicts the X-ray diffraction patterns of pure ZnO and Cts/PEG/Ag-ZnO blend nanocomposite. Two crystal forms existed in mixture Cts/PEG form I and form II depicted the major crystalline peaks at 10.3° and 20.3°, respectively [13]. However, these two main peaks (Figure 7.2 (c–e)) of Cts/PEG at 10.3° and 20.3° were affected by doping Ag and ZnO. This indicated that the incorporation of Ag and ZnO particles disrupted the regular order of polymer chains [14]. Compared with Figure 7.2 (a), the diffraction pattern of the Cts/PEG/Ag-ZnO nanocomposite exhibited six additional peaks at 31.9°, 34.6°, 36.4°, 56.8°, 62.9° and 68.2° (Figure 7.2 (c–e)), which were assigned to the (100), (002), (101), (110), (103) and (112) planes of hexagonal zinc oxide (JCPDS No. 36–1451). Besides, the peak at 38.3° indicated the presence of Ag as shown in Figure 7.2 (c–e). These data revealed the formation of Ag and ZnO in blend by the method of chemical reduction.

ZnO can be incorporated into Ag lattice as a substituent for Ag^+ or as an interstitial atom [15]. If the Zn^{2+} is substituted for Ag^+, a corresponding peak shift would be expected in the XRD. Lack of such shifts in the recorded XRD indicates the segregation of ZnO particles in the grain boundaries of Ag or only an insignificant quantity has been incorporated at the substitution silver.

However, the latter possibility is due to the difference in ionic radii between Zn^{2+} (0.72Å) and Ag^+ (1.22Å); the ZnO particles pre-frenetically choose to segregate around the Ag grain boundaries. Compared with Figure 7.2 (a), (b) the diffraction pattern of the Cts/PEG/Ag/ZnO nanocomposite exhibited four additional peaks at 31.9°, 34.6° and 56.8°, (Figure 7.2 (c–e)), which were assigned to the (100), (002) and (110) planes of zinc oxide [16]. Besides, the peak at 38.3°, 44.5° indicated the presence of Ag as shown in Figure 7.2. (c–e). The size of the ZnO doped Ag nanoparticles was estimated by applying the Scherer equation

$$D = \frac{0.9\lambda}{\beta \cos\theta} \tag{7.4}$$

where D is the crystalline size, λ is the wavelength of the incident X-ray (1.54Å), θ is the Bragg's angle and β is the full width at half maximum (FWHM). From the X-ray line broadening the crystalline sizes of ZnO, Cts/PEG/Ag and ZnO-doped Cts/PEG/Ag are estimated, which clearly shows the average particle size is reduced from 40.0 to 42.8 nm. It clearly shows the presence of nano-sized particles in the samples.

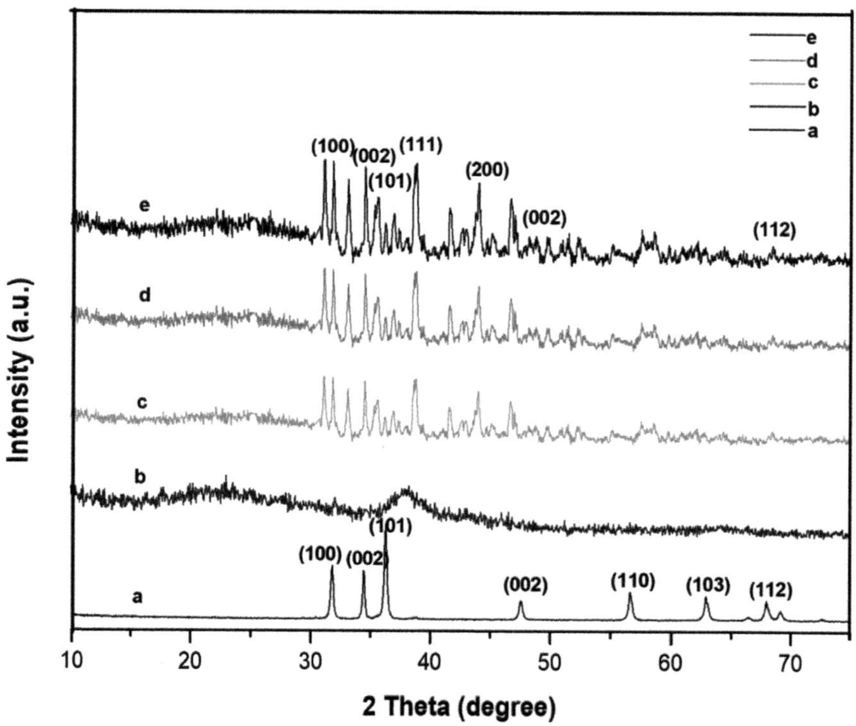

FIGURE 7.2 XRD patterns of (a) pure ZnO and (0.04% w/v) Cts/PEG assisted Ag doped with ZnO: (b) 0.0 (c) 0.03 (d) 0.06 (e) 0.09 mM.

7.3.4 SPECTRAL ANALYSIS (FT-IR)

The FT-IR spectra of ZnO, Cts/PEG/Ag, and Cts/PEG/Ag/ZnO nanoparticles were measured and presented in Figure 7.3 (a–e). In the spectrum of pure ZnO (3 (a)), the absorption band at 500 cm^{-1}, 548 cm^{-1} indicated the presence to ZnO. It was observed that ZnO stretching band (560 cm^{-1}) is present for ZnO doped Ag, indicate the presence of ZnO in the Ag nanocomposite and this could be due to a small concentration of ZnO. Figure 7.3 (b) shows the broad bands at high wavenumber range which resulted from the vibration of O-H bonds. The peaks are 1651 cm^{-1} (amide II band, N-H stretch), 1450–1411 cm^{-1} (asymmetric C-H bending of CH_2 group).

Figure 7.3 (c–e) shows the Cts/PEG/Ag-ZnO nanocomposites have shown absorption peaks at 1641 cm^{-1}, 1583 cm^{-1}, and 1325 cm^{-1} relating to amide I, amide II of C-O stretching vibrations, and N-H bending vibrations, respectively. The bands due to N-H and O-H stretching vibrations overlapped in the absorption peak at 3346 cm^{-1}. The bands corresponding to OH/NH_2 groups have shifted to 3328 cm^{-1} and became broader and stronger for Cts/PEG/Ag-ZnO nanocomposite [17]. The favorable intermolecular interactions between these groups and nanoparticles caused the variations in the characteristic absorption bands.

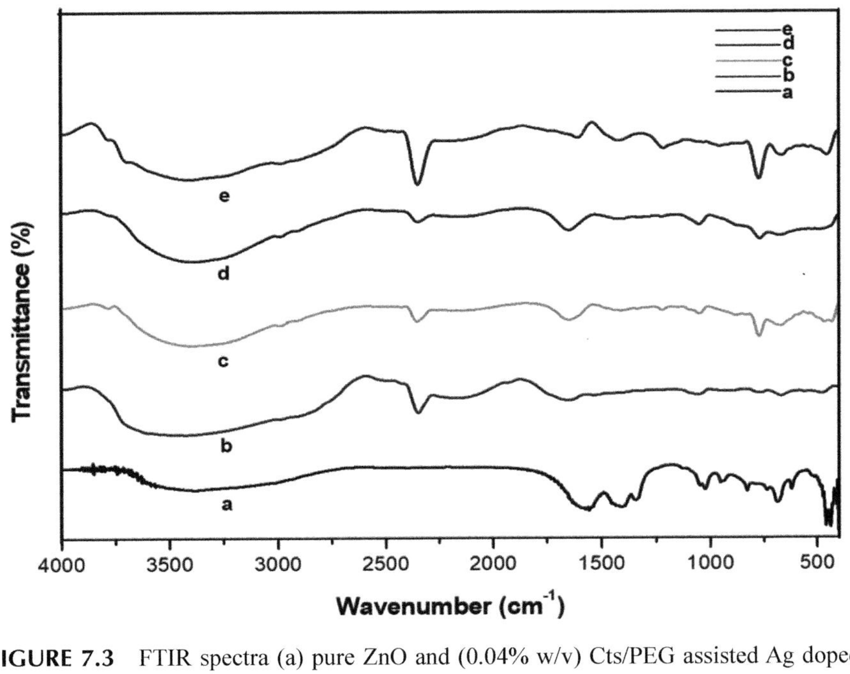

FIGURE 7.3 FTIR spectra (a) pure ZnO and (0.04% w/v) Cts/PEG assisted Ag doped with ZnO: (b) 0.0 (c) 0.03 (d) 0.06 (e) 0.09 mM.

Figure 7.3 (b) shows the FTIR spectra of ZnO doped Cts/PEG-Ag nanocomposite with various concentration of ZnO. The product has shown absorption peaks at 1641 cm^{-1} and 1583 cm^{-1} relating to amide I, amide II of C=O stretching vibrations, and N-H bending vibrations respectively. The bands due to N-H and O-H stretching vibration overlapped in the absorption peak at 3346 cm^{-1}. The bands corresponding to OH/NH_2 groups have shifted to 3328 cm^{-1} and became broader and stronger for ZnO, On Cts/PEG/Ag nanocomposite. The favorable intermolecular interactions between these groups and nanoparticles cause the variations in the characteristic absorption bands.

7.3.5 SURFACE MORPHOLOGICAL ANALYSIS

7.3.5.1 SEM/EDS STUDIES

The morphology of pure ZnO, Cts/PEG/Ag and ZnO (x=0, 0.03, 0.06 and 0.09) doped Ag nanoparticle was studied by SEM as shown in Figure 7.4 (i) (a-e). The surface morphology of the ZnO nanoparticles is as shown in Figure 7.4 (i) (a). The entire SEM picture clearly shows the average size of the nanoparticles is of the order of nanometer size. The ZnO nanoparticles show homogeneous, uniformly distributed over the surface and good connectivity between the particles. Cts/PEG/Ag (Figure 7.4 (i) (b)) showed unevenly distributed silver cubic nanoparticles on the surface of Cts/PEG.

The morphology of the particles is granular and the particles are loosely agglomerated. The sizes of undoped ZnO nanoparticles are in the range of 20–150 nm. In the case of ZnO doped Ag (Figure 7.4 (c-e)) the size of nanoparticles has a tendency to decrease firstly in size as the ZnO content increases but become almost constant thereafter. For all the samples, the distribution of particles reflects similar flake-like structures. Examination of the SEM shows extensive agglomeration of both ZnO and Ag-ZnO particles. Figure 7.4 (f), the energy dispersive X-ray spectrum of the doped Ag, shows a faint ZnO line at 3.0keV. The EDS analysis is generally accurate up to 0.1% but it provides only the surface composition. Figure 7.4 (f) illustrates the EDS spectrum of Cts/PEG/Ag-ZnO nanocomposite with 15 wt% Ag and 5 wt% ZnO. As shown in Figure 7.4 (f), that C, Zn, O, and Ag elements were identified. This result agreed well with XRD analysis.

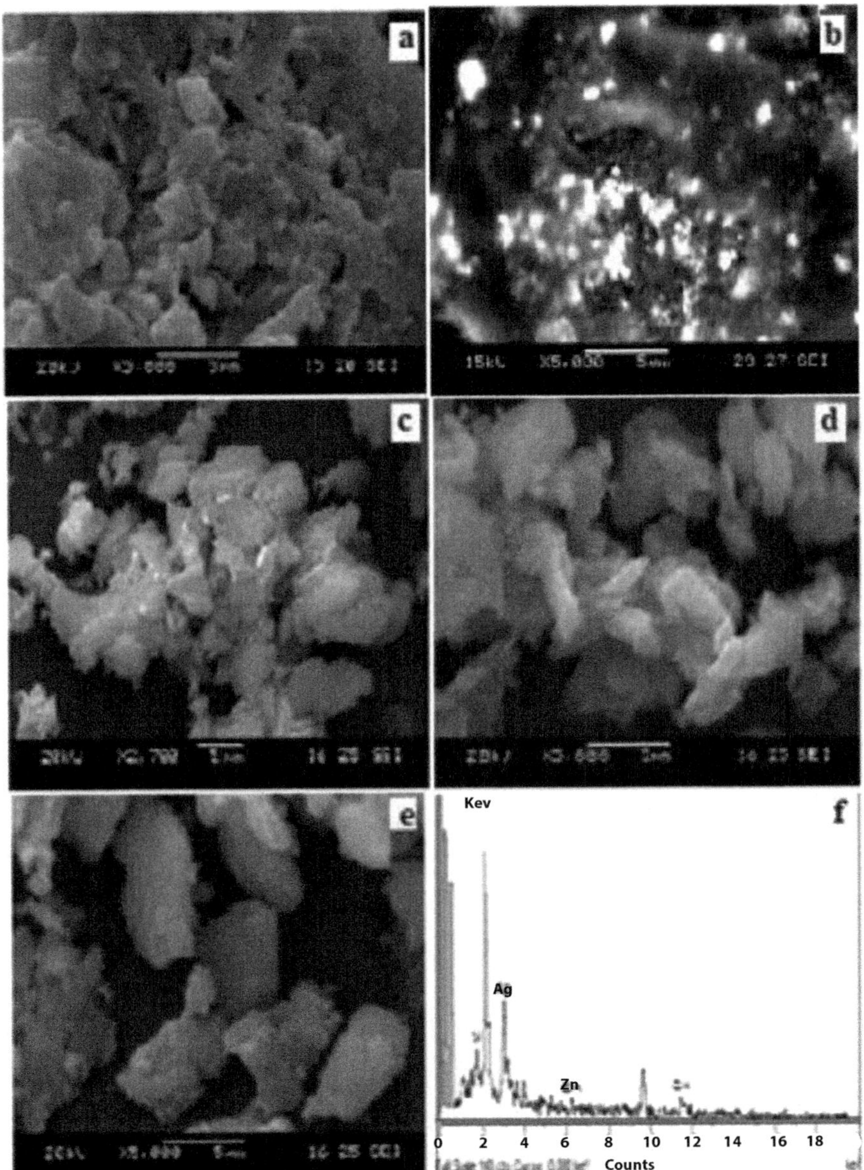

FIGURE 7.4 SEM/EDX spectrum of (a) pure ZnO and (0.04% w/v) Cts/PEG assisted Ag doped with ZnO: (b) 0.0 (c) 0.03 (d) 0.06 (e) 0.09 (f) EDS spectrum of 0.09 mM Ag nanocomposite at 80°C.

7.3.4.2 AFM STUDIES

Figure 7.5 shows an AFM image of the ZnO (0.09 mM) doped Ag nanostructures. Based on the AFM image, the root means square (RMS) roughness of the ZnO-doped Ag nanoparticle was 48.6 nm over an area of 1 μm^2. The analysis revealed that at a precursor ZnO concentration 0.09, the prepared sample exhibited very smooth surface and good crystal structures. AFM is a characterization technique for examining the nanomaterials in contact mode. The AFM has the significant advantage of probing in high details the surface topography qualitatively (by surface image) due to its nanometer-scale spatial resolution, both lateral and vertical. AFM has proved to be very helpful for the determination and verification of various morphological features and parameters.

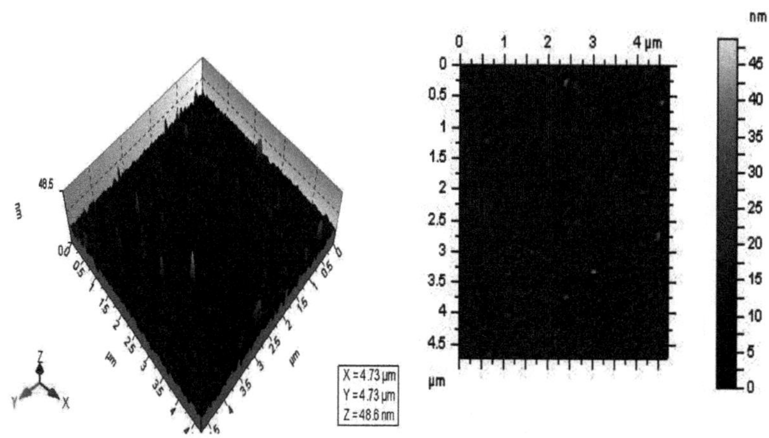

FIGURE 7.5 (See color insert.) Atomic force microscope image of Cts/PEG assisted Ag-doped with ZnO nanocomposites at 80°C.

7.3.6 DYNAMIC LIGHT SCATTERING (DLS)

DLS measurements are taken into account the whole composite nanoparticles. Finally, because of the resolution threshold of the used particle-size, particles with a hydrodynamic radius and smaller can be considered as the same fraction close to the higher range.

The measurements were performed on diluted samples, which contain elementary particles with higher particles dimensions, also measured by DLS

(129 nm, Figure 7.6). The well-dispersed nanoparticles (Cts/PEG/Ag-ZnO) were observed in the wide area of the photograph and the formation of silver nanoparticles was noticed for sulfonated derivative used in the works for other water-soluble polymers (Cts/PEG). The particle size is 129 nm.

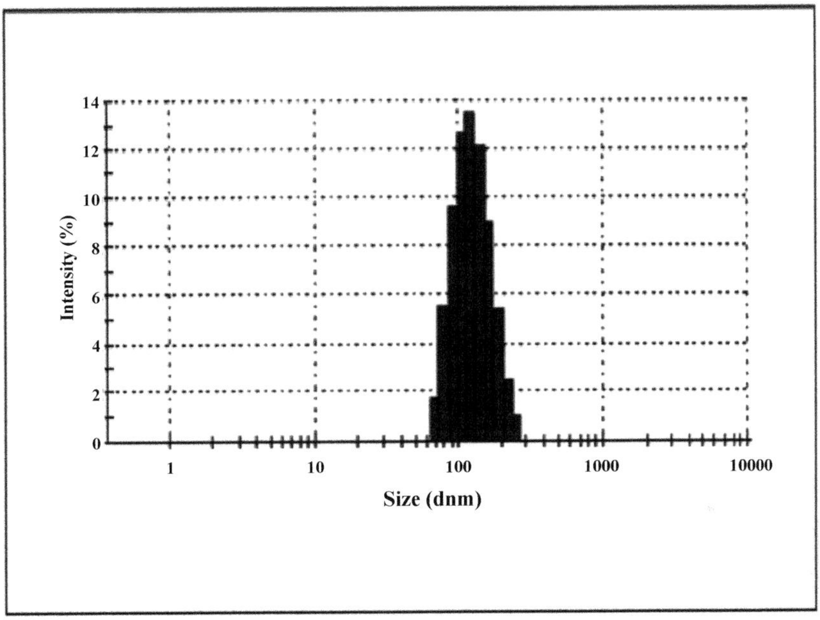

FIGURE 7.6 Particle size analyzer spectra of Cts/PEG assisted Ag-doped with ZnO nanocomposites at 80°C.

7.3.7 ZETA POTENTIAL

In the studied Cts/PEG/Ag-ZnO nanocomposites, these results indicate that the as-prepared fluid contains a fairly unimodal distribution of mono-dispersed clusters. This also means that positively charged nanospheres turn in to be more favorable to the "salting in" effect than negatively charged ones. This is consistent with the fact that the entities are typically dispersed on the positive side (Figure 7.7).

Another fundamental reason is the lower mobility of OH^- ions than protons, which renders the former to possess a higher tendency to be adsorbed on the particle surface. For the obtained nanoparticles, zeta values were measured and found to fall in +24 mV.

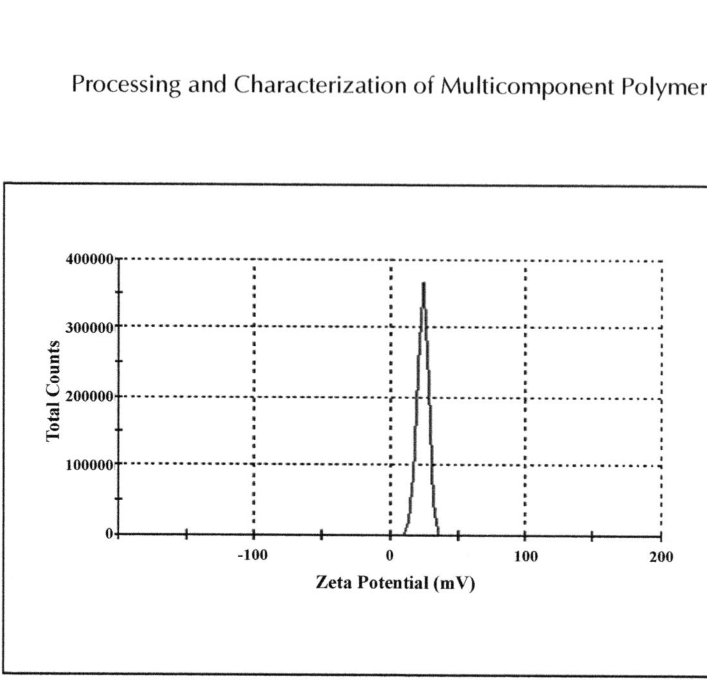

FIGURE 7.7 Zeta potential distribution of Cts/PEG assisted Ag-doped with ZnO nanocomposites at 80°C.

7.3.8 PHOTOCATALYTIC ACTIVITY MEASUREMENTS

Figure 7.8 demonstrates the enhanced photocatalytic activity of synthesized Cts/PEG/Ag-ZnO nanocomposite. The photocatalytic efficiencies of the prepared nanocomposite have been compared by employing MB. The dye used was heterocyclic dye methylene blue. The degradation profiles of dye, photocatalyzed by ZnO on Cts/PEG/Ag. The time profiles show enhancement in photocatalytic degradation of dye MB. This is in good agreement with the enhanced photocatalytic activity of ZnO doped Ag, reported by many researchers [18–22].

Increased surface area due to the small crystallite size may also contribute to the enhanced photocatalytic activity. A possible explanation for the observed enhancement of photocatalytic activities to a different extent is that the photocatalysis is also substrate specific. Photoactivity of the prepared samples was evaluated by the photocatalytic oxidation reaction of ZnO doped Cts/PEG-Ag nanocomposite under natural sunlight irradiation. The photocatalytic efficiency increased with ZnO doped Cts/PEG-Ag when compared to undoped Cts/PEG-Ag. This is possibly due to the visible light absorption ability of ZnO – Cts/PEG-Ag as evidenced by UV-Vis spectra. An increased photocatalytic activity obtained in the case of ZnO doped sample has also been ascribed to ZnO particles acting as an electron trap, retarding the electron-hole recombination process, and thereby, promising the photocatalytic activity.

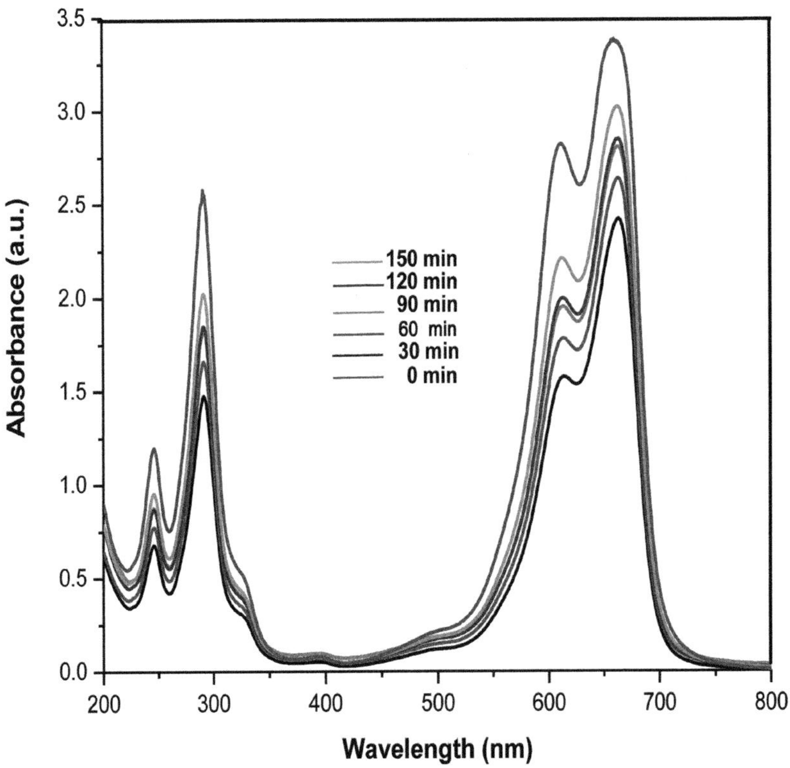

FIGURE 7.8 **(See color insert.)** UV absorption spectra of MB solution in the presence of Cts/PEG/Ag-ZnO: Synthesis with 80mg/L^{-1} under different UV irradiation time.

It has been clearly established that the photocatalyzed organic compound in solution is initiated by photoexcitation of the metal oxide semiconductor, followed by the generation of an electron-hole pair on the surface of the catalyst. Direct oxidation can also be occurring due to the high oxidative potential of the generated hole in the catalyst. Very reactive hydroxyl radicals can also be formed either by the decomposition of water or by the reaction of the hole with OH^-. The hydroxyl radical known as an extremely strong, non – selective oxidant can lead to the partial or complete reduction of several organic chemicals.

$$\left.\begin{array}{l} ZnO + hv \longrightarrow ZnO\,(e^-_{CB} + h^+_{VB}) \\ h^+_{VB} + \text{organic} \longrightarrow \text{organic} + \text{oxidation of the organic} \\ h^+_{VB} + H_2O \longrightarrow H^+ + OH^- \\ h^+_{VB} + OH^- \longrightarrow OH^- \end{array}\right\} \quad (7.5)$$

An electron in the conduction band in the catalyst surface can reduce molecular oxygen to superoxide anion. This radical, in the presence of an organic scavenger, may form organic peroxide or hydrogen peroxide.

$$\left.\begin{array}{l} e^-_{CB} + O_2 \longrightarrow O_2^- \\ O_2^- + \text{Organic} \rightarrow \text{Organic}\ ^-OO \\ O_2^- + HO_2 + H^+ \longrightarrow H_2O_2 + O_2 \end{array}\right\} \quad (7.6)$$

Electrons in the conduction band are also responsible for the production of hydroxyl radicals, species which have been indicated as the primary cause of organic compound degradation.

$$OH^{\cdot} + \text{Organic compound} \longrightarrow \text{Degradation of the organic compound}$$

The Ag nanoparticles as electron absorbent prevent electrons and holes from recombination, and the holes efficiently oxidize organic compounds, and thus the photocatalytic reaction is enhanced greatly. The major role of ZnO doping is attributed to the enhancement of photocatalytic activity. In fact, introducing ZnO nanoparticles to silver biopolymer matrix facilitates longer charge separation by trapping photogenerated electrons and photocatalytic activity under sunlight irradiation. The photocatalytic activity of Cts/PEG/Ag–ZnO nanocomposites demonstrated that by increasing sunlight illumination time the maximum absorption peak and concentration of MB decreases the presence of Ag–ZnO nanocomposites.

7.4 CONCLUSIONS

Novel ZnO doped Cts/PEG-Ag nanocomposites were prepared via the method of reduction at 80°C without chemicals. Synthesized nanocomposite has flake-like morphology had a uniform distribution within Cts/PEG copolymer. The sample consists of a mixture of nanoparticles with a particles size of 20–150nm obtained by SEM. The morphology of the particle can be controlled by selecting the precursor, temperature, the starting composition and the microstructure of the precursors. ZnO doping enhances the photocatalytic activity; the result shows that nanocomposites (ZnO doped Ag) were proven to have better properties and original size of particles after dispersion into the polymer matrix. ZnO doping on Cts/PEG/Ag decreases the average crystal size sharpens the band gap absorption and decreases the charge-transfer resistance. The formation of AgNps was confirmed in the UV-Vis absorption spectra, which showed the SPR band characteristic of AgNps in the range of 415–430 nm. The XRD result confirmed that the AgNps possess an FCC cubic crystal structure. The image of scanning electron microscopy revealed that with the ZnO concentrates large flake surfaces presented in the synthesized product. Also which shows the better compatibility was produced under high ZnO concentration on Cts/PEG-Ag nanocomposite. Photocatalytic activities of Cts/PEG-Ag and ZnO doped Cts/PEG-Ag samples were investigated by a natural sunlight-induced reduction in aqueous solution. The results clearly showed that doping of ZnO could greatly enhance the photocatalytic activity of ZnO – Cts/PEG-Ag nanocomposite. This enhanced photocatalytic activity is possibly due to the visible light absorption ability of ZnO doped Cts/PEG-Ag as observed from UV- Vis spectra.

KEYWORDS

- **methylene blue**
- **nanocomposites**
- **photocatalytic activity**

REFERENCES

1. Cohen, M. L., (2000). The theory of real materials. *Annual Review of real materials Science, 30*, 1–26.
2. Wang, Z. L., (2004). Zinc oxide nanostructure: Growth, properties, and applications. *Journal of Physics: Condensed Matter, 16*, 829–858.
3. Muzzarelli, R. A. A., (2009). Genipin crosslinked chitosan hydrogels as biomedical and pharmaceutical aids. *Carbohydrate Polymers, 77*, 1–9.
4. Adams, L. K., Lyon, D. Y., & Alvarez, P. J. J., (2006). Comparative eco-toxicity of nanoscale TiO_2, SiO_2, and ZnO water suspensions. *Water Research, 40*, 3527–3532.

5. Chen, R. Q., Zou, C. W., Bian, J. M., Sandhu, A., & Gao, W., (2011). Microstructure and optical properties of Ag-doped ZnO nanostructures prepared by a wet oxidation doping process. *Nanotechnology, 22*, 105706–105773.
6. Zhang, M., Li, X. H., Gong, Y. D., Zhao, N. M., & Zhang, X. F., (2009). *Properties and Biomedical Engineering Conference/FMBE Proceedings*, 119–120.
7. Colmenares, J. C., Aramendia, M. A., Arinas, A. M., Marinas, J. M., & Urbanr, F. J., (2006). Synthesis, characterization, and photocatalytic activity of different metal doped titania system, *Appl. Catl. A., 306*, 120–127.
8. Perrin, D. D., (1962). The hydrolysis of metal ions. *Part III. Zinc, J. Chem. Soc.*, 4500–4502.
9. Spanhel L., & Anderson, M. A., (1991). Semiconductor clusters in the sol-gel process: Quantized aggregation, gelation, and crystal growth in concentrated ZnO colloids, *J. A.m. Chem. Soc., 113*, 2826–2833.
10. Sepulveda-Guzman S., Reeja-Jayan, B., De la Rosa, E., Torres-Castro, A., Gonzalez-Gonzalez, V., & Jose-Yacaman, M., (2009). Synthesis of assembled ZnO structures by precipitation method in aqueous media, *Mater. Chem. Phys., 115*, 172–178.
11. Murugadoss, A., & Chattopadhyay, A., (2008). A 'green' chitosan–silver nanoparticles composite as a heterogeneous as well as a micro-heterogeneous catalyst, *Nanotechnology, 19*, 015603/1–015603/9.
12. Huang H. Z., Yuan, Q., & Yang, X. R., (2004). Preparation and characterization of metalchitosan nanocomposites, *Colloids Surf. B., 39*, 31–37.
13. Samuels, R. J., (1981). Solid state characterization of the structure of chitosan films, *J. Polym. Sci.: Polym. Phys. Ed., 19*, 1081–1105.
14. Dong Z. F., Du, Y. M., Fan, L. H., Wen, Y., Liu, H., & Wang, X. H., (2004). Preparation and properties of chitosan/gelatin/nano-TiO_2 ternary composite films, *J. Funct. Polym., 17*, 61–66.
15. Georgekutty, R., Seery, M. K., & Pillai, S C., (2008). A highly efficient Ag-ZnO photocatalyst: Synthesis, properties, and mechanism, *J. Phys. Chem. C., 112*, 13563–70.
16. Li-Hua, L., Jian-Cheng, D., Hui-Ren, D., Zi-Ling, L., & Xiao-Li, L., (2010). "Preparation, characterization, and antimicrobial activities of chitosan/Ag/ZnO blend films" *Chemical Engineering Journal, 160*, 378–382.
17. Yangshuo, L., & Hyung-Il, K., (2012). Characterization and antibacterial properties of genipin-crosslinked chitosan/poly (ethylene glycol)/ZnO/Ag nanocomposites. *Carbohydrate Polymers, 89*, 111–116.
18. Gu, C., Cheng, C., Huang, H., Wong, T., Wang, N., & Zhang, T. Y., (2009). Growth and photocatalytic activity of dendrite-like ZnO@Ag heterostructure nanocrystals. *Cryst. Growth Design, 9*, 3278–3285.
19. Liqiang, J., Dejun, W., Baiqi, W., Shudan, L., Baifu, X., Honggang, F., & Jiazhong, S., (2006). Effects of noble metal modification on surface oxygen composition, charge separation and photocatalytic activity of ZnO nanoparticles. *J. Mol. Catal. A., 244*, 193–200.
20. Zheng, Y., Zheng, L., Zhan, Y., Lin, X., Zheng, Q., & Wei, K., (2007). Ag/ZnO heterostructure nanocrystals: Synthesis, characterization, and photocatalysis. *Inorg. Chem., 46*, 6980–6.
21. Lin, D., Wu, H., Zhang, R., & Pan, W., (2009). Enhanced photocatalysis of electrospun Ag-ZnO heterostructured nanofibers, *Chem. Mater, 21*, 3479–84.

CHAPTER 8

DEVELOPMENT OF HIGH-PERFORMANCE *IN-SITU* POLYPROPYLENE/NYLON 6 MICROFIBRILLAR COMPOSITES

B. D. S. DEERAJ[1], K. JAYANARAYANAN[2], and KURUVILLA JOSEPH[1]

[1]*Department of Chemistry, Indian Institute of Space Science and Technology, Kerala–695547, India*

[2]*Department of Chemical Engineering and Materials Science, Amrita University, Coimbatore, India*

ABSTRACT

Microfibrillar in-situ composites of polypropylene (PP)/nylon 6 (N6) blends were prepared in the presence and absence of carbon nanotubes (CNTs). The N6/PP/CNT blends were prepared by melt blending in a twin screw extruder. These extruded strands were elongated in a lab scale cold stretching unit followed by consolidation in a compression molding machine. The scanning electron microscope (SEM) micrographs of extruded blends displayed immiscible blend morphology which became aligned and oriented after fibrillation step. The mechanical performance of the extrudates affirms the contribution of microfibrils and CNTs in reinforcing PP matrix. The storage modulus of the MFCs was found to be superior to the individual polymers throughout the temperature range of analysis.

8.1 INTRODUCTION

Microfibrillar composites (MFCs) are the advanced class of fiber-reinforced composites, formed in-situ from immiscible and incompatible polymer blends. These blends offer many advantages like excellent mechanical performance, superior storage modulus, high creep and fatigue resistance, good impact strength and low gas permeability [1–3]. The main applications and end users of MFC technology include the plastics recycling industry, automobile industry, and packaging industry. The choice of blend components, blend composition, stretching ratio, the morphology of fibrils formed and consolidation temperature are the main factors that determine the performance of these composites [4–6].

Kiss et al. [7], proposed the concept of in-situ composites in which the reinforcements, matrix, and interphase are formed intrusively with reduced processing steps. This approach leads to the development of composites with better mechanical properties and eliminates cumbersome steps associated with conventional composites processing. Initial work on in-situ composites was reported on blends of liquid crystalline polymers (LCPs) and thermoplastics [8–10]. But reinforcing thermoplastics with LCPs was not economical and the resultant composite manifested poor comprehensive properties. This led to the development of thermoplastic/thermoplastic MFCs which are cheap and easily processable.

Many thermoplastic combinations were considered for preparing MFCs, which include polyesters/N6 [11], poly (ethylene terephthalate) (PET)/PP [5, 12–18], PET/N6 [19], polyamide/polyethylene (PE) [20], polycarbonate (PC)/high density polyethylene (HDPE) [21], poly (phenylene sulfide)/PP [22], PET/low density polyethylene (LDPE) [23], PET/PE [24–27], etc., The main steps involved in fabrication of MFCs are: (1) melt mixing of component polymers, (2) stretching of the blended extrudate uniformly and (3) consolidation step to form in-situ composites. In the initial step, the two constituent polymers are compounded at a processing temperature above the melting point of both selected polymers to achieve a homogeneous melt. In the second step, the blended extrudates are extended in a stretching unit and in the next step; the drawn extrudates were isotropized at a temperature in between the melting points of both the polymers. This result in the melting of low-temperature polymer to form the continuous phase (matrix) incorporated with high-temperature polymer fibrils. Here, the fibril diameter is found to be in the micron range and are formed in-situ during the process. Hence, the composites developed from this technique were termed as microfibrillar composites.

The conversion of dispersed phase droplets to stable fibril form is a fundamental requirement to form in-situ MFCs. In the case of incompatible blend systems, the morphology obtained after blending depends on rheological properties, viscosity ratio, mixing mode and parameters [28, 29]. Viscosity ratio is termed as the ratio of dispersed phase viscosity to the matrix viscosity [30]. Smaller viscosity ratio and elongation type of flow assist the fibrillation of the dispersed phase [31]. The morphology of the MFC prepared from a low viscosity pair PET/PE was observed at ratio 15/85 w% by SEM and variation in dispersed phase geometry was reported [24]. The effect of stretching on the dispersed phase morphology in the case of PP/PET at 85/15 w/w% was reported [5]. The impact of consolidation temperature (molding temperature) on the dispersed phase morphology of PET/PP composites was also studied and low-temperature molding was found to be better than that carried out at higher temperature [32].

At the same blend composition, the mechanical strength of in-situ composites was found to be much higher than that of normal blends due to the fibrillar morphology attained by the dispersed phase. The high aspect ratio of fibrils and fibril/matrix interface interactions are the important factors that contribute to the enhanced mechanical properties of these composites. The effect of blend ratio in PET/LDPE blends by varying PET content was analyzed and they reported that the tensile strength increased up to 25% of PET loading and deteriorated on further loading. The optimum blend composition of PET/PE system was predicated and experimentally investigated [24]. The influence of stretch ratio in PET/PE composites was studied and it was observed that with an increase in stretch ratio the strength and modulus of the MFC increases, reaches a maximum and decrease beyond a certain level. The reason for the aforementioned decrease in strength was attributed to the ductile to the brittle transition of the composite [26]. The similar trend was observed in the case of PET/PP composites, where the decrease in mechanical properties at stretch ratios was explained as fibril breakage which was supported by the scanning electron micrographs [13]. The impact strength of microfibril composites was found to be higher than normal polymer blends [13, 33]. The Dynamic mechanical analysis of PP/PET systems was performed to evaluate the effect of temperature on the storage modulus, loss modulus and tan δ at different stretch ratios.

It was found that the storage modulus of PP drastically decreased at elevated temperatures. At the stretch ratio of 5 and 8, these fibrils helped to improve the modulus especially at high temperature and loss modulus also exhibited a positive shift with stretch ratio 5, showing good energy dissipation [14]. The effect of nanofillers incorporated in MFCs has not been widely reported. It is expected that nano-sized inclusions in MFCs improve the performance of the combined system due to the high filler aspect ratio and surface area. When these incorporations are conductive, in addition to the mechanical property the conductivity of the system also increases. So, the conversion of MFC to conducting polymer composites (CPCs) is a novel thought. These CPCs enjoy the benefit of high strength and conductivity because of the conductive path created with low percolation threshold. Conducting MFCs were developed by incorporating carbon black (CB) in the PET/PE blends and. These MFCs were found to have percolation threshold at 3.7 vol.% carbon black while it is 5 vol.% and 7.3 vol.% for CB/PET/PE normal blends and CB/PE composites respectively [34]. MFCs were developed by incorporating CNTs into the dispersed phase or compatibilizer (in the interface) of MFCs and the electrical resistivities were evaluated. in the CNT were loaded in the interface and in the dispersed phase of MFCs and significant property improved was evaluated [35]. Another work on PP/ Poly (butylene terephthalate) (PBT)/MWCNT reported the improvement in mechanical properties and electrical conductivity of the system [36].

In this work, the optimum blend ratio and stretching ratio for developing MFCs of PP/N6 blend system were investigated and presented. The effect of CNTs (concentration w%) on the morphology and mechanical property was examined. The dynamic mechanical analysis of PP, N6, and MFC was performed and reported.

8.2 EXPERIMENTAL WORK

8.2.1 PREPARATION OF PP/N6 MFCS

The polymers selected for the study were: Nylon 6 (as fiber forming a polymer, procured from Sigma Aldrich) and PP (as the matrix, grade: H110MA). The melting temperatures of PP and Nylon were around 170°C and 220°C, respectively. Both the polymers were dried overnight at 80°C for removing the moisture content. The blend components were melted mixed in a twin screw extruder at specified combinations required. The melt blending was carried out at an optimized temperature profile of 240°C, 245°C, 250°C, 250°C, and 255°C from the feed zone to die zone. The speed of the screw was maintained at 10 rpm. Subsequently, the extrudates coming out of the die were cooled in a water bath and were taken to stretching unit. The stretching unit consists of two pairs of grip rollers and an oven maintained at 105°C. The speed of the first roller set was maintained the same as the extrudate velocity (S_1), but the speed of second roller set (S_2) was varied to attain different stretch ratios. The stretch ratio can be defined as the ratio of velocities (S_2/ S_1), which can be varied. These extrudates obtained during the orientation step were referred to as microfibrillar blends (MFBs). In a compression molding machine, these MFBs were consolidated to form MFCs at a temperature of 200°C. This resulted in the generation of the transformation of PP as a matrix filled with N6 microfibrils.

8.2.2 PREPARATION OF PP/N6/CNT MFCS

The CNTs (both multi-walled and single-walled) were procured from Sigma Aldrich. The preparation of PP/N6/CNT MFCs was similar to the method mentioned earlier except that CNTs were mixed with N6 prior to blending with PP. The scheme for the preparation of these composites and the fibril morphology was presented in Figure 8.1.

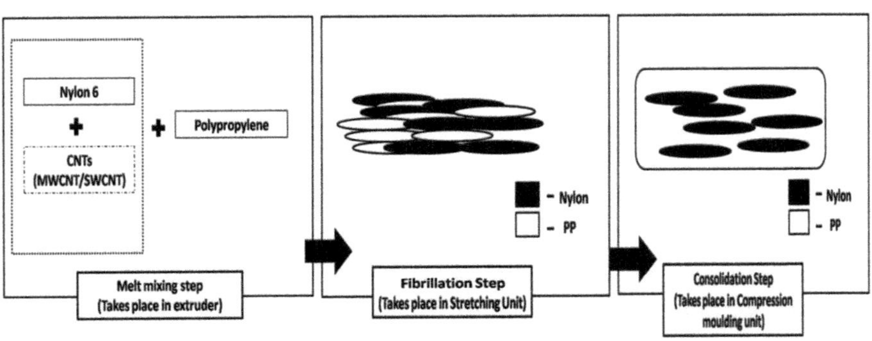

FIGURE 8.1 Scheme for the preparation of PP/N6/CNT MFCs.

8.3 CHARACTERIZATION METHODS

8.3.1 MORPHOLOGY

The PP phase from PP/N6 blend extrudates was extracted by hot xylene and the morphology of N6 fibrils was observed. A high-resolution SEM with an acceleration voltage of 20 KV was used for this purpose. Prior to observation, the samples were given a gold coating to make them conductive.

8.3.2 MECHANICAL PROPERTIES

The tensile strength of MFBs at different blend compositions and stretch ratios was evaluated by TINIUS OSLEN Universal testing machine at room temperature as per ASTM C1557 standard. For testing, a gauge length of 100 mm and a crosshead speed of 100 mm/min was maintained.

8.3.3 DYNAMIC MECHANICAL ANALYSIS

The Dynamic storage modulus of PP, N6, and MFC was found using a dynamic mechanical thermal analyzer (DMA 800, Perkin Elmer, USA). The sample dimensions were maintained as 55mm x 15 mm x 3.2 mm and the analysis was carried out in 3 points bending mode. At a frequency of 1 Hz, the samples were tested in the temperature range –20°C to 120°C, at a heating rate of 2°C/min.

8.4 RESULTS AND DISCUSSION

8.4.1 OPTIMIZATION OF BLEND RATIO

8.4.1.1 MORPHOLOGY

The performance of MFCs strongly depends on the morphology, aspect ratio and volume fraction of fibrils formed which in-turn depends on the blend composition and stretch ratio. The morphologies of PP/N6 MFBs at different compositions by varying N6 (from 10 w% to 40 w%) was at stretch ratio 2 observed and presented in Figure 8.2.

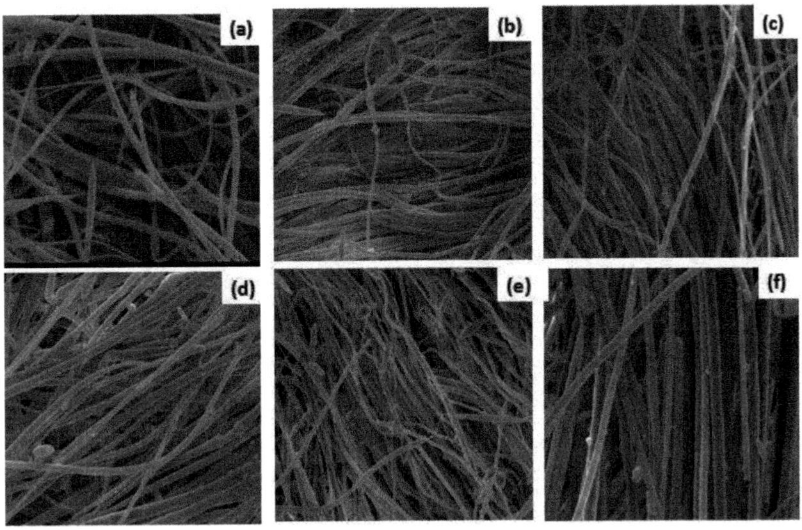

FIGURE 8.2 SEM images of PP/N6 blends at (a) 90/10, (b) 85/15, (c) 80/20, d)75/25, (e) 70/30 and f) 60/40 (w%/w%) at stretch ratio 2.

From the Figure 8.2, it can be realized that the N6 phase (dispersed phase) forms fibril morphology at stretch ratio 2. At all blend compositions, well-defined fibril morphology was obtained but volume fraction and fibril diameter were distinctive. The average N6 fibril diameter was found and reported in Figure 8.3, which reveals that the avg. fibril diameter increased with increase in the N6 ratio. The volume fraction of the fibrils at blend ratios 70/30, 75/25 and 80/20 tends to be high. The dependence of blend ratio when the stretch ratio is fixed was well conveyed in Figure 8.3.

8.4.1.2 MECHANICAL PERFORMANCE

The aspect ratio of the N6 phase and the interface between the PP and N6 microfibrils determine the mechanical properties of PP/N6 MFCs. The effect of blend composition on the static mechanical properties of PP/N6 MFBs was studied by varying N6 content and reported in Figure 8.4. It can be seen the tensile strength of the extrudates increased with N6 content until 30 w% beyond which it decreased. In the former case, the addition of N6 reinforced the PP matrix, whereas at 40 w% of N6 the brittle nature of the system increased and strength decreased.

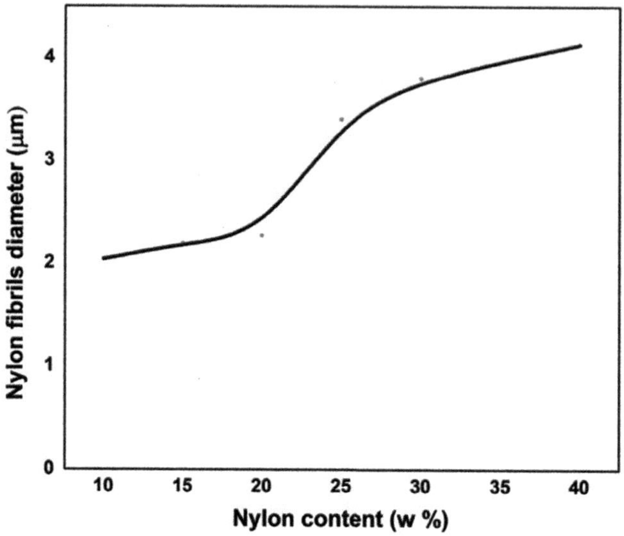

FIGURE 8.3 Variation of the average diameter of N6 fibrils with N6 content (w%).

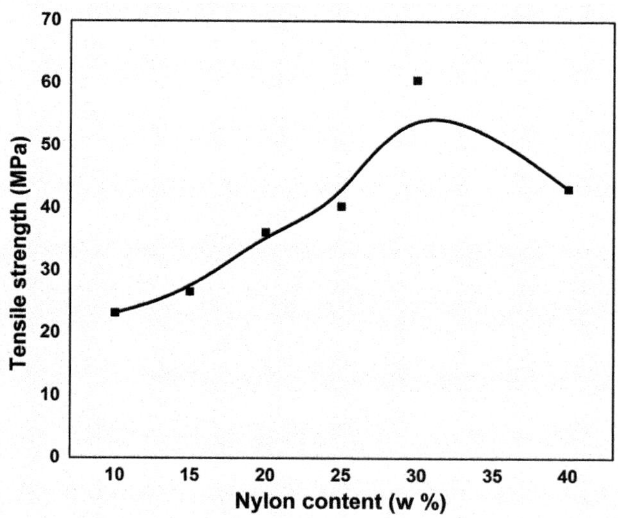

FIGURE 8.4 Variation of tensile strength PP/N6 extrudates with N6 content (w%).

By considering the mechanical strength and morphology, the PP/N6 blend ratio at 70/30 was considered as optimum and further studies were continued at this composition.

8.4.2 OPTIMIZATION OF STRETCH RATIO

At a fixed composition of PP/N6 (70/30 w/w%), the extrudates at different stretch ratios were prepared. Figure 8.5 shows the SEM images of PP/N6 extrudates at stretch ratios of 2, 5 and 8 respectively.

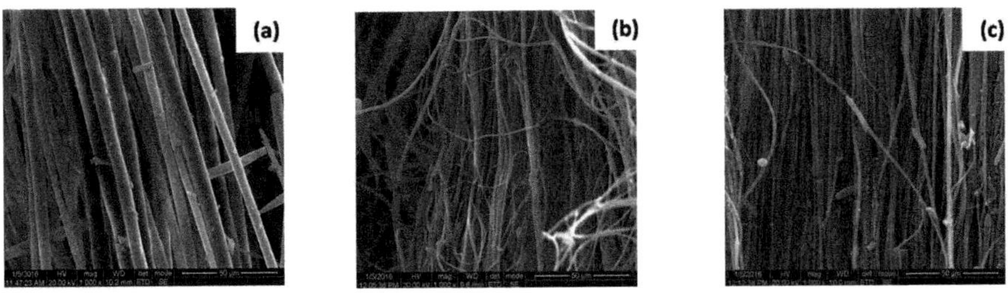

FIGURE 8.5 SEM image of PP/ N6 (70/30 w/w%) blends at stretch ratios of (a) 2, (b) 5 and (c) 8 respectively.

From the Figure 8.5(a), one can observe that the microfibrils with an average diameter of 3.7 microns are aligned, but as the stretch ratio increases the diameter of the microfibers decreases drastically to 1.1 microns. As a result, its aspect ratio elevates which can be observed in Figure 8.5 (b). Further increase in the stretch ratio leads to the increase of the fibril diameter to 1.67 microns, which can be attributed to the breakage of microfibers due to high elongational stresses. From Figure 8.5(c) the attrition of fibers can be evidenced. In Figure 8.6, the variation of the microfibril diameter at different stretch ratios is presented.

8.4.2.1 MECHANICAL PERFORMANCE

The tensile properties of the MFBs were significantly enhanced with increasing stretch ratio from 1 to 5, indicating that the microfibrils act as good reinforcement for the isotropic matrix. Figure 8.7 shows the tensile strength variation of PP/N6 extrudates with respect to stretch ratio. The increase in strength of MFBs from stretch ratios 1 to 5 is because of the improvement in the nylon fibril aspect ratio and its distribution in the PP matrix. As the stretching forces increase the fibrils become more aligned and oriented towards the flow field of the extruder. The decrease in the tensile strength at stretch ratios beyond 6

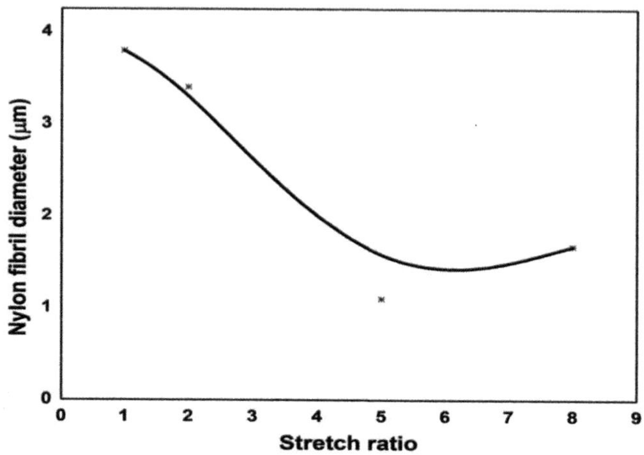

FIGURE 8.6 Variation of the average diameter of nylon 6 microfibril with the stretch ratio.

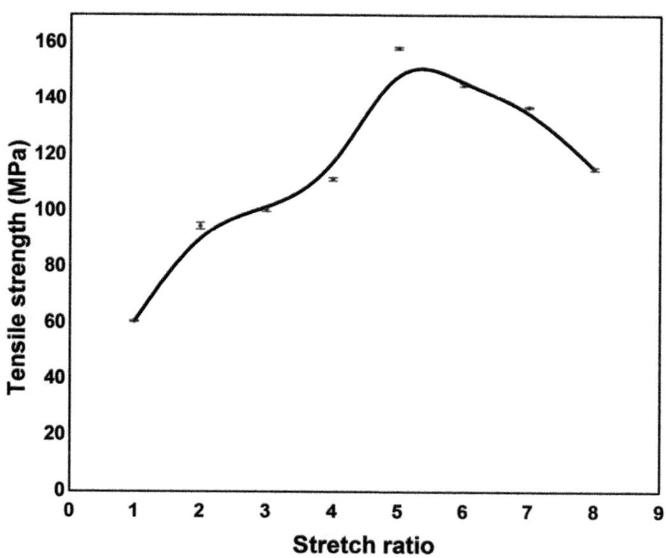

FIGURE 8.7 Variation of tensile strength of PP/N6 extrudates with respect to stretch ratio.

is due to the ductile to brittle transition in the failure of the extrudates. For stretch ratios up to 5, the fibrils have a ductile failure with necking but from 5 to 8 due to increased fibrillation, the extrudates became brittle and abrupt failure took place. Further, the decrease in ratio 8 can be attributed to the breakage or attrition of the nylon fibrils due to high applied forced during stretching. So, the morphology supports the increased fibrillation and fineness of fibrils at stretch ratio 5. The breakage of fibrils and increment in fibril diameter at 8 affirms the breakage occurred during drawing.

An improvement of around 160% in tensile strength is noted at stretch ratio 5 and it decreased to 90% at stretch ratio 8. It has to be noted that the strength of stretched samples is always greater than the un-stretched samples. So, above a critical ratio, the nylon microfibrils cannot provide effective reinforcement to the PP matrix. By, considering morphology and tensile performance the stretch ratio of 5 is found to be optimum of PP/N6 blend ratio 70/30 w/w%.

8.4.3 EFFECT OF CNT LOADING

8.4.3.1 MORPHOLOGICAL DEVELOPMENT OF PP/NYLON 6/ CNT MFBS

The SEM micrographs of CNT (MWCNT & SWCNT, respectively) loaded (0.2 w%) MFCs and those prepared in its absence were observed and are presented in Figure 8.8. The Nylon fibril average diameter was 1.01 micron in the case of extrudates, while it increased to 2.2 and 2.05 microns for CNT loaded samples. The CNTs added during the processing entered the interface region and provides a compatiblizing effect, thereby increasing the diameter of the fibrils.

FIGURE 8.8 SEM image of PP/Nylon 6/CNT extrudates at (a) 0% CNT loading, (b) 0.2 w% MWCNT loading, c/) 0.2 w% SWCNT loading at stretch ratio 5.

8.4.3.2 MECHANICAL PERFORMANCE

To the PP/Nylon 6 system, nanofillers CNTs (both MWCNTs and SWCNTs) were added, and MFCs were prepared. The extrudates made from PP/N6/CNT at compositions of 0.2%, 0.3%, 0.35% and 0.4% are tested at stretch ratio 5 and the tensile strength values are presented in Figures 8.9 and 8.10. The extrudates with CNTs showed a significant improvement in the mechanical performance and the strength increased with nanofiller loading. This is because the CNTs incorporated in the system acts good reinforcements and have a constructive effect in strengthening PP. The CNTs added to the system enters the interface of the fibrils contribute synergistically in improving the mechanical properties. Both the type of CNTs gave almost the same values when filler loading remained the same. The MWCNT loaded samples gave slightly better property than the SWCNT loaded extrudates owing to the presence of more number of walls and forces of attraction between them.

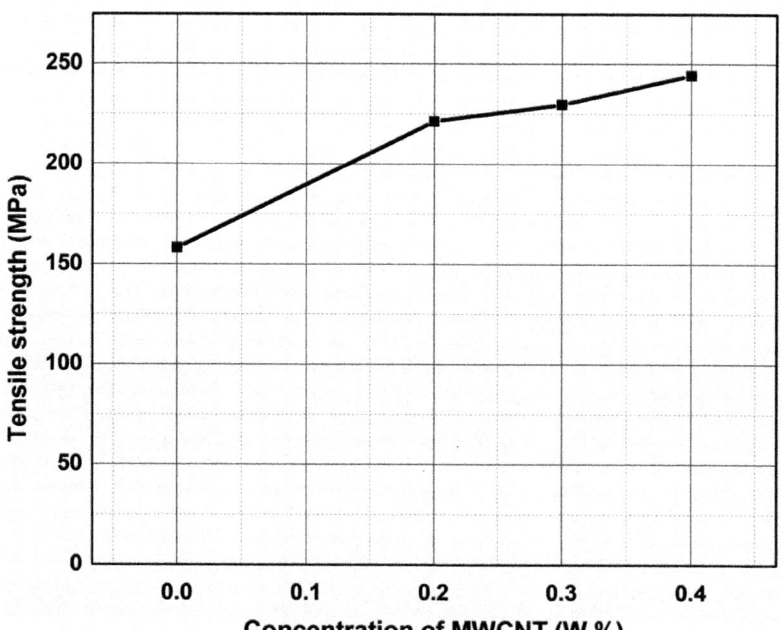

FIGURE 8.9 Variation of mechanical strength of PP/N6 extrudates with a concentration of MWCNTs.

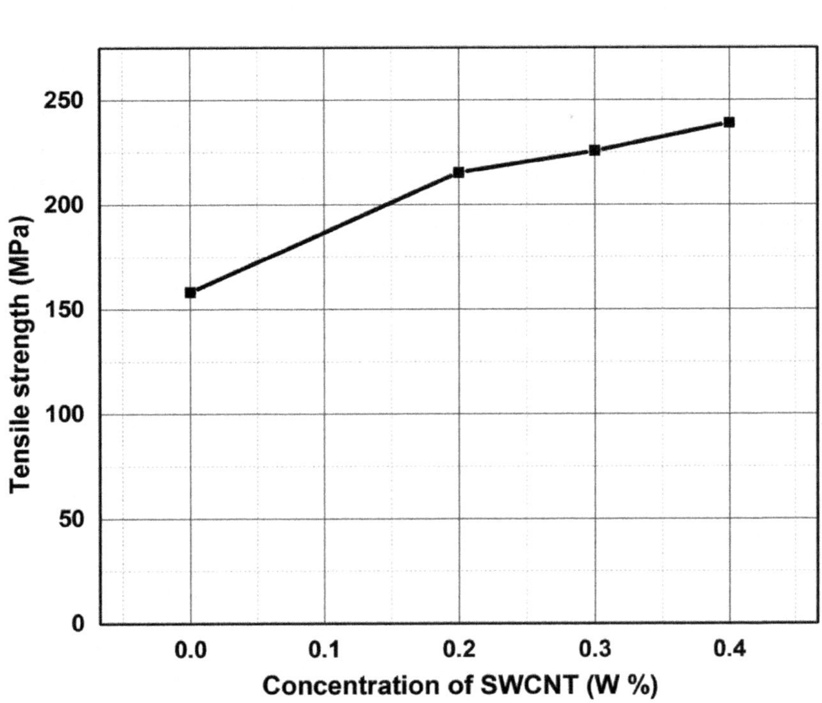

FIGURE 8.10 Variation of mechanical strength of PP/N6 extrudates with a concentration of SWCNTs.

The improvement in tensile strength in the CNT loaded MFBs was found and it was compared with those prepared in the absence of nanotubes. The tensile strength improvement for the MFBs prepared at stretch ratio is reported in Table 8.1. It can be clearly observed that the MWCNT loaded MFBs manifest better properties than the corresponding SWCNT loaded ones. It can also be inferred that the concentration of filler has a positive effect on MFB performance.

TABLE 8.1 Percentage Improvement (in Tensile Strength) in CNT Loaded MFBs with CNT Concentration

Filler concentration (w%)	MWCNT Loaded samples	SWCNT loaded samples
0.2	39.9%	36.04%
0.3	45.2%	42.5
0.4	54.43%	50.9%

8.4.3 DYNAMIC MECHANICAL ANALYSIS (DMA)

The dynamic mechanical analysis is usually employed to investigate the changes that occur in polymer systems over a broad range of temperature and frequency. This method is the most effective way to find the performance of polymeric matrices. It helps us to know about the link between the molecular parameters and the mechanical properties.

In this study, an attempt is made to analyze the dynamic mechanical properties like storage modulus of in-situ microfibrillar composites of PP/N6 system (70/30 w/w% and stretch ratio 5) with that of the pure polymers PP and N6. The dynamic analysis of molded samples was carried out from –20°C to 120°C at 1 Hz. The storage modulus (G') obtained from the DMA determines the load-bearing capacity of the visco-elastic material. In the Figure 8.11, the variation of the storage modulus with the temperature for PP, N6 and MFC system at 1 Hz frequency is shown.

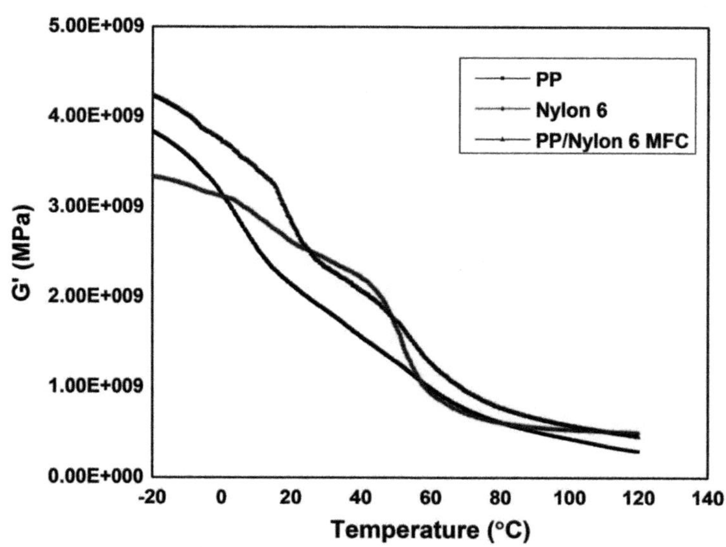

FIGURE 8.11 Variation of the storage modulus with temperature for PP, Nylon 6 and MFC.

In the case of pure PP, there is a steady fall in its storage modulus after its glass transition temperature. The increased molecular mobility of the polymer chains at this temperature may be the reason for that behavior. Throughout the temperature range, the MFCs is having high modulus compared to PP, indicating its efficiency in inducing stiffness to the system due to the presence of nylon fibrils (Figure 8.12).

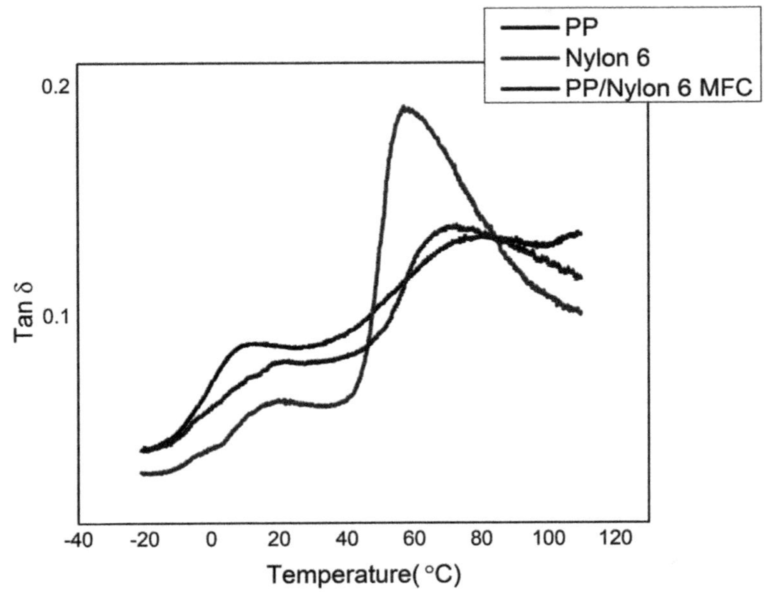

FIGURE 8.12 Variation of the mechanical damping factor with temperature.

The loss modulus (E") is the viscous response of viscoelastic materials. It is a measure of the energy dissipated or lost as heat per cycle of sinusoidal deformation when different systems are compared at the same strain amplitude. PP exhibits three relaxations localized in a temperature range of –80°C, 10°C and 100°C. There is a steady fall of the moduli beyond the glass transition for PP. The loss moduli are found to be maximum for MFC due to the improved energy dissipation, which could contribute to the better impact properties

The ratio of the loss modulus to the storage modulus is measured as the mechanical loss factor (tan δ). The damping properties will give the balance between the elastic phase and viscous phase in a polymeric structure. The main factors that determine the damping behavior are: (i) relaxation between the matrix and fiber, (ii) the interface between the fiber and matrix, (iii) fiber loading and length. In this study, the variation of the mechanical loss factor with the temperature at 1 Hz is represented in

Figure 8.12. The lowering and broadening of the tan delta peak indicate the restriction offered for the relaxation of the polymer chains. In comparison with PP it could be mentioned that the tan delta peak of MFCs lowered which can be assigned to the presence of N6 fibrils arresting the mobility of PP chains.

8.5 CONCLUSION

From the thermoplastic combinations of Nylon and polypropylene, microfibrillar composites were successfully developed in the presence and absence of CNTs. The blend composition of 70/30 (PP/N6) was found as the ideal blend composition for developing MFCs. At this optimized blend ratio, the effect of stretching was analyzed and found that stretch ratio 5 tends to give MFCs with superior mechanical properties because of the fine fibril morphology. The influence of CNTs on MFCs was evaluated and found that its incorporation tends to further improve the performance of the system with filler loading. The dynamic storage modulus of MFCs developed from extrudates of stretch ratio 5 was found superior to the component polymers. Hence, the careful choice of blend composition, stretching ratio and choice of filler & its loading, can help in developing high-performance thermoplastic composite materials (Figure 8.13).

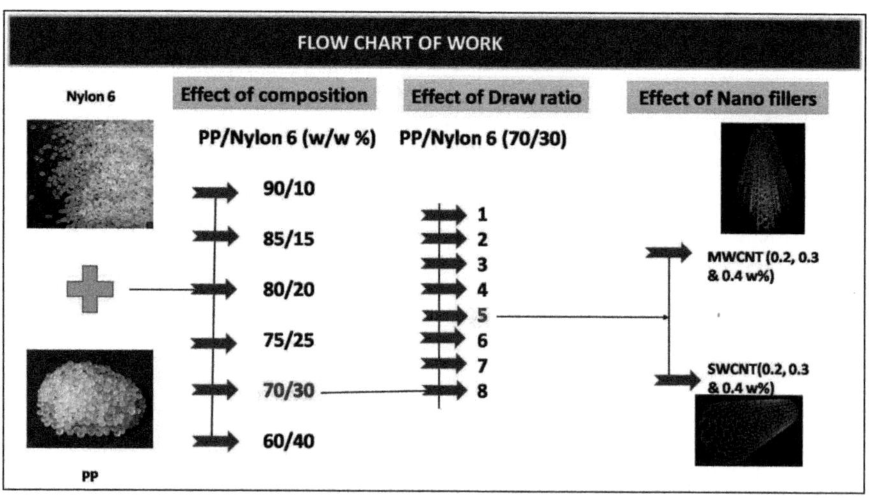

FIGURE 8.13 (See color insert.) The pictorial conclusion of work.

ACKNOWLEDGMENT

The authors deeply acknowledge Indian Institute of Space Science and Technology (IIST) for funding and SAIF, IIT Madras for the SEM images.

KEYWORDS

- **dynamic mechanical thermal analyzer**
- **mechanical properties**
- **microfibrillar composite**
- **morphology**
- **scanning electron microscope**

REFERENCES

1. Fakirov, S., (2013). Nano- and microfibrillar single-polymer composites: A review. *Macromolecular Materials and Engineering, 298*(1), 9–32.
2. Shields, R., Bhattacharyya, D., & Fakirov, S., (2008). Fibrillar polymer-polymer composites: Morphology, properties, and applications. *Journal of Materials Science 43*(20), 6758.
3. Evstatiev, M., Fakirov, S., & Friedrich, K., (2005). Manufacturing and characterization of microfibrillar reinforced composites from polymer blends. *Polymer Composites,* 149–167.
4. Denchev, Z. Z., & Dencheva, N. V., (2008). Transforming polymer blends into composites: A pathway towards nanostructured materials. *Polymer International, 57*(1), 11–22.
5. Jayanarayanan, K., Jose, T., Thomas, S., & Joseph, K., (2009). Effect of draw ratio on the microstructure, thermal, tensile, and dynamic rheological properties of in-situ microfibrillar composites. *European Polymer Journal, 45*(6), 1738–1747.
6. Jayanarayanan, K., Thomas, S., & Joseph, K., (2012). Effect of blend ratio on the mechanical and sorption behavior of polymer-polymer microfibrillar composites from low-density polyethylene and polyethylene terephthalate. *Journal of Reinforced Plastics and Composites, 31*(8), 549–562.
7. Kiss, G., (1987). In situ composites: Blends of isotropic polymers and thermotropic liquid crystalline polymers. *Polymer Engineering & Science, 27*(6), 410–423.
8. Postema, A., & Fennis, P., (1997). Preparation and properties of self-reinforced polypropylene/liquid crystalline polymer blends. *Polymer, 38*(22), 5557–5564.

9. Saengsuwan, S., Bualek-Limcharoen, S., Mitchell, G. R., & Olley, R. H., (2003). Thermotropic liquid crystalline polymer (Rodrun LC5000)/polypropylene in-situ composite films: Rheology, morphology, molecular orientation and tensile properties. *Polymer, 44*(11), 3407–3415.
10. Chiou, Y. P., Chiou, K. C., & Chang, F. C., (1996). In situ compatibilized polypropylene/liquid crystalline polymer blends. *Polymer, 37*(18), 4099–4106.
11. Fakirov, S., Evstatiev, M., & Petrovich, S., (1993). Microfibrillar reinforced composites from binary and ternary blends of polyesters and nylon 6. *Macromolecules, 26*(19), 5219–5226.
12. Jayanarayanan, K., Bhagawan, S., Thomas, S., & Joseph, K., (2008). Morphology development and nonisothermal crystallization behavior of drawn blends and microfibrillar composites from PP and PET. *Polymer Bulletin, 60*(4), 525–532.
13. Jayanarayanan, K., Thomas, S., & Joseph, K., (2008). Morphology, static, and dynamic mechanical properties of in situ microfibrillar composites based on polypropylene/poly (ethylene terephthalate) blends. *Composites Part A: Applied Science and Manufacturing, 39*(2), 164–175.
14. Jayanarayanan, K., Thomas, S., & Joseph, K., (2009). Dynamic mechanical analysis of in situ microfibrillar composites based on PP and PET. *Polymer-Plastics Technology and Engineering, 48*(4), 455–463.
15. Jayanarayanan, K., Thomas, S., & Joseph, K., (2011). In situ microfibrillar blends and composites of polypropylene and poly (ethylene terephthalate): Morphology and thermal properties. *Journal of Polymer Research, 18*(1), 1–11.
16. Li, Z. M., Lu, A., Lu, Z. Y., Shen, K. Z., Li, L. B., & Yang, M. B., (2005). *In-situ* microfibrillar PET/iPP blend via a slit die extrusion, hot stretching and quenching process: Influences of PET concentration on morphology and crystallization of iPP at a fixed hot stretching ratio. *Journal of Macromolecular Science, Part B: Physics, 44*(2), 203–216.
17. Li, Z. M., Li, L. B., Shen, K. Z., Yang, W., Huang, R., & Yang, M. B., (2004). Transcrystalline morphology of an in situ microfibrillar poly (ethylene terephthalate)/poly (propylene) blend fabricated through a slit extrusion hot stretching-quenching process. *Macromolecular Rapid Communications, 25*(4), 553–558.
18. Li, Z. M., Li, L. B., Shen, K. Z., Yang, M. B., & Huang, R., (2004). In-situ microfibrillar PET/iPP blend via slit die extrusion, hot stretching, and quenching: Influence of hot stretch ratio on morphology, crystallization, and crystal structure of iPP at a fixed PET concentration. *Journal of Polymer Science Part B: Polymer Physics, 42*(22), 4095–4106.
19. Fakirov, S., Evstatiev, M., & Schultz, J., (1993). Microfibrillar reinforced composite from drawn poly (ethylene terephthalate)/nylon–6 blend. *Polymer, 34*(22), 4669–4679.
20. Li, Z., Yang, M., Huang, R., Lu, A., & Feng, J., (2002). In-situ composite based on poly(ethylene terephthalate), polyamide, and polyethylene with microfibres formed through extrusion and hot stretching. Cailiao Kexue Yu Jishu (*Journal of Materials Science and Technology*)(China)(USA) *18*, 419–422.
21. Xu, H. S., Li, Z. M., Pan, J. L., Yang, M. B., & Huang, R., (2004). Morphology and rheological behaviors of polycarbonate/ high-density polyethylene in situ microfibrillar blends. *Macromolecular Materials and Engineering, 289*(12), 1087–1095.

22. Quan, H., Zhong, G. J., Li, Z. M., Yang, M. B., Xie, B. H., & Yang, S. Y., (2005). Morphology and mechanical properties of poly (phenylene sulfide)/isotactic polypropylene in situ microfibrillar blends. *Polymer Engineering & Science, 45*(9), 1303–1311.
23. Fakirov, S., Kamo, H., Evstatiev, M., & Friedrich, K., (2004). Microfibrillar reinforced composites from PET/LDPE blends: morphology and mechanical properties. *Journal of Macromolecular Science, Part B., 43*(4), 775–789.
24. Li, Z. M., Yang, W., Xie, B. H., Shen, K. Z., Huang, R., & Yang, M. B., (2004). Morphology and tensile strength prediction of in situ microfibrillar poly (ethylene terephthalate)/polyethylene blends fabricated via slit-die extrusion-hot stretching-quenching. *Macromolecular Materials and Engineering 289*(4), 349–354.
25. Li, Z. M., Yang, M. B., Feng, J. M., Yang, W., & Huang, R., (2002). Morphology of in situ poly (ethylene terephthalate)/polyethylene microfiber-reinforced composite formed via slit-die extrusion and hot-stretching. *Materials Research Bulletin, 37*(13), 2185–2197.
26. Li, Z. M., Yang, M. B., Xie, B. H., Feng, J. M., & Huang, R., (2003). In-situ microfiber-reinforced composite based on PET and PE via slit die extrusion and hot stretching: Influences of hot stretching ratio on morphology and tensile properties at a fixed composition. *Polymer Engineering & Science, 43*(3), 615–628.
27. Jayanarayanan, K., Ravichandran, A., Rajendran, D., Sivathanupillai, M., Venkatesan, A., Thomas, S., & Joseph, K., (2010). Morphology and mechanical properties of normal blends and in-situ microfibrillar composites from low-density polyethylene and poly (ethylene terephthalate). *Polymer-Plastics Technology and Engineering, 49*(5), 442–448.
28. Chapleau, N., & Favis, B., (1995). Droplet/ fiber transitions in immiscible polymer blends generated during melt processing. *Journal of Materials Science, 30*(1), 142–150.
29. Tsebrenko, M., (1983). Fibrillation of the mixtures of crystallizable, amorphous, and poorly crystalline polymers. *International Journal of Polymeric Materials, 10*(2), 83–119.
30. Taylor, G. I., (1932). The viscosity of a fluid containing small drops of another fluid. *Proceedings of the Royal Society of London Series A, Containing Papers of a Mathematical and Physical Character, 138*(834), 41–48.
31. Elmendorp, J., & Van der Vegt, A., (1986). A study on polymer blending microrheology: Part IV. The influence of coalescence on blend morphology origination. *Polymer Engineering & Science, 26*(19), 1332–1338.
32. Kuzmanović, M., Delva, L., Cardon, L., & Ragaert, K., (2016). The effect of injection molding temperature on the morphology and mechanical properties of PP/PET blends and microfibrillar composites. *Polymers, 8*(10), 355.
33. Friedrich, K., Evstatiev, M., Fakirov, S., Evstatiev, O., Ishii, M., & Harrass, M., (2005). Microfibrillar reinforced composites from PET/PP blends: Processing, morphology, and mechanical properties. *Composites Science and Technology, 65*(1), 107–116.
34. Li, Z. M., Xu, X. B., Lu, A., Shen, K. Z., Huang, R., & Yang, M. B., (2004). Carbon black/poly (ethylene terephthalate)/polyethylene composite with electrically conductive in situ microfiber network. *Carbon, 42*(2), 428–432.

35. Panamoottil, S., Pötschke, P., Lin, R., Bhattacharyya, D., & Fakirov, S., (2013). The conductivity of microfibrillar polymer-polymer composites with CNT-loaded microfibrils or compatibilizer: A comparative study. *eXPRESS Polymer Letters, 7*(7)1-30.
36. Fakirov, S., Rahman, M. Z., Pötschke, P., & Bhattacharyya, D., (2014). Single polymer composites of poly (butylene terephthalate) microfibrils loaded with carbon nanotubes exhibiting electrical conductivity and improved mechanical properties. *Macromolecular Materials and Engineering, 299*(7), 799–806.

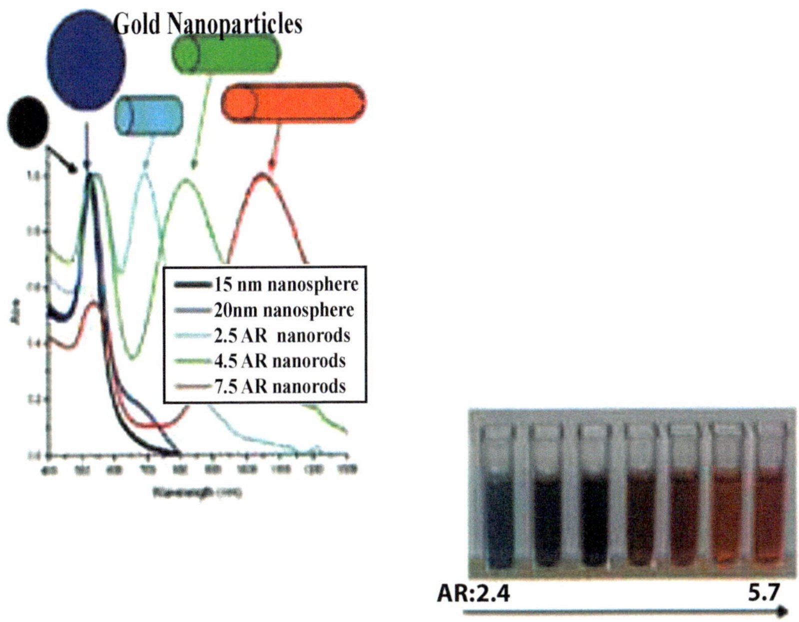

FIGURE 2.3 LSPR plasmon band of nanoparticles with different shapes. It can be seen that the LSPOR band corresponding to different shapes is centered at different wavelengths.

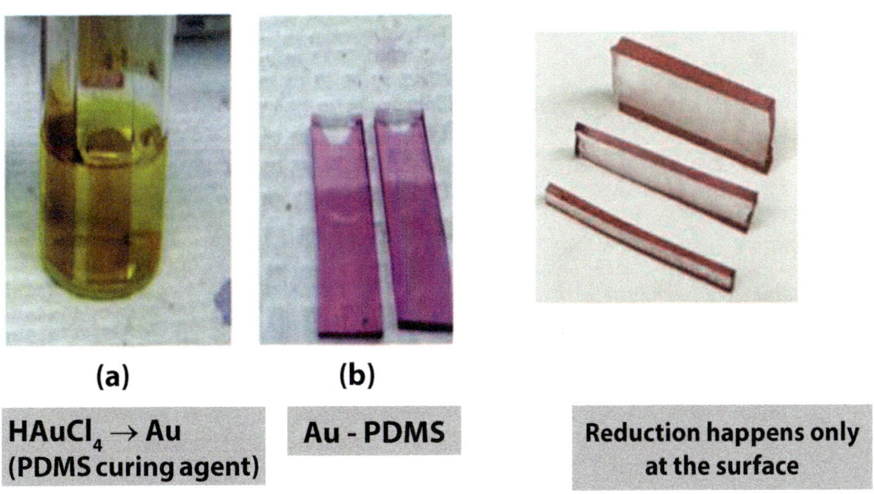

FIGURE 2.4 *In-situ* synthesis of Au – PDMS nanocomposites. The composite is prepared by immersing a PDMS sample in the solution of chloroauric acid. The change of color shows the presence of AuNPs on the surface of the polymer.

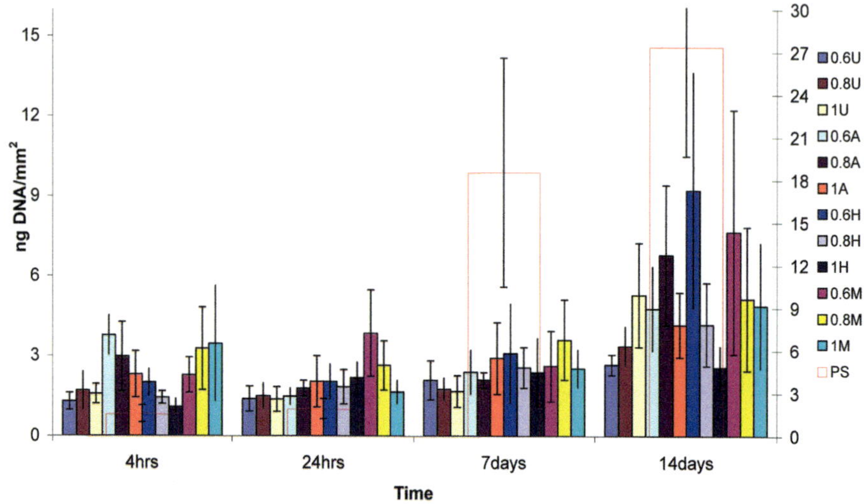

FIGURE 3.5 DNA quantification in cell lysates at 4 h, 24 h, 7 days, and 14 days. A – aminolyzed; H – hydrolyzed; M – multilayered; U – untreated film; PS – polystyrene.

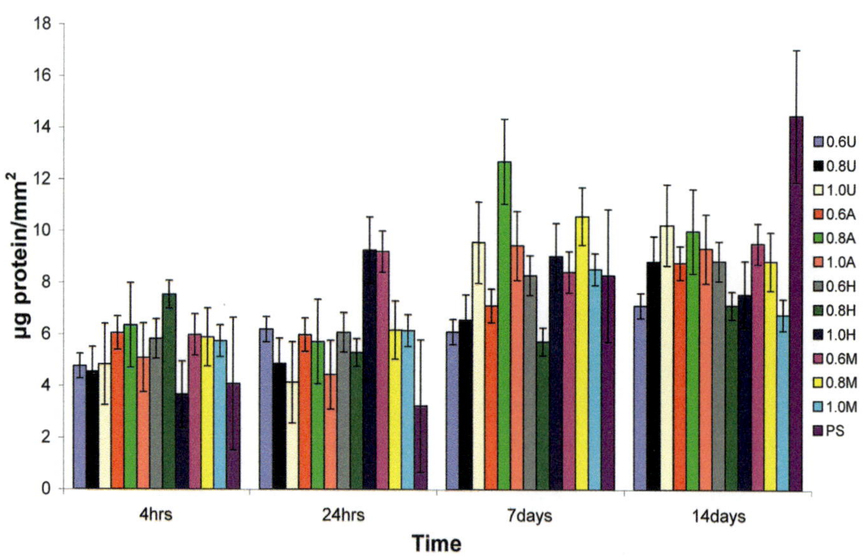

FIGURE 3.8 Quantification of protein in MSC cell lysates at 4 h, 24 h, 7 days, and 14 days time- points. A – aminolyzed, H – hydrolyzed, M – multilayered, U – untreated film, PS – polystyrene.

FIGURE 4.5 Samples after chemical treatment.

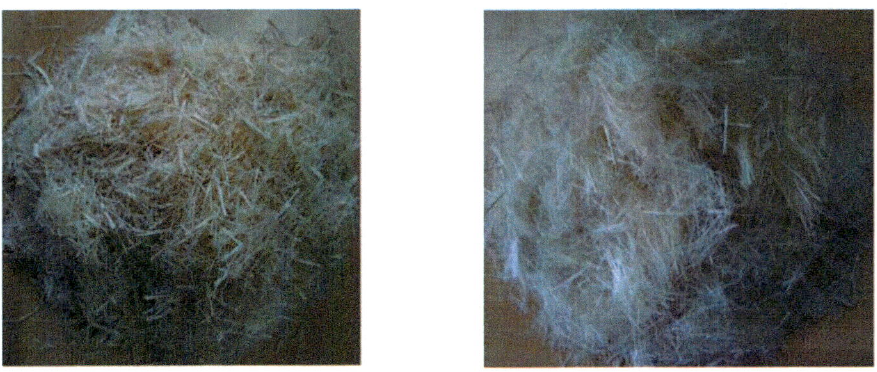

(a) (b)

FIGURE 5.1 Short (a) banana fiber and (b) jute fiber.

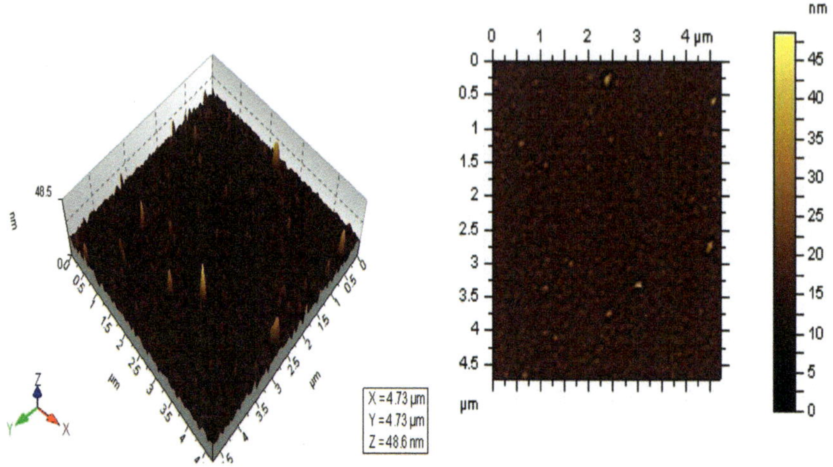

FIGURE 7.5 Atomic force microscope image of Cts/PEG assisted Ag-doped with ZnO nanocomposites at 80°C.

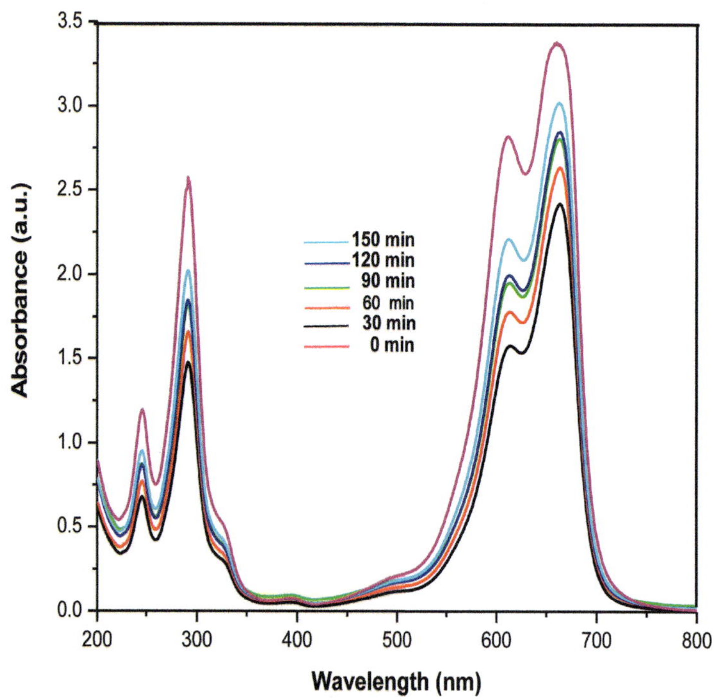

FIGURE 7.8 UV absorption spectra of MB solution in the presence of Cts/PEG/Ag-ZnO: Synthesis with 80mg/L^{-1} under different UV irradiation time.

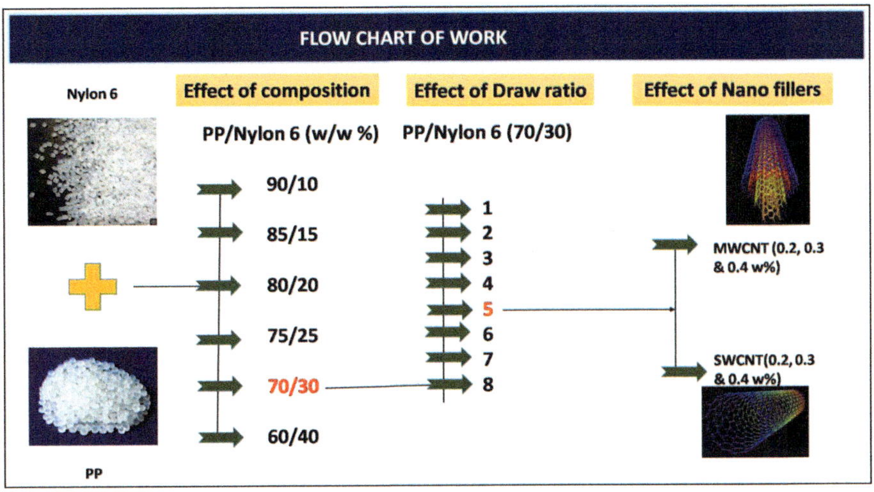

FIGURE 8.13 The pictorial conclusion of work.

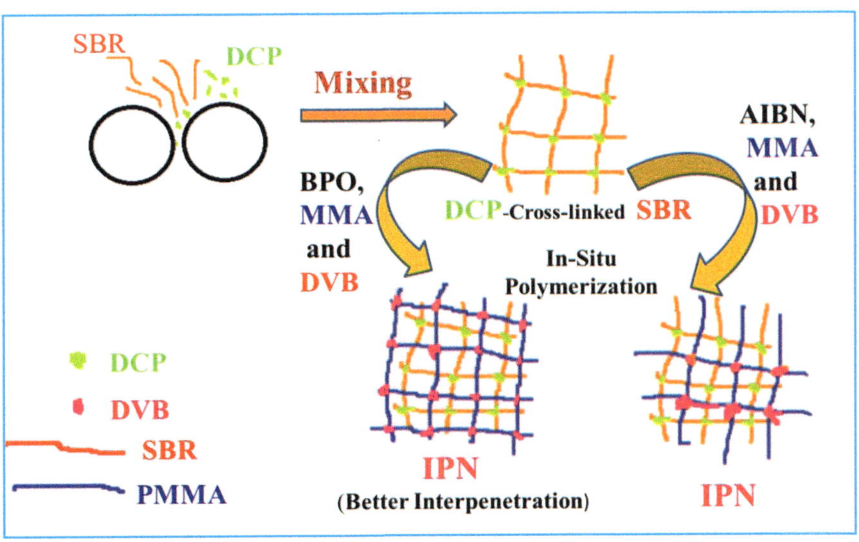

FIGURE 10.1 Scheme of preparation of IPN.

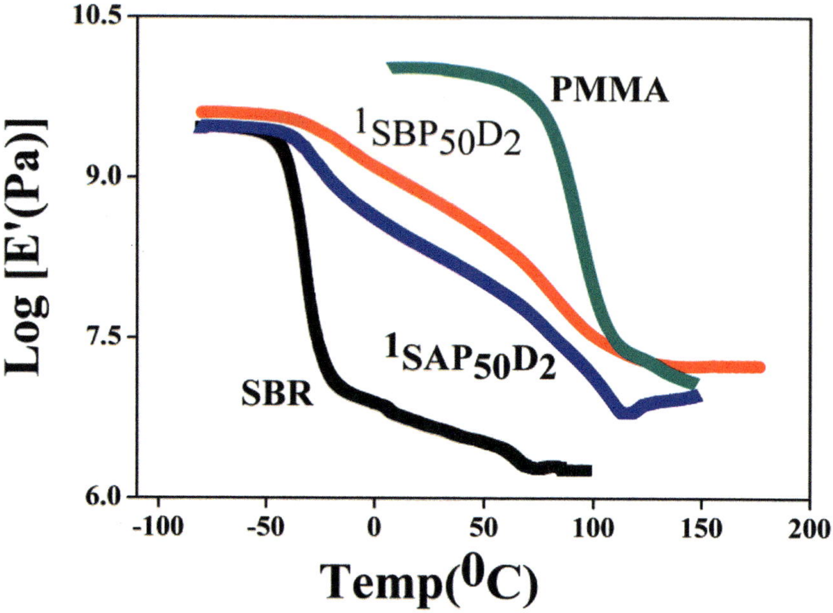

FIGURE 10.2 Effect of initiator on *E*' for SBR, PMMA, $^{1}SBP_{50}D_2$ and $^{1}SAP_{50}D_2$ IPNs at 1Hz.

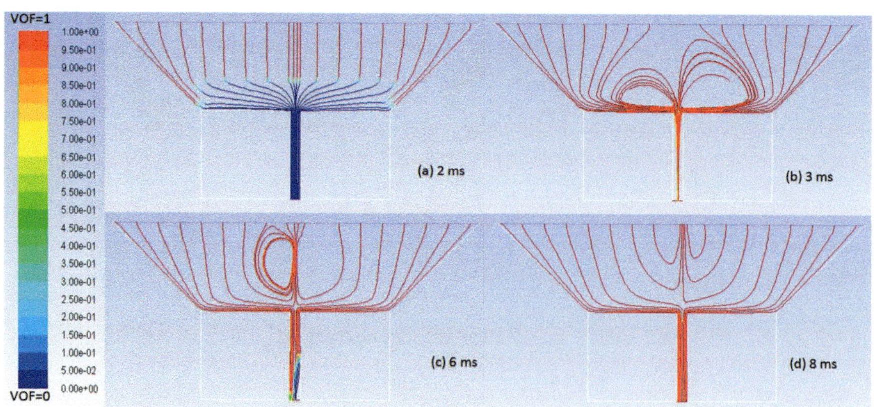

FIGURE 15.8 Pathlines in the crucible after (a) 2 ms (b) 3 ms (c) 6 ms and (d) 8 ms.

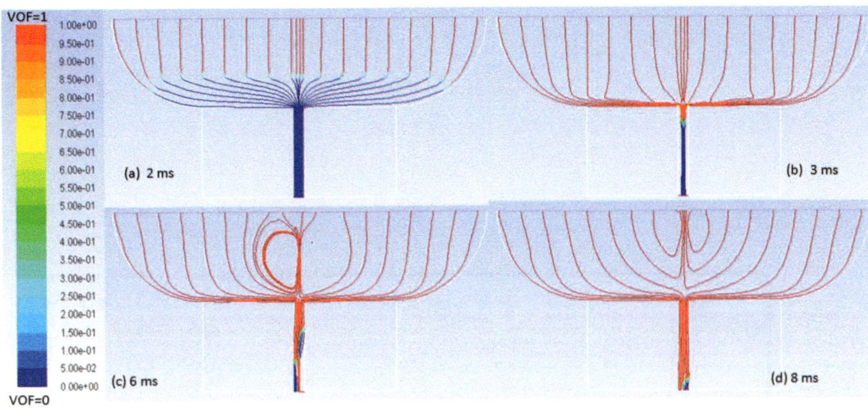

FIGURE 15.10 Pathlines in the crucible after (a) 2 ms (b) 3 ms (c) 6 ms and (d) 8 ms.

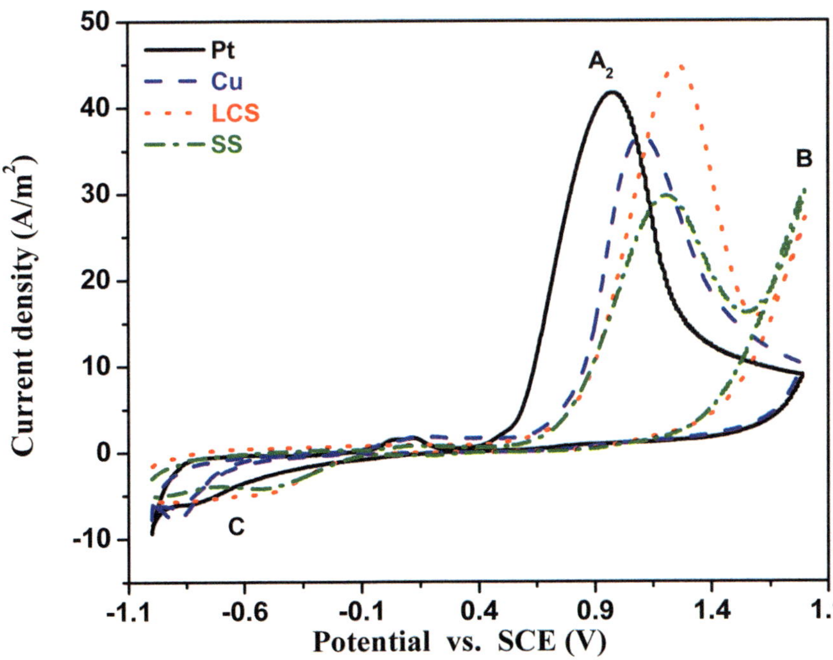

FIGURE 16.2 The first scan of cyclic voltammograms of POT coating in salicylate solution on Pt, LCS, SS, and Cu.

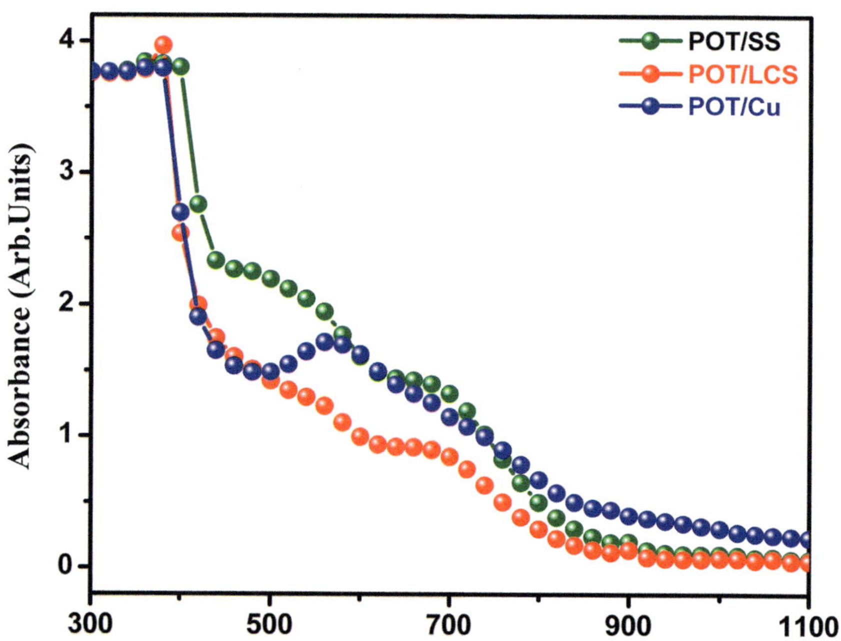

FIGURE 16.4 UV-Vis. spectra of the POT (salicylate) coatings on LCS, SS, and Cu.

CHAPTER 9

EFFECT OF CO-PRECIPITATION ON THE PHYSICO-MECHANICAL AND DIELECTRIC PROPERTIES OF NICKEL FERRITE EMBEDDED NATURAL RUBBER COMPOSITES

SANKAR S. MENON[1], RADHU KRISHNA[1], LIDA WILSON[1], SREEDHA SAMBHUDEVAN[1], BALAKRISHNAN SHANKAR[2], ANSHIDA MAYEEN[3], and NANDAKUMAR KALARIKKAL[3]

[1]*Department of Chemistry, Amrita School of Arts and Sciences, Amritapuri, Amrita Vishwa Vidyapeetham, Amrita University, India*

[2]*Department of Mechanical Engineering, Amrita School of Engineering, Amritapuri, Amrita Vishwa Vidyapeetham, Amrita University, India*

[3]*School of Pure and Applied Physics, Mahatma Gandhi University, Kottayam, Kerala*

ABSTRACT

Spinel structured nickel ferrite has been prepared using the co-precipitation method. The ferrite particles prepared were characterized using XRD, FTIR, and TEM and was confirmed to be in the nano-regime. Natural rubber composites were prepared with different loadings of nickel ferrite like 5, 15, 25, 50, 75 (in part per hundred rubber, phr). The mechanical, swelling, and magnetic properties were analyzed using standard methods. Dielectric measurements show that permittivity decreases with increase in frequency and increases with increase in ferrite loading. Tan delta value also was found to increase with filler loading which may be attributed to the presence of interfacial polarization.

9.1 INTRODUCTION

Studies on magnetic properties based on introducing ferrite into elastomers are significant from fundamental as well as application viewpoint [1, 2]. This class of materials is perfect models to probe phenomenon that could produce an impact on the physicomechanical properties of composite structures. Magnetic behavior of ferrites with spinel structure rests on numerous aspects like particle dimension, a method of preparation, thermal procedures and the microstructure [3, 4]. Nickel ferrite is a representative of a wide range of combinations named spinels. Nickel ferrite ($NiFe_2O_4$) is having an inverse spinel structure and is likely to display fascinating magnetic stuff in the nanosystem whose insertion into systems like natural rubber will definitely impart magnetism to the elastomer matrix.

The utmost significant property linked with the ferrite nanoparticles is the practical lessening in saturation magnetization (Ms) related to the bulk counterpart [5, 6]. The reduction in Ms is linked with various ins and outs like surface effects [7], spin canting [8] and dead layer formation [9]. To establish a thick association between the structure and magnetic behavior of various spinel ferrites, several types of research were testified [10, 11]. The dispersal of the metal ions in octahedral and tetrahedral lattice points is the main factor which decides the ferrimagnetism possessed by spinel ferrites. Modification of rubber ferrite composites is highly crucial as far as end-uses are considered. Simple mixture equation is able to articulate to foretell the magnetic characteristics of the rubber ferrite composites. Estimation of magnetic properties, as well as modeling with identified equations, are also important.

The dielectric properties of ferrites too rest on other parameters like applied voltage frequency, humidity, pressure, and temperature. A detailed learning of the dielectric properties of spinel ferrites is of prime importance because, the general behavior of ferrites are highly reliant on the microstructure, molecular arrangement, and the processing parameters. Like other dielectric materials, the dielectric properties of composite materials are also generally articulated by its complex permittivity, conductivity, and resistivity. All the above-said terms are functions of temperature and the kind (ac/dc) and quantity of the voltage applied. Since the dielectric behavior of rubber ferrite composites is highly dependent on temperature and frequency, it helps us to understand the polarization pattern and the conduction mechanism [12, 13]. In order to explain the experimental dielectric properties of rubber composites, numerous theories and equations can be used. Such detailed observations reveal important information such as filler- matrix collaboration, distribution of filler in the matrix and separation limit of the filler in the matrix.

Vibrating sample magnetometer of model EG and G PAR 4500 was used to measure magnetic characteristics of prepared nickel ferrite fillers and the rubber composites containing these fillers. Characterizations were done at a normal temperature and all the hysteresis curve factors like saturation magnetization, remanent magnetization, and coercivity were calculated. Natural rubber based ferrite composites or rubber ferrite composites (RFCs) were synthesized by including pre characterized $NiFe_2O_4$ nanoparticles. The change of dielectric properties with frequency and temperature of the filler and rubber composites were examined. Diverse theoretical equations were utilized to compare the experimental and pre-calculated dielectric data of the nickel ferrite and the analogous composites.

9.2 EXPERIMENTAL DETAILS

9.2.1 PREPARATION OF NICKEL FERRITE ($NIFE_2O_4$)

Nickel ferrite powders were prepared by a co-precipitation method using sodium hydroxide (NaOH) as the precipitating agent. About 0.3 g of polyvinylpyrrolidone (PVP) is dissolved in 100 ml of deionized water taken in a round bottom flask (RB). Then 16.16 g iron nitrate and 5.8158 g nickel nitrate were added into the PVP solution. NaOH solution was added dropwise to attain a p^H of around 12 with continuous stirring for 2 hours at 80°C. The solution containing nanopowders of $NiFe_2O_4$ is centrifuged and washed several times with water and ethanol to remove excess PVP. The obtained powder was dried in the oven for 24 hours at 100C. The resulting dark brown powder was crushed and calcined at different temperatures say 400°C, 500°C, 600°C, and 700°C for 3 hours. The resultant compounds were characterized using FTIR and XRD to confirm the modification and particle size.

9.2.2 PREPARATION OF NANO-COMPOSITES BY SOLUTION MIXING METHOD

The filler was first dispersed in an organic solvent like toluene using sonicator and then mixed with NR swollen in the same solvent by stirring. The nanocomposites thus obtained are casted on a Petri dish at room temperature to obtain a thin film. The solvent was allowed to evaporate at room temperature until there was no weight variation. The casted film and curing system are added on a two-roll mill and compounded for 15 minutes by carefully controlling the nip gap and temperature [14]. The samples were milled for sufficient time to uniformly disperse the fillers in the matrix at a mill opening of 1.25 mm. The fillers were added at the end of the mixing process. After complete mixing, the mix was passed through a tight nip gap and finally sheeted at a preset nip gap. The composites were left for a day in a desiccator before vulcanization. A schematic representation of solution mixing method is shown in Figure 9.1.

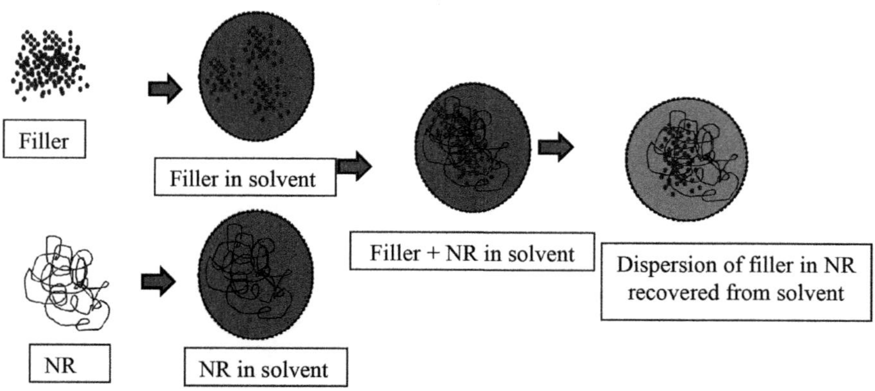

FIGURE 9.1 Schematic representation of solution mixing method.

The mixes were kept in a desiccator for 24 hours to avoid moisture content and then vulcanized at 150°C in an electrically heated hydraulic press up to their corresponding cure times (t_{90}) at a pressure of 150 kgf/cm^2 to get sheets of 2 mm thickness and the compositions are listed in Table 9.1. These samples were kept in a desiccator for 24 hours at room temperature to make sure that all the samples are having uniform thermal history when exposed to testing.

TABLE 9.1 The composition of NR/Nickel Ferrite Nano Composites

Ingredients in phr	Gum S	NFN5	NFN15	NFN25	NFN50	NFN75
NR	100	100	100	100	100	100
Nickel ferrite	0	5	15	25	50	75
Stearic acid	1	1	1	1	1	1
Zinc oxide	5	5	5	5	5	5
TDQ	1	1	1	1	1	1
CBS	1	1	1	1	1	1
Sulfur	2.5	2.5	2.5	2.5	2.5	2.5

9.3 RESULTS AND DISCUSSION

9.3.1 X-RAY POWDER DIFFRACTION

The X-ray patterns of the samples milled for 2 hours and annealed for 3 hours in various temperatures are shown in Figure 9.2. The diffraction pattern shows reflection planes at (220), (311), (400), (422), (511) and (440) which clearly indicates the presence of spinel cubic phase. It is also evident that by raising the annealing temperature, the diffraction peaks turn out to be finer and strident, because of the rise in crystallinity and particle size.

The average particle size has been calculated using the Debye Scherer equation

$$t = \frac{0.9\lambda}{\beta Cos\theta} \tag{9.1}$$

where t is the crystallite size, β is the full width of the diffraction line at half of the maximum intensity measured in radians, λ is x-ray wavelength (Cu $K\alpha 1$ radiation, 0.154 nm) and θ is the Bragg angle.

Results show that with the rise in hardening temperature, the particle sizes also enhanced and touch the values of 10, 12, 17 and 25 nm for temperatures 400, 500, 600 and 700°C, respectively. A possible explanation for this increase can be the supremacy of the activation energy at the time of particle formation course as well as temperature assisted crystal growth.

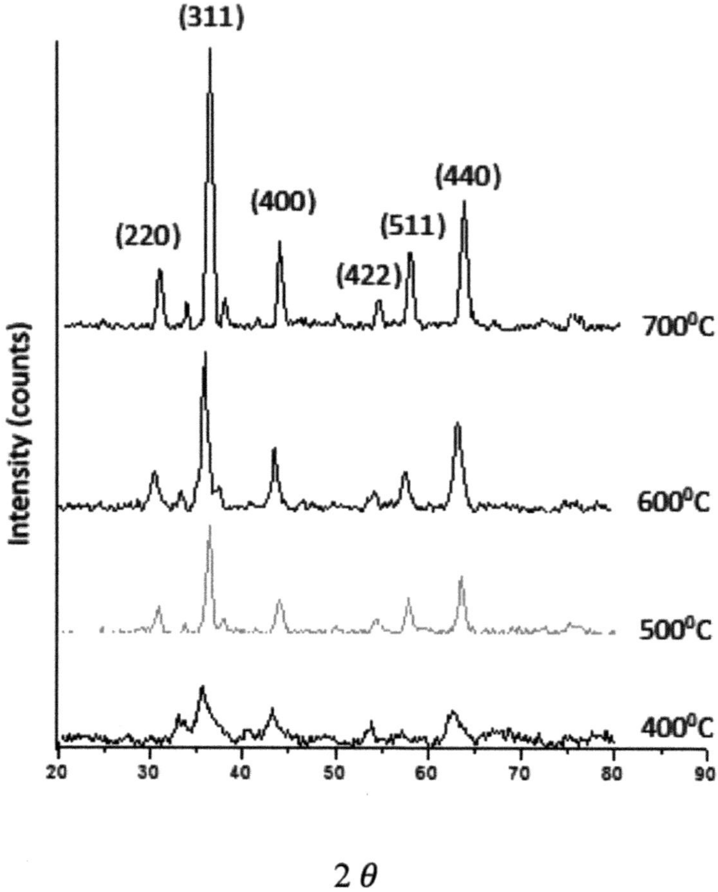

FIGURE 9.2 XRD spectra of $NiFe_2O_4$ sintered at different temperatures.

9.3.2 FTIR ANALYSIS

FTIR spectra of nickel ferrite prepared by co-precipitation method and thereafter sintered at various temperatures are shown in Figure 9.3. The band at 551 cm^{-1} corresponds to intrinsic stretching vibrations of the metal at the tetrahedral site [15]. The band at 470 cm^{-1} corresponds to octahedral stretching. The bands at 3410 cm^{-1} and at 1631 cm^{-1} are attributed to the stretching modes and H-O-H bending vibrations of the free or absorbed water [16].

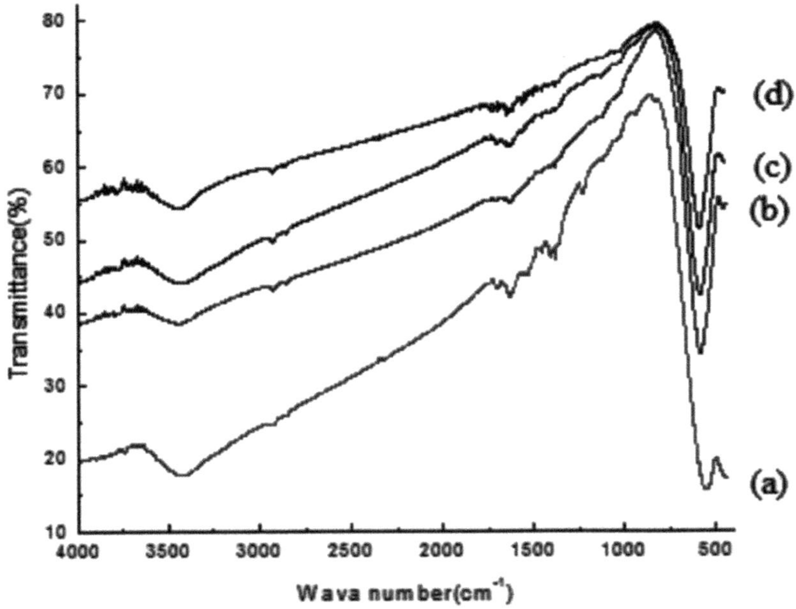

FIGURE 9.3 FTIR spectra of $NiFe_2O_4$ sintered at (a) 400°C, (b) 500C, (c) 600°C and (d) 700°C.

9.3.3 TEM ANALYSIS

TEM images in Figure 9.4 show that the nanoparticles are of a nearly uniform shape and have a moderate tendency to agglomerate. The average particle size was found to be 23 nm which is in good agreement with XRD studies. Earlier studies show that the agglomeration tendency could be reduced by increasing the PVP concentration [17].

9.3.4 MECHANICAL PROPERTIES OF NICKEL FERRITE-NATURAL RUBBER COMPOSITES

The reinforcing ability of the fillers is chiefly decided by the size, shape, specific surface area and a synergistic effect of the size dispersal and shape [18, 19]. It was found that ferrite fillers exhibit a semi-reinforcing effect when used in natural rubber based rubber composites [20–21].

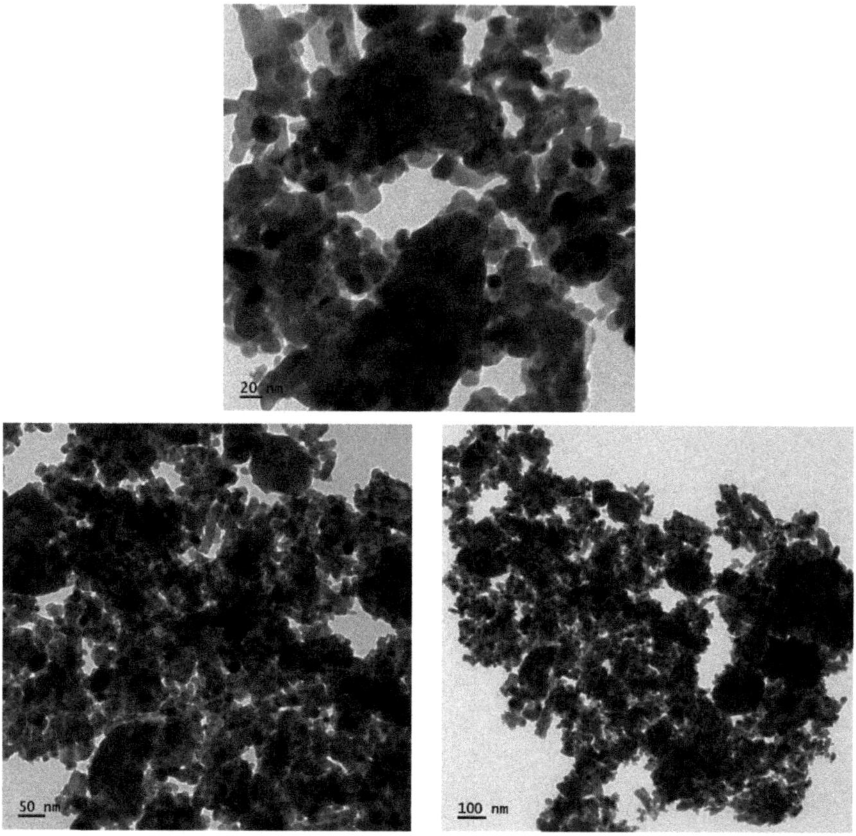

FIGURE 9.4 TEM images of Nickel ferrite particles at different magnifications sintered at 700°C.

The increase in tensile strength of the composites as given in Table 9.2, reveals the reinforcing property of the ferrite filler. The increase in tensile strength value is attributed to the enhanced surface area offered by the fillers because the filler is in the nano region (size 23 nm).

The tensile strength decreased with the increase in the amount of ferrites. One reason for this is the inferior interfacial bonding among the filler and the matrix. Due to the reduction in the effective cross section of the matrix in the composites, there is a rise in internal stress, for every point of external loading when compared with the bare rubber matrix. Another reason for diminishing property is the dilution effect, which may be attributed to the decreasing volume fraction of rubber in the composite.

TABLE 9.2 Mechanical Properties of NR Based Composites Containing Different Ladings of Nickel Ferrite

Sample	Tensile Strength (MPa)	Elongation at break (%)	Modulus 100% (MPa)	Modulus 200% (MPa)	Modulus 300% (MPa)	Hardness (Shore (a)
Gum	11.78	710.07	0.47	0.71	0.96	35.4
NFN5	6.0	678.12	0.48	0.81	1.24	37.8
NFN15	6.5	652.5	0.55	0.88	1.27	39.4
NFN25	6.72	625.9	0.58	0.91	1.76	40.8
NFN50	7.33	603.09	0.61	1.23	2.15	42.1
NFN75	6.26	570.58	0.77	1.82	2.48	43.3

There is a slow increase in modulus value with filler loading, which is the characteristic of the reinforcing filler. The elongation at break showed a decreasing trend with increasing loading of ferrite fillers in the case of natural rubber based RFCs. The hardness of these magnetic materials showed a steady increase with the filler.

9.3.5 MAGNETIC MEASUREMENTS OF $NIFE_2O_4$

The hysteresis loop for $NiFe_2O_4$ filler obtained at normal conditions is shown in Figure 9.5. Magnetic parameters measured from the hysteresis curve are given in Table 9.3. The saturation magnetization of the nickel ferrite filler is estimated as 44.21 emu/g, which is a smaller value when compared with the reported M_s value of the bulk counterpart [22]. The abnormality in magnetic properties may be due to numerous reasons. The arrangement of the particles possibly will be altered by the occurrence of lattice imperfections or may be due to a variation in the dispersal of the component ions among different vacant crystal sites.

The reduction in M_s value for the prepared $NiFe_2O_4$ when compared with the bulk is attributed to the small dimension influence and surface effects. The saturation magnetization of fillers falls with the size of particles which may be attributed to surface spin disorder. In nano regime structures, as the surface to volume ratio is comparatively high, the number of surface spins will be higher than the overall count of spins. This may cause disorder of surface spins leading to surface anisotropy and deviance from the usual bulk properties. The alternative reason of surface spin disorder found in nano-sized nickel ferrite may be due to shattered exchange bonds that weaken the magnetic order leading to spin-obstruction [23].

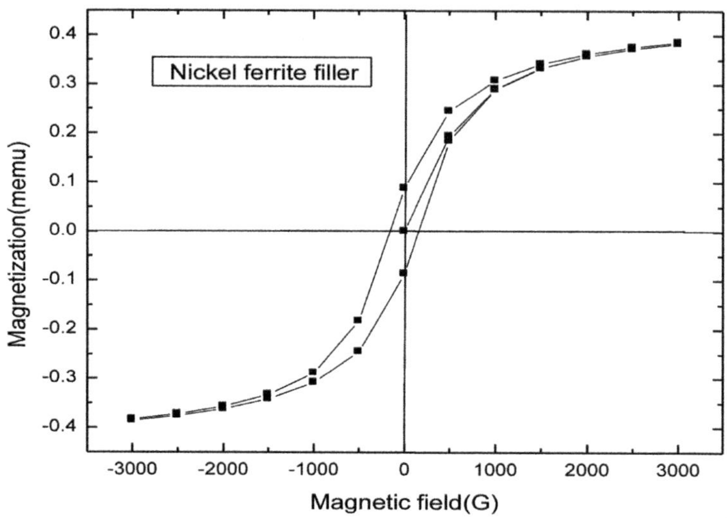

FIGURE 9.5 Hysteresis loop for $NiFe_2O_4$.

Literature shows that as the particle size comes down, coercivity rises touches an extreme value and then comes down when the particle size reaches a specific value [24]. The magnetization/demagnetization produced by sphere domain wall displacement needs less energy than that required for domain spinning. The number of domain walls falls with a reduction in particle size and the impact of wall movement to magnetization/ demagnetization is lesser when compared to domain rotation. Hence the particles with smaller grain size are established to possess superior coercivity. In this specific case, the particle size of the as-prepared $NiFe_2O_4$ sample is found to be lower than the critical diameter and therefore coercivity was found to be higher than that of its bulk counterpart.

TABLE 9.3 Magnetic Characteristics of $NiFe_2O_4$

Sample	Coercivity H_c(G)	Magnetic Remanence M_r(emu/g)	Saturation magnetization M_s(emu/g)	M_r/M_s
Nickel Ferrite	162	8.74	44.21	0.1977

9.3.6 MAGNETIC MEASUREMENTS OF $NIFE_2O_4$ BASED NR COMPOSITES

Rubber ferrite composites (RFC) loaded with various concentration of $NiFe_2O_4$ built on NR were examined and magnetic measurements were undergone using VSM as in the case of ferrite particles (Figure 9.6 and Table 9.4).

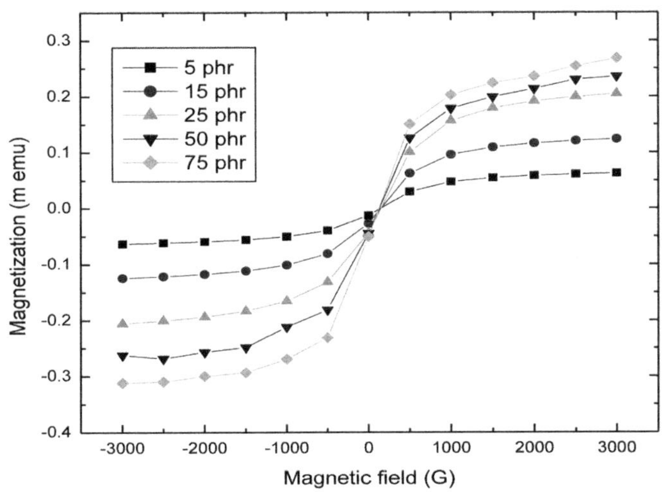

FIGURE 9.6 Hysteresis loop for nickel ferrite natural rubber composites.

TABLE 9.4 Magnetic Characteristics of Differently Loaded $NiFe_2O_4$ Containing NR Composites

Sample	Coercivity H_c(G)	Magnetic Remanence M_r(emu/g)	Saturation magnetization M_s(emu/g)	Mr/Ms
NFN5	155.01	1.32	7.2	0.18
NFN15	155	2.69	14.18	0.19
NFN25	155	3.48	18.46	0.19
NFN50	155	4.48	23.53	0.19
NFN75	155	5.08	26.78	0.19

In RFCs, as the Ms value solely depends on filler properties, the saturation magnetization values can be enhanced by properly selecting the nature and quantity of the magnetic filler within the matrix. If the Ms values of the ferrite are known, a simple mixture formula of the common form including the weight fractions of the filler can be applied to determine the Ms of the composites as follows.

$$M_{RFC} = W_1M_1 + W_2M_2 \tag{9.2}$$

where M_{RFC} is the saturation magnetization of the composite, W_1 and W_2 are the weight fractions of the filler and polymer, M_1 and M_2 are the saturation magnetization of the filler and the polymer respectively. As the matrix taken for the synthesis of rubber ferrite composites is non-magnetic, Eq. (9.1) can be condensed into the resulting formula [25]

$$M_{RFC} = W_1M_1 \tag{9.3}$$

Figure 9.6 depicts the discrepancy in magnetization tendency of $NiFe_2O_4$ containing NR composites with ferrite loading. The resultant intended values of Ms also scheme.

From Figure 9.7, it is evident that the premeditated values are in decent covenant with the experimental data, particularly at lower filler concentration. The small disparity witnessed at higher filler concentration can be attributed to the deviance from the perfect distribution of the ferrite particles inside the matrix. Grouping of the filler particles happens at upper loadings as evinced from the physicomechanical properties of the composites.

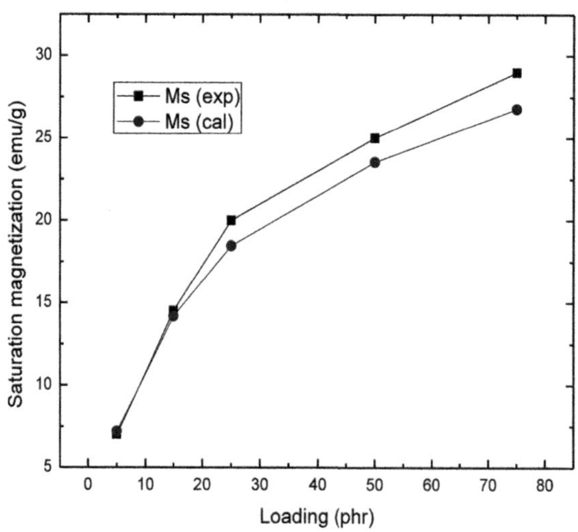

FIGURE 9.7 Experimental and calculated values of saturation magnetization of $NiFe_2O_4$ filled NR composites.

9.3.7 DIELECTRIC PROPERTIES

9.3.7.1 DIELECTRIC PERMITTIVITY

It is usually observed that dielectric data is regarded as the combination of two progressions, a conductivity involvement that results in an increase of both real part ε' and imaginary part ε" of the dielectric function on lessening frequency and a relaxation contribution showing a maximum in ε" that changes upper-frequency part with rise in temperature. Figure 9.8 shows the variation of imaginary part of dielectric permittivity ε" with frequency for various loadings of $NiFe_2O_4$. The advanced value of dielectric loss at low frequency may be attributed to the free charge motion inside the materials (Figure 9.9).

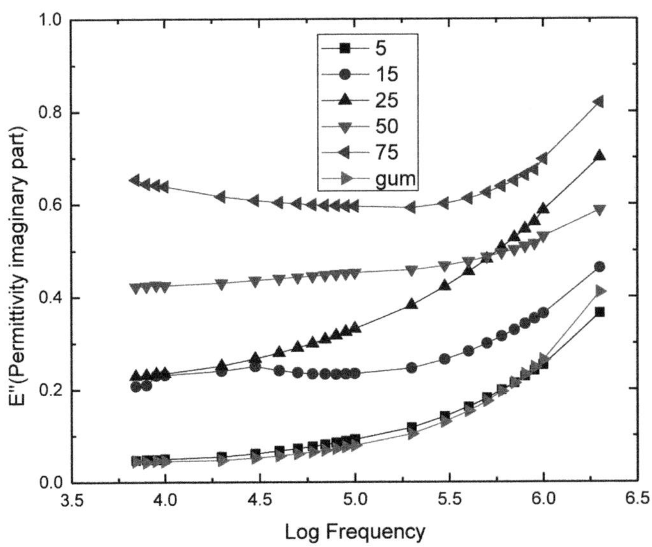

FIGURE 9.8 Variation of permittivity (imaginary) with frequency in $NiFe_2O_4$ filled NR composites.

Figure 9.9 shows the real part of dielectric permittivity ε" with frequency at room temperature for different loadings of $NiFe_2O_4$ in NR matrix. Dielectric properties of a polymeric material underneath the impact of an external electric field rely on the polarization influence that takes place inside the sample.

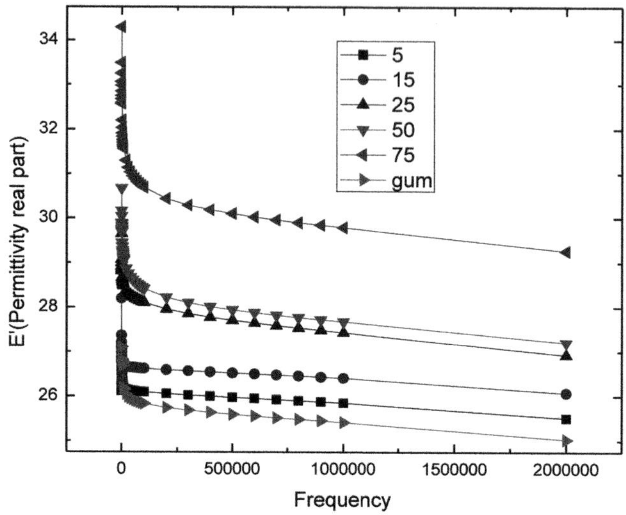

FIGURE 9.9 Variation of permittivity (real) with frequency in $NiFe_2O_4$ filled NR composites.

Overall polarization is the totality of the deformational polarization and orientational polarization. At small given frequencies all the above-mentioned aspects donate to the total polarization. With the increase in frequency, the orientational polarization turns out to be out of phase by the applied field, to be exact, the dipolar motion cannot further track the fast vibration in the electric field. Hence the dielectric permittivity of the matrix constituents declines with the rising in applied frequency. The polarization and therefore the dielectric permittivity can be considered as an intricate amount.

$$\in^{*} = \in' - i \in'' \tag{9.4}$$

The complex dielectric permittivity $\in^{*}$ is specified by the Debye equation

$$\in^{*} = \in_{\alpha} + \frac{\in_0 - \in_{\alpha}}{1 + i\omega\tau} \tag{9.5}$$

where ω is the angular frequency, $\in_0$ is the static dielectric permittivity and $\in''$, is the dielectric permittivity at infinite frequency. The real and imaginary components $\in'$ and $\in''$ are given by [26]

$$\in' = \in_{\alpha} + \frac{\in_0 - \in_{\alpha}}{1+(\omega\tau)^2} \tag{9.6}$$

$$\in'' = \frac{(\in_0 - \in_{\alpha})\omega\tau}{1+(\omega\tau)^2} \tag{9.7}$$

As per Eq. (9.6), dielectric permittivity declines as frequency increases.

The actual permittivity in nanocomposites is governed by the dielectric polarization mechanism in the bulk of the material. For composites, they are polarizations connected with the rubber matrix and the filler and interfacial polarizations at the matrix-filler interface. Apart from nanoparticles, nanocomposites take a huge volume of interfaces where interfacial polarization could occur. It is clear from Figure 9.9 that at higher frequencies, interfacial polarization is not occurring. Interfacial or space charge polarization happen because of the gathering of space charges at interface boundaries. When a high-frequency field is given, the chance of these space charges to gather at the interface is more probable.

The impact of the interface on the dielectric properties of nanofiller embedded composite materials was studied using theoretical. A multi-core model was established by Tanaka *et al.* to explore the dielectric characteristics of rubber nanocomposites [27]. According to them, the polymer matrix embedded with sphere-shaped filler possess interface which contains three different regions: a bonded level (1st level), a bound level (2nd level) and a loose level (3rd level). In the 1st level polymer is in near connection with the filler, and the 2nd level resembles the region of the interface. Lastly, at the 3rd level, it represents polymer bulk properties.

According to the above-mentioned multi-layer model, Smith et al. suggested a systematic theory for the interface organization [28]. With the intention of preserving charge neutrality on the interface, a rearrangement of charge happens at the interface which results in a Helmholtz or Stern layer. This diffusible binary layer exists in the polymer matric farther as of the interface. This double layer, consecutively, rests on the charge in the Stern layer. Appropriate change of the interface leads to variations in mobility, free volume and trap points for charge carriers (here, electrons). Titanium dioxide was reformed with a coupling group and a reduction in the movement of charge carriers was observed [29]. The occurrence of filler in small concentrations eradicates overlying of native conductive areas and consequently stops early total breakdown in samples.

Figure 9.10 gives the variation of tan delta value with frequency for various filler loadings of nickel ferrite in NR composites. It is found that tan delta value increases with increase in filler loading. The occurrence of interfacial phenomena in a system is usually associated with a sharp increase of tan delta and effective permittivity with regard to frequency, particularly at higher filler loadings. Depending on the interaction between the polymer and the nanofiller, a nanolayer will be formed on the filler surface which is obviously immobile due to the interaction with a polymer as well as nanofiller. When these immovable nanolayers spread over all the nanofillers in the polymer matrix, the movements of the polymer chain are also restricted.

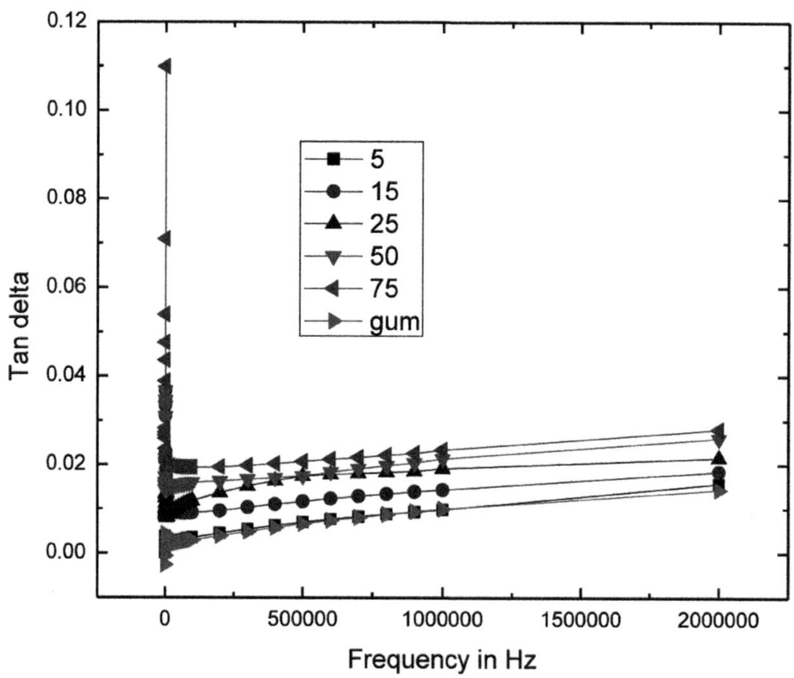

FIGURE 9.10 Variation of dissipation factor (tan delta) with frequency.

It is also proved that when the length scale of the polymer chain and nanofillers becomes comparable, the interface wall-wall distance becomes shorter and a second polymer chain network develops which leads to intermingled structures [30]. This would lead to an additional reduction in the polymer mobility and the degree of entanglement and reduction in mobility depends on filler quantity. The same concept can be applied to the NR composite in the present study where the NR rubber segments interact with the ferrite nanofillers leading to a restriction in the movement of NR segments (Figure 9.11).

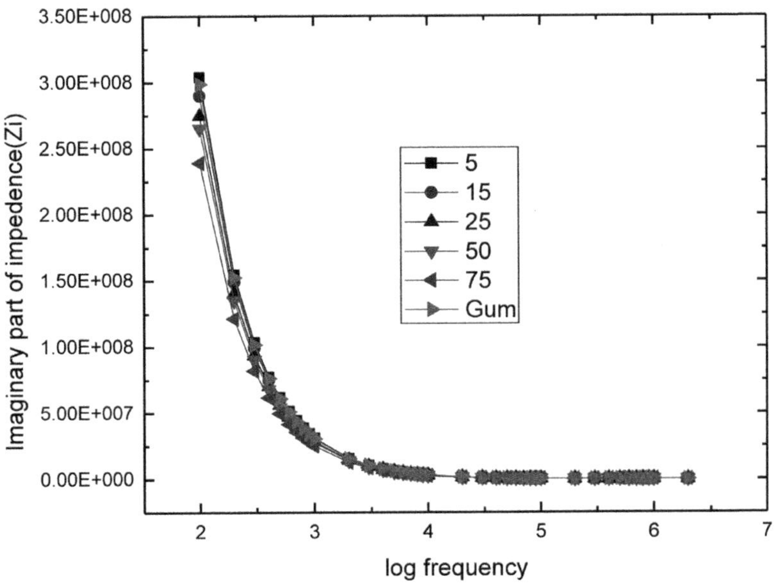

FIGURE 9.11 Variation of Impedance (imaginary) with frequency in $NiFe_2O_4$ filled NR composites.

The large increase of $\in'$ and $\in''$ to a large extent may be attributed to the formation of a percolation structure of nanoparticles. The low-frequency losses may be either due to the Maxwell-Wagner effect [31] due to alternating current in phase with the applied potential or the direct current conductivity as a result of an increase in ion mobility, or both.

Variation of impedance shows an opposite trend to that of permittivity as we expected (Figures 9.11 and 9.12). Negative values for the real part of the input impedance also means negative values for the real part of the input admittance. If we assume that the input terminals are connected with a very large resistor (an open circuit in our case), the situation does not change very much. Therefore, we are considering a circuit that is stable while exhibiting a negative input conductance.

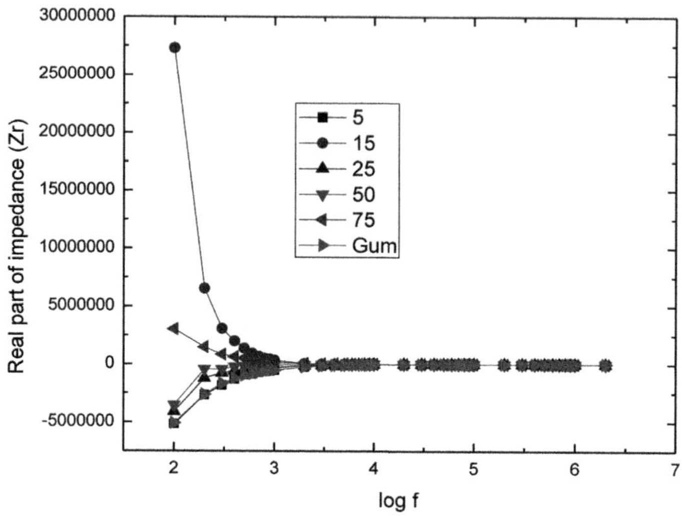

FIGURE 9.12 Variation of Impedance (real) with frequency in $NiFe_2O_4$ filled NR composites.

9.3.7.2 MODELING OF EFFECTIVE DIELECTRIC CONSTANT

Numerous hypothetical methods have been established to forecast the actual dielectric constants of rubber- composite systems. Among them, the volume-fraction average method is a modest one to evaluate the effective dielectric constant of a polymer composite system.

$$\epsilon_{eff} = \varnothing_1 \epsilon_1 + \varnothing_2 \epsilon_2 \tag{9.8}$$

Here the numbers 1 and 2 signify the rubber and the filler system, and Ø is the volume fraction of the ingredients. On the basis of this model (Eq. 9.7), the effective dielectric constant of the composite rises suddenly for a small fraction of filler. But several types of research including both experiments [32] and theory [33] invalidate the tendency expected by Eq. (9.7).

Supplementary accurate models are centered on mean field theory. The Maxwell equation:

$$\epsilon_{eff} = \epsilon_1 \frac{\epsilon_2 + 2\epsilon_1 - 2(1-\varnothing_1)(\epsilon_1 - \epsilon_2)}{\epsilon_2 + 2\epsilon_1 + (1-\varnothing_1)(\epsilon_1 - \epsilon_2)} \tag{9.9}$$

is created on a mean-field rough calculation of a single spherical insertion enclosed by an uninterrupted matrix of the polymer [34].

One more theory, called the Bruggeman model, considers the binary mixture to be consist of repetitive unit portions of the matrix polymer with sphere-shaped insertions in the focal point. The actual dielectric constant of the binary mixture is given by:

$$\varnothing_1\left[\frac{\in_1-\in_{eff}}{\in_1+2\in_{eff}}\right]+\varnothing_2\left[\frac{\in_2-\in_{eff}}{\in_2+2\in_{eff}}\right]=0 \tag{9.10}$$

In two-phase models, each component of the composite structure is well thought-out as separate phases more than seeing one component of the composite as an insertion in another continuous phase. On the basis of effective-medium theory (EMT), Rao et al. prepared mathematical calculations of the effective dielectric constant for rubber-filler composites [35].

Accordingly, when the filler size is lesser compared to that of the composite, it is possible to calculate the dielectric permittivity of the composite based on an active source whose dielectric permittivity is averaged above the dielectric permittivity of the two components. Figure 9.13 depicts the contrast between the theoretical and experimental values of effective dielectric constant versus frequency for a 50 phr loaded natural rubber nanocomposites. The experimental values are found to be in decent agreement with the Maxwell model as suggested by the literature.

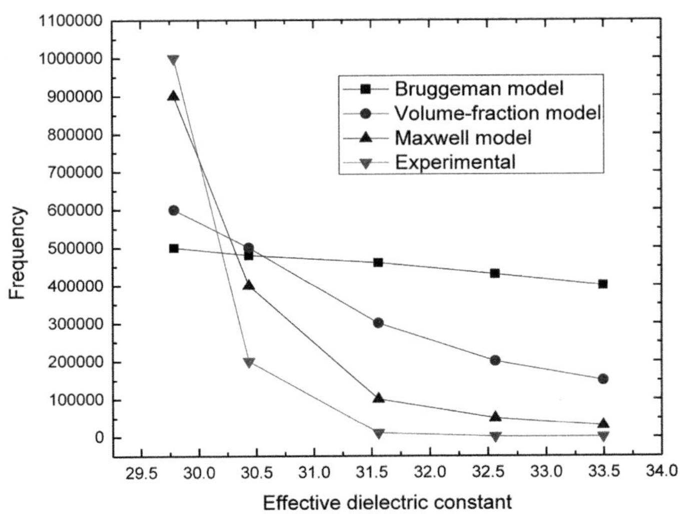

FIGURE 9.13 Comparison of experimental and theoretical values of the effective dielectric constant with frequency for 50 phr $NiFe_2O_4$ filled natural rubber nanocomposite.

9.4 CONCLUSIONS

Nano-sized nickel ferrite particles were prepared successfully using co-precipitation method and were characterized. The nanoparticles show excellent magnetic properties when compared with their bulk counterparts. Co-precipitation method was successful for preparing magnetic nickel ferrite particles with M_s value of 44 emu/g. The calculated values of magnetic properties are in good agreement with theoretical values at lesser loadings, but at upper loadings, there is a deviance due to agglomeration of particles which is in good agreement with other physicomechanical properties. Maximum dielectric permittivity is found at lower frequencies and higher filler loadings. Modeling of dielectric properties shows that experimental values are in good agreement with the Maxwell model.

KEYWORDS

- **dielectric permittivity**
- **interfacial polarization**
- **natural rubber**
- **nickel ferrite**

REFERENCES

1. Aliahmad, M., & Noori, M., (2013). *Indian Journal of Physics*, *5*(87), 431–434.
2. Kamellia, N., & Rezvanh, Z., (2012). *Chemistry Central Journal*, *6*, 23.
3. Kinnari, P., (2010). *Indian Journal of Pure and Applied Physics*, *48*, 581–585.
4. George, M., John, A. M., Nair, S. S., Joy, P. A., & Anantharaman, M. R., (2006). *J. Magn. Magn. Mater*, *302*, 190–195.
5. Atif, M., & Nadeem, M., (2014). *Journal of Sol-Gel Science and Technology*, *72*(3), 615–626.
6. Saso, G., Miha, D., & Darko, M., (2012). *Materials Chemistry and Physics 133*, 515–522.
7. Sousa, E., Alves, R., Aquino, R., Sousa, M. H., Goya, G. F., Rechenberg, H. R., Tourinho, F. A., & Depeyrot, J., (2005). *J. Magn. Magn. Mater*, *289*, 118–121.
8. Pankhurst, Q. A., & Pollard, R. J., (1991). *Physical Review Letters*, *67*(2), 248.
9. Kodama, R. H., Berkowitz, A. E., McNiff, E. J., & Foner, S., (1996). *Phys. Rev. Lett.*, *77*, 394.

10. Patil, R. P., Delekar, S. D., Mane, D. R., & Hankare, P. P., (2013). *Results in Physics*, *3*, 129–133.
11. Mamilla, L., Katrapally, V. K., & Krishnan, T., (2016). *Advances in Nanoparticles*, *5*, 103.
12. Aly, M., Badr, H., Elshaikh, A., & Ibraheim, M. A., (2011). *Journal of Modern Physics*, *2*, 12–25.
13. Aliuzzaman, M., Manjurul, M. H., Jannatul, M. F., Manjura, S. H., & Abdul, H. M., (2014). *Scientific Research, 4-28.*
14. Saritha, A., Kuruvilla, J., Sabu, T., & Muraleekrishnan, R., (2012). *Journal of Applied Polymer Science*, *124*(6), 4590–4597.
15. Rahimi, M., Kameli, P., Ranjbar, M., Hajihashemi, H., & Salamati, H., (2013). *Journal of Materials Science*, *48*(7), 2969–2976.
16. Ibraheem, O. A., (2014). *Journal of Thermal Analysis and Calorimetry*, *116*(2), 805–816.
17. Mahmoud, G. N., Elias, S., & Nazrin, K. Z., (2013). *International Nano Letters*, *3*, 19.
18. Thomas, H., & Dorothée, V. S., (2010). *Materials*, *3*, 3468–3517.
19. Gerard, J. F., (2001). *Fillers and Filled Polymers*, Wiley-VCH: Weinheim, Germany.
20. Ananthika, V., Sreedha, S., & Balakrishnan, S., (2015). *International Journal of Applied Engineering Research*, ISSN 0973–4562, *10*, 91.
21. Malini, K. A., Mohammed, E. M., Sindhu, S., Joy, P. A., Date, S. K., Kulkarni, S. D., Kurian, P., & Anantharaman, M. R., (2001). *J. Mater. Sci.*, *36*, 5551.
22. Kooti, M., & Naghdi, A. S., (2013). *J. Mater. Sci. Technol.*, *29*(1), 34–38.
23. Muscas, G., Concas, G., Cannas, C., Musinu, A., Ardu, A., Orrù, F., et al., (2013). *J. Phys. Chem. C.*, *117*(44), 23378–23384.
24. Arati, G., Kolhatkar, A., Jamison, C., Dmitri, L., Richard, C. W., & Randall, T. L., (2013). *Int. J. Mol. Sci.*, *14*(8), 15977–16009.
25. Anantharaman, M. R., Malini, K. A., Sindhu, S., Mohammed, E. M., Date, S. K., Kulkami, S. D., Joy, P. A., & Philip, K., (2001). *Bull. Mater. Sci.*, *24*, 623.
26. Cole, K. S., & Cole, R. H., (1941). "Dispersion and absorption in dielectrics-I, Alternating current characteristics," *J. Chem. Phys.*, *9*, 341.
27. Tanaka, T., Kozaka, M., Fuse, N., & Ohki, Y., (2005). *IEEE Trans Dielectr. Electr. Insul.*, *12*, 669–681.
28. Smith, R. C., Liang, C., Landry, M., Nelson, J. K., & Schadler, L. S., (2008). *IEEE Dielect. El. In.*, *15*, 187–196.
29. Ma, D., Hugener, T. A., Siegel, R. W., Christerson, A., Martensson, E., Onneby, C., & Schadler, L., (2005). *Nanotechnology*, *16*, 724.
30. Mrinal, B., (2016). *Materials, 9*, 262.
31. Hamon, B. V., (1953). *Australian Journal of Physics*, *6*.304.
32. Wang, S., Zhang, Y., Peng, Z., & Zhang, Y., (2005). *J. Appl. Polym. Sci.*, *98*, 227.
33. Fritzsche, J., Das, A., Jurk, R., Stöckelhuber, K. W., Heinrich, G., & Klüppel, M., (2008). *eXPRESS Polymer Letters*, *2*(5), 373–381.
34. Das, A., Jurk, R., Stöckelhuber, K. W., Engelhardt, T., Fritzsche, J., Klüppel, M., & Heinrich, G., (2008). *Journal of Macromolecular Science, Part A: Pure and Applied Chemistry*, *45*, 144.
35. Yang, R., Jianmin, Q., Tom, M., & Wong, C. P., (2001). *IEEE Transactions on Components and Packaging Technologies*, *23*(4), 680–683.

PART II

Interpenetrating Polymeric Networks and Nanostructured Materials

CHAPTER 10

EFFECT OF INITIATING SYSTEMS ON THE VISCOELASTIC BEHAVIOR OF SBR-PMMA INTERPENETRATING POLYMER NETWORKS

JOSE JAMES[1–3], GEORGE V. THOMAS[1], and SABU THOMAS[2,3]

[1]*Research and Post-Graduate Department of Chemistry, St: Joseph's College, Moolamattom, Kerala, India*

[2]*International and Interuniversity Center for Nanoscience and Nanotechnology, Mahatma Gandhi University, Kottayam, 686560, Kerala, India*

[3]*School of Chemical Sciences, Mahatma Gandhi University, Kottayam, 686560, Kerala, India*

ABSTRACT

A series of interpenetrating polymer networks (IPNs) based on styrene butadiene rubber (SBR) and poly [methyl methacrylate] (PMMA) have been synthesized by Sequential polymerization technique. The effect of two initiating systems (Benzoyl peroxide (BPO) and Azobisisobutyronitrile (AIBN)) are employed in the polymerization of MMA during the fabrication of IPN. The viscoelastic properties of these IPNs were investigated in detail in the temperature range of –80 to 200°C and at a frequency of 1Hz. IPNs with 50 wt% of PMMA has broad transitions arising from β- and α-relaxations in PMMA. The morphology of the IPNs was analyzed with TEM images and tried to correlate these results with viscoelastic behavior and their damping properties.

10.1 INTRODUCTION

Multi-component polymeric systems provide an apt protocol for the modifications of properties to meet specific needs in material science. Interpenetrating polymer networks (IPNs) are one of the important hybrid polymeric systems [1] and is one of the rapidly growing areas of polymeric material science.

IPNs consist of two or more polymeric systems, out of which, one is in network form, and others are dissolved in the first system on a molecular scale [2]. IUPAC Compendium of chemical terminology defines IPNs as polymers comprising two or more networks that are at least partially interlaced on a molecular scale but not covalently bonded to each other and cannot be separated unless chemical bonds are broken. IPNs can be prepared by simultaneous polymerization, sequential polymerization or latex blending technique [3]. In sequential polymerization technique for IPN involves the sequential addition of selective crosslinkers to a homogenous mixture of two polymers in solution or in melt form. They are in fact synthesized by a- two-step process. In the first step, polymerization of the first mixture (consisting of monomer, crosslinking agent, and initiator or catalyst) forms a network I. This network is swollen with the second combination of monomer and cross-linking agent and polymerized to form an IPN, that is the polymer–2 is polymerized and cross-linked in situ in network I.

In simultaneous polymerization method, a polymer is synthesized (from the monomer) and simultaneously cross-linked within the network of another polymer to give rise to an interpenetrating network. Here an IPN is formed by polymerizing two different monomers and cross-linking agent pairs together in one step. The key to the success of this process is that the two components must polymerize by reactions that will not interfere with one another. This is often accomplished by polymerizing one network by a condensation reaction, while the other network is formed by a free radical reaction.

Systems with no cross-linked phases are actually polymer blends. IPN in which both the polymeric systems are cross-linked is referred to as Full-IPN [4]. In semi-IPN, one of the two components has a linear structure [5]. IPN do not interpenetrate on a monomer scale but has a microheterogeneous morphology. IPN synthesis is the only way of mixing two cross-linked polymers intimately. IPN formation not only retards the process of phase separation but also results in enforced miscibility.

Numerous techniques have been explored for finding the most effective and efficient IPN material that displays the best possible compatibility with minimal phase separation. If the phase that is synthesized first is too heavily cross-linked, there may not be enough room for the second polymer to swell and penetrate the first network, thus eventually creating two separate phases. The existence of interpenetration is generally judged through the combination of DMA analysis and morphology characterization [6].

IPNs are a special class of polymer blends with special features and potential applications. IPN as a combination of an elastomer and a glassy polymer show properties ranging from high-impact plastics to reinforced elastomers. The analysis of the glass-transition temperatures (*T*g's) of virgin polymers and their IPNs through dynamic mechanical analysis is the most dependable characterization tool for determining the miscibility of IPNs. Homogeneity in the composite is indicated as a single *T*g and two separate *T*g's are reflections of immiscibility. IPN synthesis is generally accompanied by widening the loss peak and as a result, IPN manufactures have been widely used as tailoring route for the synthesis of damping materials. In this mode of synthesis, most IPNs show microphase separation with considerable molecular-level mixing. IPNs with partial miscibility usually exhibit broad glass transitions. The mechanical vibration is damped to a maximum extent near the glass-transition region. The damping behavior of an IPN over a wide range of temperatures and frequencies is the common expectation in this field [7]. IPNs have a variety of applications and they are in fact a commercially successful form of polymer blends, probably owing to the crosslinked structure that provides better thermal stability, mechanical properties, chemical resistance and so on. IPNs are traditionally used as damping materials, impact-resistant materials, adhesives, and so on.

The aim of this research is to create thermoplastic-elastomer blend materials with comparable mechanical performance, damping properties and as a sensor like separation membrane based on polarity. The full-IPN system in this work was prepared by combining a highly stiff PMMA phase with soft SBR phase through sequential interpenetration followed by in-situ polymerization.

One of the major factors affecting the viscoelastic properties of IPN is the initiating system in its formation. Here, this factor is analyzed in full detail. The temperature range of analysis is –80 to 200°C. The studies were carried out at a frequency of 1 Hz. The morphology of the IPNs was analyzed with transmission electron microscopy. An attempt was made to relate the viscoelastic behavior and their predictions to the morphology of the IPNs.

10.2 EXPERIMENTAL

10.2.1 MATERIALS

Styrene-butadiene rubber (Synaprene 1502) with 25% styrene content was used for this study was supplied by Indian Synthetic Rubber Limited (ISRL). Dicumyl peroxide (DCP 99%-) is the crosslinker for SBR system. Methyl methacrylate (MMA, Aldrich) was freed from inhibitor and used. Benzoyl peroxide (BPO) and Azobisisobutyronitrile (AIBN, Aldrich) were employed as an initiator for the polymerization of MMA to PMMA. The cross-linker for MMA, Divinylbenzene (DVB, Aldrich) was distilled under vacuum prior to use.

10.2.2 PREPARATION OF IPNS

SBR rubber was mixed with DCP (1phr) in a two-roll mixing mill at room temperature as per ASTM standards. The curing behavior of SBR compounds was studied on a Rheometer and the optimum cure time was determined. The mixture was vulcanized at 150°C on a hydraulic press and as a result, cross-linked SBR sheet was obtained.

In the preparation of IPN using BPO as the initiator for PMMA phase (B Series), the following method has been adopted. Vulcanized SBR sheets of definite weight and of thickness (2 mm) were immersed in a homogeneous mixture of MMA, DVB, and BPO (the concentration of DVB and BPO were based on the MMA content) for different time intervals. As a result, we get swollen samples with different weight percentages of MMA. To get an equilibrium distribution of MMA monomer in the matrix, swollen samples were kept at 273 K for three hours.

The SBR sheet with equilibrium distribution of MMA was wrapped with aluminum foil to minimize the evaporation of monomer during in-situ polymerization in an oven. This SBR sheet was kept in between two stainless steel plates inside an oven at 353K for 16 h for polymerization. Afterward, the rubber sheet was vacuum dried to constant weight. The unreacted MMA monomers were removed by treatment under reduced pressure. By this technique, we could synthesize a series of flexible IPN sheets. The second set of IPNs (A series) using AIBN as the initiator for the PMMA phase, the above-mentioned method for the synthesis of IPN has been utilized, where AIBN replace the position of BPO.

The composition of the IPN samples was determined on the basis of their final weights. Preparation of IPN can be schematically represented as follows (Figure 10.1).

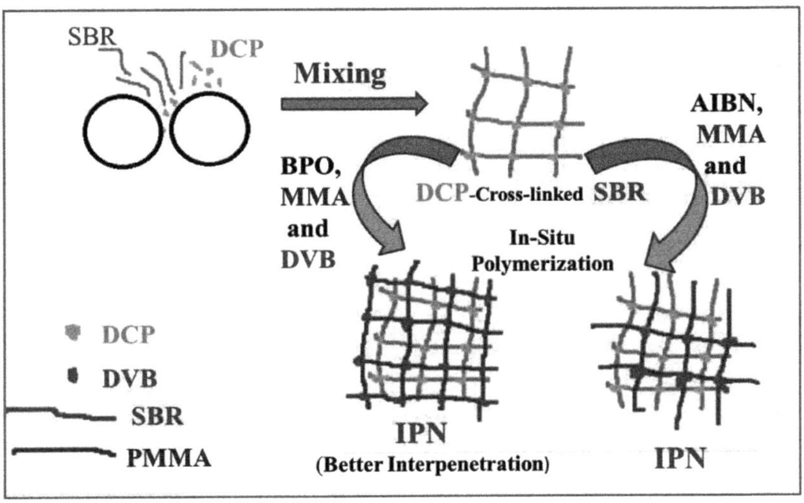

FIGURE 10.1 **(See color insert.)** Scheme of preparation of IPN.

The IPNs using can be represented as **^{a}S A/B P_c D_d** (Here 'A' stands for AIBN and 'B' for BPO. 'a' indicates the weight of DCP per 100 gram of SBR rubber, S indicates SBR rubber, P represents PMMA, c indicates weight percentage of PMMA, D stands for Divinyl benzene [DVB], d corresponds to wt.% of DVB content) (Table 10.1).

TABLE 10.1 Composition of IPNs

Type of IPN	DCP (Phr)	Weight% of PMMA and SBR in the IPN	DVB (wt.%)	The initiator for the Polymerization of MMA
$^{1}SBP_{50}D_2$	1	50:50	2	BPO
$^{1}SAP_{50}D_2$	1	50:50	2	AIBN

10.2.3 VISCOELASTIC MEASUREMENTS

The viscoelastic properties were determined on TA Instruments Q800 DMA in Single cantilever mode at frequencies of 1Hz and Amplitude 20 µm in the temperature range of –80 to 200°C. The heating rate was 3°C/min. The samples for testing were taken as rectangular bars (17.5 mm x 12.6 mm x 2mm).

10.2.4 MICROSCOPIC ANALYSIS

The morphology of IPNs was studied using JOEL-JEM 2010 model high-resolution transmission electron microscope. The samples were cryogenically fractured and examined under the microscope. Ultrathin sections of bulk specimens (Cryo cut specimens ~100nm thickness) prepared using an ultramicrotome (Leica, Ultracut UCT) were placed on 300 mesh Cu grids (35mm diameter) and were analyzed without staining.

10.3 RESULTS AND DISCUSSIONS

Dynamic mechanical analysis (DMA) is widely used for the measurement of the viscoelastic behavior of IPNs. DMA is a technique used to characterize IPNs properties as a function of temperature, time, frequency, stress or a combination of these parameters. It measures stiffness and damping (dissipation of energy) in a material under cyclic loading.

Dynamic mechanical thermal analysis (DMTA) is the most dependable technique employed to explain the miscibility between the components in IPNs. The common parameters usually employed to determine the extent of miscibility or phase separation in IPN are the height of the loss tangent peak and any shift of the glass transition temperature of one or both of the polymer components. Here, we consider the 50:50 ratio composition (SBR: PMMA ratio in IPN) of BPO and AIBN initiated systems. i.e., $BP_{50}D_2$ and $AP_{50}D_2$ IPN samples are compared in this study. Crosslink content in SBR and PMMA phases are kept constant in this analysis. The effect of initiating system in the polymerization of second component (PMMA) in IPN fabrication can be summarized as follows.

In Figure 10.2, the effects of the initiating system on *E*' (storage modulus) for P_{50} samples are shown ($BP_{50}D_2$ and $AP_{50}D_2$) in comparisons with neat

SBR and PMMA. B and A corresponds to "BPO" and "AIBN." These are the initiator for the polymerization of MMA in the course of IPN formation. In the case of neat IPNs, relaxation process in SBR has usually observed around –29°C and a sudden drop in the modulus of PMMA at around 100°C is due to its glass transition. Here two IPN samples show a glassy phase below –28.8°C. Upon increasing the temperature from –28.8°C to around 65°C, the samples soften and the modulus decreases dramatically. The intermediate plateau value of the elastic modulus is reached at around 65°C and extends up to around 100°C which corresponds to the beginning of the PMMA phase mechanical relaxation. The IPN materials do not flow at higher temperatures and keep a stable modulus value up to 200°C. *E'* was in the following order: BPO > AIBN IPN system.

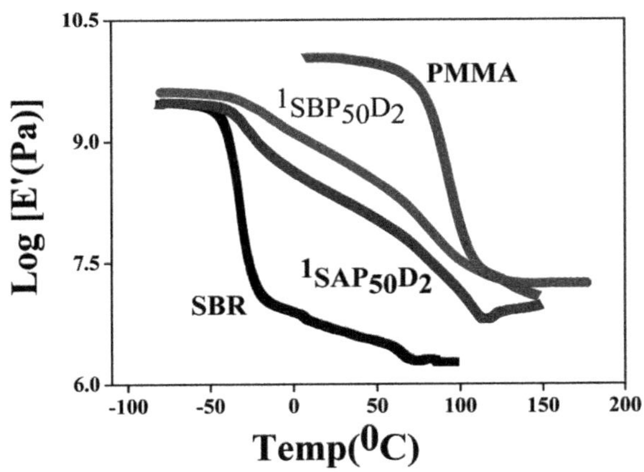

FIGURE 10.2 (See color insert.) Effect of initiator on *E'* for SBR, PMMA, $^1SBP_{50}D_2$ and $^1SAP_{50}D_2$ IPNs at 1Hz.

The degree of entanglement of the polymer blends can be obtained from the dynamic mechanical analysis. We can use the storage modulus data for determining the entanglement density (N) using the Eq. (10.1): [8]

$$N = E'/6RT \tag{10.1}$$

where *E'* is the storage modulus obtained from the plateau region of *E'* versus temperature curve, *R* is the universal gas constant and *T* is the absolute temperature. The value of entanglement density of IPNs with different composition of PMMA at 298K is given in Table 10.2.

TABLE 10.2 Entanglement Density of IPNs

IPN	Entanglement Density (moles/m^3)
$^1SBP_{50}D_2$	44035
$^1SAP_{50}D_2$	39685

From the Table 10.2, it is clear that $^1SBP_{50}D_2$ possess high entanglement density. This is because by the addition of BPO as the initiator for the polymerization of MMA in IPN formation, the entanglement between the homopolymers increases and a better adhesion is achieved as a result of a decrease in the interfacial tension compared with corresponding AIBN initiated IPN [9].

The following Table 10.3 summarizes the viscoelastic results of our IPNs and neat samples.

TABLE 10.3 T_g and Related Values of SBR (Peroxide Vulcanized [1phr]), PMMA, and IPNs

Sample	Half peak width of SBR transition (°C)	T_g SBR (°C)	Half peak width of PMMA transition (°C)	T_g-PMMA (°C)
SBR	36	–28.8	-	-
$^1SBP_{50}D_2$	58	–23.52	53	96.8
$^1S\ AP_{50}D_2$	59	–24.43	45	115.1
PMMA	-	-	40	109.7

In Figure 10.3, the effects of the initiating system on tanδ for P_{50} samples are shown ($BP_{50}D_2$ and $AP_{50}D_2$) at low and high-temperature regions, respectively. The low-temperature transition (SBR transition) was in the following order: AIBN initiated IPN ($AP_{50}D_2$) > BPO initiated IPN ($BP_{50}D_2$). Therefore, the damping property was higher for $AP_{50}D_2$ and lower for $BP_{50}D_2$. It proves that the phase mixing was efficient in the Benzoyl peroxide (BPO) series. The efficient mixing of the two phases in BPO-initiated system restricted the mobility of the SBR phase and reduced the damping behavior of that system. The peak width at half height is higher for 'B' series than that for 'A' series. IPN belongs to BPO initiated system show an inward shift in Tg values with respect to PMMA by 12.9°C, whereas AIBN initiated IPN shows an outward shift by 5.4°C with respect to neat PMMA. These two results clearly indicate the existence of better interpenetration in BPO initiated IPNs.

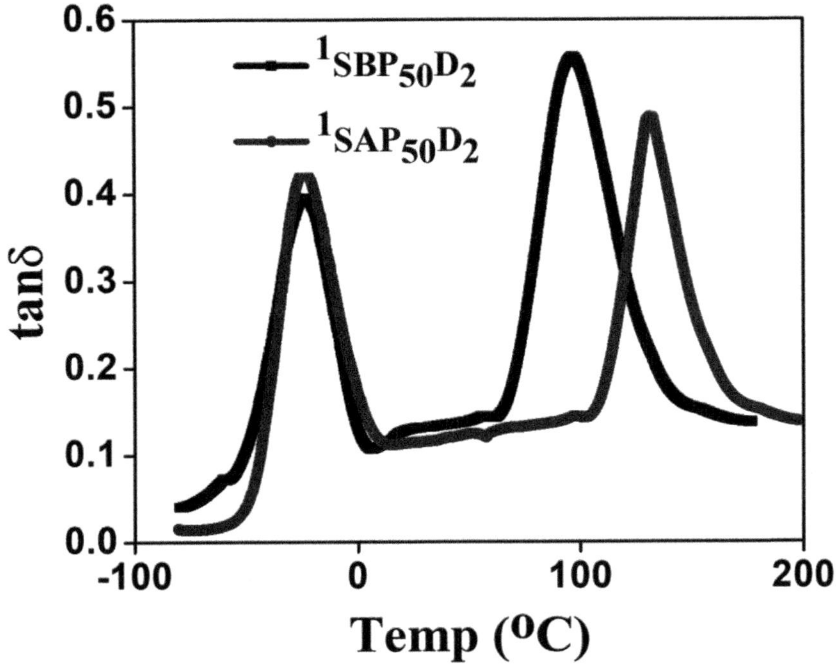

FIGURE 10.3 Effect of initiator on tanδ for $^{1}SBP_{50}D_2$ and $^{1}SAP_{50}D_2$ IPNs at 1Hz.

The morphology of IPNs depends on the method of synthesis, on the compatibility of the polymer systems employed, and on the relative rates of formation. In sequential IPNs, the network first formed is most likely to be the continuous network. Its crosslink density is the controlling factor in determining the morphology of the system of each network. The morphological properties of IPNs are best characterized by using the Scanning Electron Microscopy (SEM) and Transmission Electron Microscopy (TEM).

The second component in IPN has an important role in designing the morphology of the resultant system [10]. As a result, an initiator for the polymerization of the second phase has a deceive role in the morphology of the resultant system. In our system, BPO initiated IPNs possess a higher level of entanglement than those of AIBN system. The better interpenetration and co-continuous morphology of IPN ($^1SBP_{50}D_2$) over with that of $^1SAP_{50}D_2$ is proved through the following TEM images (Figure 10.4).

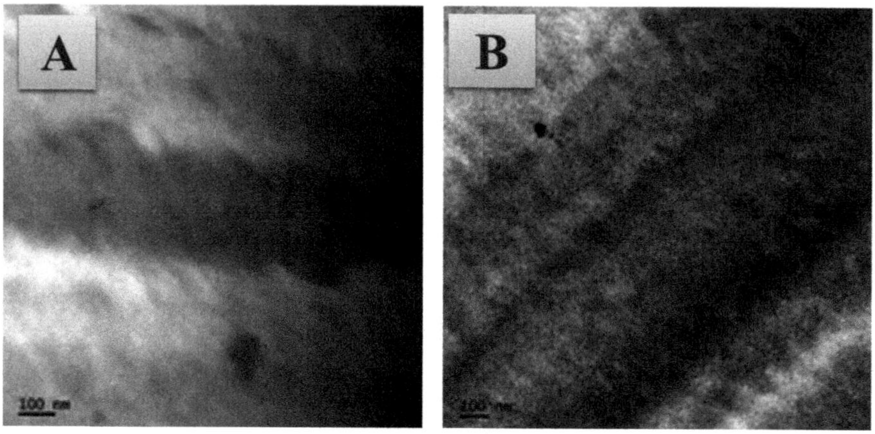

FIGURE 10.4 Transmission electron micrographs of IPNs (a) $^1SAP_{50}D_2$ and (b) $^1SBP_{50}D_2$.

Initiators play an important role in designing the morphology and in engineering the properties of resultant IPNs. In our system, BPO is the effective initiator for the polymerization of MMA during the IPN synthesis, which provides better interpenetration in resultant IPN system.

10.4 CONCLUSION

The sequential method has been employed for the synthesis of full IPNs using SBR and PMMA. All IPN show two separate glass transitions and is an indication of extensive phase separation. But a certain extent of component mixing was indicated by the significant shifting of glass transition of the SBR and PMMA component in the IPNs. The initiator for the polymerization of the second component in IPN performs a decisive role in the properties of resultant IPNs. The BPO initiated IPN system, which had maximum phase mixing with better interpenetration, showed lower damping properties than the AIBN initiated system. This was unambiguously proved through TEM images. *E*' was highest for the BPO series because of intimate and effective crosslinking of the phases.

KEYWORDS

- **azobisisobutyronitrile**
- **benzoyl peroxide**
- **interpenetrating polymer networks**
- **poly [methyl methacrylate]**
- **sequential polymerization technique**
- **styrene butadiene rubber**

REFERENCES

1. Sperling, L. H., (2012). *Interpenetrating Polymer Networks and Related Materials*. Springer Science & Business Media.
2. James, J., Thomas, G. V., Pramoda, K., & Thomas, S., (2017). Transport behavior of aromatic solvents through styrene butadiene rubber/poly [methyl methacrylate](SBR/PMMMA) interpenetrating polymer network (IPN) membranes. *Polymer*, *116*, 76–88.
3. Hanoosh, W. S., & Saleh, H. M., (2016). Tensile strength and dynamic mechanical analysis of new IPNs based on epoxy resin-polysulphide elastomer. *European Journal of Chemistry*, *7*(3), 352–356.
4. Trakulsujaritchok, T., & Hourston, D. J., (2006). Damping characteristics and mechanical properties of silica filled PUR/PEMA simultaneous interpenetrating polymer networks. *European Polymer Journal*, *42*(11), 2968–2976.
5. James, J., Thomas, G. V., Akhina, H., & Thomas, S., (2016). Micro-and nano-structured interpenetrating polymer networks: State of the art, new challenges, and opportunities. *Micro-and Nano-Structured Interpenetrating Polymer Networks: From Design to Applications*, 1.
6. Mathew, A. P., Groeninckx, G., Michler, G., Radusch, H., & Thomas, S., (2003). Visco-elastic properties of nanostructured natural rubber/polystyrene interpenetrating polymer networks. *Journal of Polymer Science Part B: Polymer Physics*, *41*(14), 1680–1696.
7. Librado, P., & Rozas, J., (2009). DnaSP v5: A software for comprehensive analysis of DNA polymorphism data. *Bioinformatics*, *25*(11), 1451–1452.
8. Singh, S. P., Kumar, C. P., Nagarjuna, P., Kandhadi, J., Giribabu, L., Chandrasekharam, M., Biswas, S., & Sharma, G., (2016). Efficient solution processable polymer solar cells using newly designed and synthesized fullerene derivatives. *The Journal of Physical Chemistry C.*, *120*(35), 19493–19503.
9. Sperling, L., (1985). Recent advances in interpenetrating polymer networks. *Polymer Engineering & Science*, *25*(9), 517–520.
10. Ahmed, S., Chakrabarty, D., Mukherjee, S., & Bhowmik, S., (2017). Characteristics of simultaneous epoxy-novolac full interpenetrating polymer network (IPN) adhesive. *Journal of Adhesion Science and Technology*, 1–16.

CHAPTER 11

INFLUENCE OF CATIONIC, ANIONIC, AND NONIONIC SURFACTANTS ON HYDROTHERMAL SYNTHESIS OF NANO CUS: STRUCTURAL, MORPHOLOGICAL, AND CAPACITANCE BEHAVIOR

D. GEETHA[1], P. S. RAMESH[2], and SUREKHA PODILI[1]

[1]*Department of Physics, Annamalai University, Annamalai Nagar – 608002, Tamilnadu, India*

[2]*Department of Physics (DDE Wings), Annamalai University, Annamalai Nagar – 608002, Chidambaram, Tamilnadu, India, E-mail: geeramphyau@gmail.com*

ABSTRACT

A different ionic surfactant was used in a typical hydrothermal process for controlling the morphology of the CuS nanostructure. Here in we demonstrate the synthesis and formation mechanism of CuS nanostructures by a simple hydrothermal route using organic surfactants anionic, cationic, and non-ionic surfactants as templates and thiourea as the sulfur source in 130°C. The effect of cationic cetyltrimethylammonium bromide (CTAB), anionic sodium dodecyl benzene sulphonate (SDBS) and nonionic (Triton X–100) surfactants for adjusting the shape/size, porosity, and electrochemical properties of CuS nanostructures was examined. The as-obtained CuS were characterized by XRD, FT-IR, UV-Vis, SEM/EDS, TEM, XPS, and cyclic voltammetry (CV). The XRD pattern reveals that the obtained nanostructures are crystalline in nature. The effect of surfactants on the morphology of the CuS shows that the diameter of the product in the range of 7–14 nm for (0.1 mM) CTAB stabilized CuS, 12–27 nm for (0.1 mM) SDBS stabilized CuS and 9–32 nm for (0.1 mM) Triton X–100 stabilized CuS. The result of electrochemical measurements by cyclic voltammetry shows that the specific capacitance value changes with the stabilizing agents and porous nature of the samples. The specific capacitance values were found to increase in the order of CuS – Triton X–100 (164.47Fg^{-1}) < CuS – SDBS (257.13Fg^{-1}) < CuS- CTAB (328.26 Fg^{-1}) at a scan rate of 5 mV per second in 2 M KOH aqueous electrolyte solution. The electrochemical measurement shows that the specific surface area and capacitance changes with the ionic nature of the surfactant. Among these electrodes, the CuS electrode show an using specific capacitance 328.26 Fg^{-1} with mass loading (1.5 mg/cm^3), good power capability, excellent cycling stability and high columbic efficiency. This exceptional performance is benefited from the almost mono dispersed morphology and high specific surface area. At the same time, the supercapacitor, employing the CuS electrode with porous nanostructure as the positive electrode and the activated carbon electrode as the negative electrode was successfully assembled.

11.1 INTRODUCTION

Copper mono Sulfide as one of the low-cost transition metal sulfides has attracted a great deal of attention in many fields owing to their unique physical and chemical properties [1]. Recently, as a promising electrode material, CuS has been investigated widely for pseudocapacitor applications on account of the excellent electrochemical properties, abundant resources and environmental compatibility [2–4]. Supercapacitors, a family of electrochemical capacitors, are considered to be one of the most promising energy-storage devices because of their many advantages, including long cycle life, high power density, faster charge-discharge processes and relatively low cost [5–9]. Pseudocapacitors along with electric double layer capacitors (EDLCs) create a supercapacitor, which can store charge by redox-based faradaic reactions [10–12] and thus can have higher capacitance values than electric EDLCs; moreover, they can have higher power densities than secondary batteries.

Chalcogenides are a significant class of materials that have found potential uses in catalysis [13], energy storage [14] and optoelectronic devices [15–17]. CuS is one of the essential chalcogenides received a great attention to its wide stoichiometric composition with porous morphology and optical property. The synthesis technique plays a crucial role in controlling the morphology which determines their structural, optical, electrical properties. Low-temperature solution-based techniques Viz., co-precipitation [18], hydrothermal [19], solvothermal [20] and electrodeposition [21] and commonly employed to synthesize of CuS. We demonstrate the synthesis of CuS using very common and inexpensive reagents, such as copper nitrate trihydrate and thiourea by the hydrothermal technique with short duration (10h) in ethanol. The reaction medium plays a significant role in controlling the shape, size, and phase of the products for practical applications in a supercapacitor. From the review reports, it can be understood that the specific capacitance will vary with stabilizing agents (surfactants) with increased self-aggregation and electron transfer. Therefore, it is important to prevent the self-aggregation and accelerate the electrolyte permeation and electron transfer to further enhance the electrochemical performance of the CuS electrode. The effect of cationic cetyltrimethylammonium bromide (CTAB), anionic sodium dodecyl benzene sulphonate (SDBS) and nonionic (Triton X–100) surfactants for adjusting the shape/size, porosity, and electrochemical properties of CuS nanostructures were examined.

Due to the unique property such as its electrical conductivity [22] resembling metals and absorption of solar energy [23] of sulfides, they have numerous potential applications. Many researchers focused on the preparation of CuS with different morphology [24, 25]. Surfactants are usually used in reaction systems and they play critical roles in the morphological control of CuS nanostructures. Inorganic nanomaterials are well-defined morphologies having peculiar properties and potential applications Viz., sensors, energy storage and energy conversion [26]. CuS has attracted significant attention as an active electrode material owing to its high theoretical specific capacitance, excellent reversibility, good electrical conductivity and environmental friendliness [27]. For practical application of high power density devices, the powder forms yield better volumetric energy density thin films. CuS powders of various morphologies have been synthesized by different routes, and their utility as electrode materials for pseudocapacitor has been reported [28]. It is well accepted that pseudocapacitance is an interfacial phenomenon which is tightly related to the specific surface area and porous structure of active electrode materials. The mesoporous sulfide system can provide a very short diffusion pathway for ion as well as high specific surface area, leading to improved electrochemical performance.

In this work, we report three kinds of surfactants; CTAB, SDBS, and Triton X–100 were used in a hydrothermal process for controlling the morphology of the CuS nanostructure for high-performance supercapacitors. The results showed that different surfactants had remarkable effects on the morphology, specific surface area and electrochemical properties of CuS electrodes. The addition of CTAB-CuS the supercapacitor exhibited low resistance, good long-term electrochemical stability, and high specific energy. Porous nanostructure of CuS had been employed in super capacitor fabrication on a glassy carbon electrode to examine the electrochemical behavior and results were discussed elaborately.

11.2 EXPERIMENTAL SECTION

11.2.1 PREPARATION OF POROUS CUS

In a typical synthesis, all reagents were analytically pure. Copper nitrate trihydrate ($Cu(NO_3)_2$.–$3H_2O$), Ethylene glycol (EG) ($(C_2H)_6O_2$), Thiourea (Tu,$Sc(NH_2)_2$), Sodium dodecyl benzene sulphonate (SDBS), Cetyl trimethyl ammonium bromide (CTAB), Triton X–100, Ethanol, and deionized water were purchased and used as such without further purification.

1 mM of copper nitrate trihydrate was dissolved into 40 ml EG, 2 mM of thiourea and 0.1 mM of CTAB, SDBS, and Triton X–100 was added separately with copper nitrate trihydrate and EG solution with stirring of 45 minutes under room temperature and then aged for 10 hrs at 130°C. Finally, the precipitate was separated and washed with ethanol and de-ionized water several times to eliminate excess chemicals. The product was dried in a vacuum at 60°C for 6 hrs.

11.2.2 MATERIALS CHARACTERIZATION

XRD pattern of the as-synthesized samples was recorded on X'Pert-PRO using Cu Kα radiation. X-ray diffractometer at a scan rate of 1° min^{-1} in the 2θ range of 10°–70°. The presence of a functional group of nanostructures was characterized by Fourier transform infrared spectrometer (as pellets in

KBr) using a Perkin Elmer spectrometer range from 400–4000 cm^{-1}. The band gap of the nanoparticles was determined by UV-Vis spectroscopy (ShimadzuUV1700). The morphology of the synthesized product was inspected through a (JEOL-JSM – 5610 LV with INCA EDS) scanning electron microscope (SEM) and the elemental compositional analysis of the products was accomplished using EDS, in combination with SEM. The morphologies were characterized on a TEM CM–200 transmission electron microscope (TEM). Chemical bonding states were investigated by X-ray photoelectron spectroscopy (Kratos Analytical). The electrochemical characterization of the CuS was carried out in an electrochemical workstation (CHI 660C, USA) by cyclic voltammetry (CV) and using a three-electrode design.

11.2.3 ELECTRODE PREPARATION

A bare glassy carbon electrode (GCE) was cleaned and washed with double-distilled water. The as-prepared CuS (1.55 mg m/L) was dissolved in ethanol to get slurry. Then it was sonicated for 10 min. 10ml of the slurry was placed onto glassy carbon electrode surface using a micropipette and dried at 70°C for 1.5h under vacuum to evaporate the solvent. The CuS modified GCE was taken as the working electrode, a saturated calomel electrode as the reference electrode and platinum foil acts as the counter electrode. The electrochemical test was conducted in 2M KOH electrolyte solution at room temperature. The CV measured in the potential window of 0.0 to 0.5 V at different scan rates. Typical EIS study observed over the frequency limit of 100 mHz to 100 kHz at open circuit potential with an AC voltage of 5 mV.

11.3 RESULT AND DISCUSSION

In this work, a pale green complex could be formed in the starting system due to the coordination between Cu $(No_3)_2 3H_2O$ and thiourea. Simultaneously the coordination disassociation equilibrium existed in the system. When the system was heated at 130°C, few thiourea molecules hydrolyzed to produce H_2S which rapidly reacted with Cu^{2+} to form CuS. The coordination dissociation equilibrium was breached and moved to produce more Cu^{2+} ions and thiourea molecules. Thus, CuS nanoparticles were obtained, when the procedure was repeated at the same temperature for various surfactants for 10 hrs.

It is notable that the composition of solvent (EG) and stabilizer plays an important role in controlling the phase structures and the morphologies of copper sulfides. The diffraction peaks of samples match well with the standard peaks. In the previous work, the reaction time plays an important role in controlling the structure and morphology of the product. The product was not obtained when the reaction time was shorter than 5hrs.

11.3.1 XRD

Figure 11.1(a–c) shows the X-ray diffraction (XRD) pattern of the samples (CuS). The main peaks are indexed at 2θ values 27.45°, 29.21°, 31.90°, 32.79°, 47.89°, 52.63° and 59.34° corresponding to the planes (101), (102), (103), (006), (110), (108) and (116), respectively, which agree well with standard powder diffraction patterns of CuS covallite structure (JCPDS card no. 06–0464) [29]. The peak intensities were lower and the peak shapes were wider, this might be caused by their porous present in the sample and the low preparation temperature of CuS. No reflections could be detected corresponding to impurities, which indicate the complete transformation of CuS.

The estimated crystallite sizes of the as-prepared samples were calculated using the Scherrer equation (11.1).

$$D = \frac{0.9\lambda}{\beta cos\theta} \tag{11.1}$$

The lattice parameters for hexagonal structured for CTAB-CuS, SDBS-CuS, and Triton X–100-CuS are calculated from the following equation (11.2).

$$\frac{1}{{d_{hkl}}^2} = \frac{4}{3}\left(\frac{h^2 + hk + k^2}{a^2}\right) \tag{11.2}$$

where a and c are the lattice constants; h, k, and l are the Miller indices; and d_{hkl} is the interplanar spacing. This interplanar spacing can be calculated from Bragg's law. The crystalline peaks of (103) and (110) planes are used for the calculation of lattice parameters. The volume of the unit cell for the hexagonal system was calculated by using the Eq. (11.3) [30].

$$V = 0.866 \times a^2 \times \mathrm{c} \tag{11.3}$$

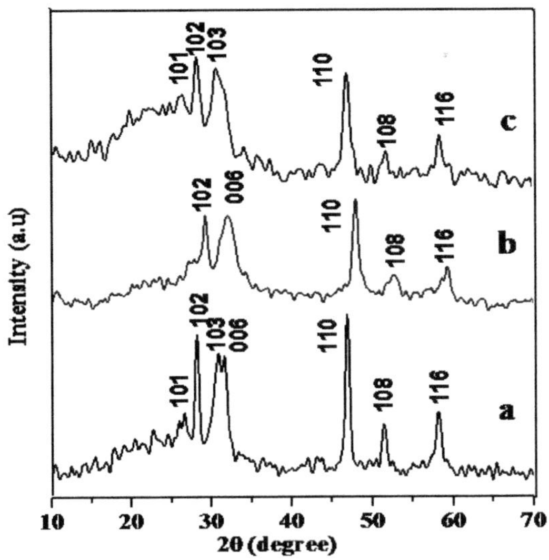

FIGURE 11.1 XRD pattern of **(a)** CTAB (0.1 mM)-CuS **(b)** SDBS (0.1 mM)-CuS **(c)** Triton X–100 (0.1 mM)-CuS.

Table 11.1 summarizes the lattice parameter a and c, average crystal size and volume of the unit cell. Both a and c values increase when CTAB stabilized with CuS nanostructures, which demonstrates the successful incorporation of dopant in lattice sites. When CTAB stabilized with CuS nanostructures the volume of the unit cell also increases compared to other stabilized surfactants.

TABLE 11.1 Doping Concentration, Lattice Parameters, Cell Volume and Average Crystal Size

Concentration of Doping (mol%)	Lattice parameters (Å)		Cell volume Å^3	Average crystal size D (nm)
	a = b	c		
CTAB-CuS	3.789	16.345	203.213	14
SDBS-CuS	3.781	16.339	202.281	20
Triton X–100-CuS	3.779	16.337	202.043	25

The influence of surfactants on the crystalline size of CuS is significant. The cationic surfactant CTAB is effective in producing smaller crystallites of CuS. The relatively broad XRD peaks reveal the small size of copper sulfide crystals, and according to the Scherrer diffraction formula, the average particle size is 14 nm (CuS-CTAB), 20 nm (CuS- SDBS) and 25nm (Cus-TritonX–100). Moreover, the variation in the diffraction intensity from the XRD spectra with changing the ionic nature of the template materials (surfactants) clearly indicates that the reaction condition plays a crucial role for obtaining porous CuS nanostructures. The results suggest that the present synthesis route is suitable for obtaining pure CuS with porous nature in a lower temperature and shorter time.

11.3.2 FTIR

Figure 11.2(a–c) shows the chemical constituents of CuS nanostructure stabilized with different surfactants presented in the range 4000–400 cm^{-1} were investigated by FTIR spectroscopy. The well-known bands at 3718 and 3855 cm^{-1} are assigned to the O-H stretching vibration indicates the existence of water molecule and the weak absorption band at 2969 cm^{-1} is attributed to the asymmetric stretching vibration of $-CH_3$. The band at 1776 cm^{-1}, which is attributed to the C=O stretching mode (spectrum (a). In all samples the strong absorption bands found at 1112, 1096 and 1108 cm^{-1} for CTAB-CuS, SDBS-CuS, and Triton X–100-CuS are assigned to the stretching vibrations C=C and C=S bonds [31]. In all synthesized samples the strong peaks observed at 617, 611 and 609 cm^{-1} indicates the vibrational peaks of Cu-S stretching modes [32, 33].

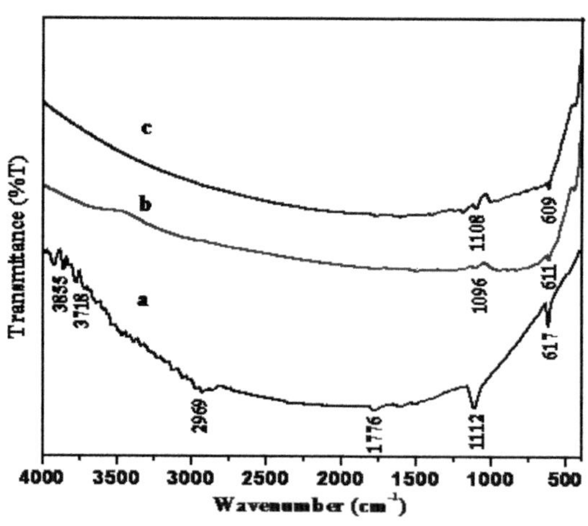

FIGURE 11.2 FTIR spectra of **(a)** CTAB (0.1 mM)-CuS **(b)** SDBS (0.1 mM)-CuS **(c)** Triton X–100 (0.1 mM)-CuS.

11.3.3 UV-VIS SPECTROSCOPY

Figure 11.3 (a–c) shows the absorption spectra in the range of 200–800 nm of CTAB-CuS, SDBS-CuS, and Triton X–100-CuS. The as-synthesized samples were taken for UV-Vis measurements and were found to be easily dispersed in methanol.

All the absorption peak positions are located at 265 nm for CTAB-CuS, 257 nm for SDBS-CuS and 251 nm for Triton X–100-CuS. Especially a broad absorption edge is observed in the IR region, which is the characteristic of covellite CuS [34] strongly blue-shifted, it indicates a decrease of the crystallite size of the samples. A broad absorption is observed from 400 to 600 nm.

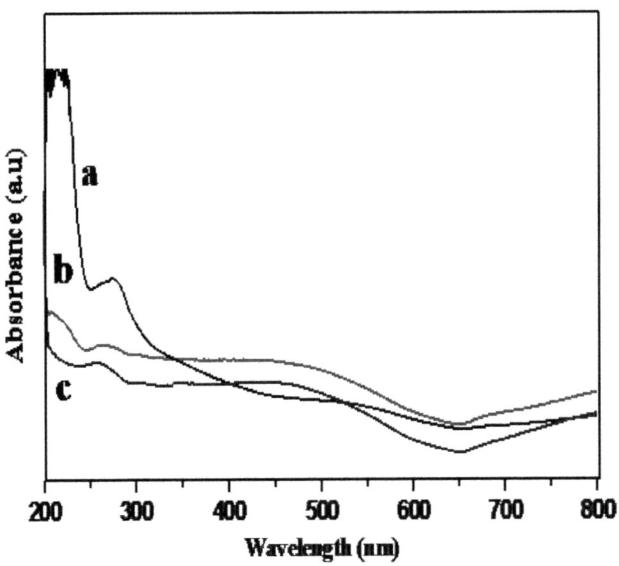

FIGURE 11.3 UV-Vis spectra of **(a)** CTAB (0.1 mM)-CuS **(b)** SDBS (0.1 mM)-CuS **(c)** Triton X–100 (0.1 mM)-CuS.

Figure 11.4 shows the band gap plots of $(\alpha h\nu)^2$ vs. $h\nu$ as per Tauc's equation for all the samples. Using the absorption value the measured band gaps are 2.61, 2.25 and 2.01 eV for CTAB-CuS, SDBS-CuS, and Triton X–100-CuS nanostructures respectively. Based on the optical absorption spectra, the band gap energy values were calculated using the following Eq. (11.4)

$$(\alpha h\nu)^n = K(h\nu - E_g) \quad (11.4)$$

where hv is the photon energy, K is a constant, E_g is band gap, α is absorbance, and n is either two for direct transition or 1/2 for an indirect transition [35].

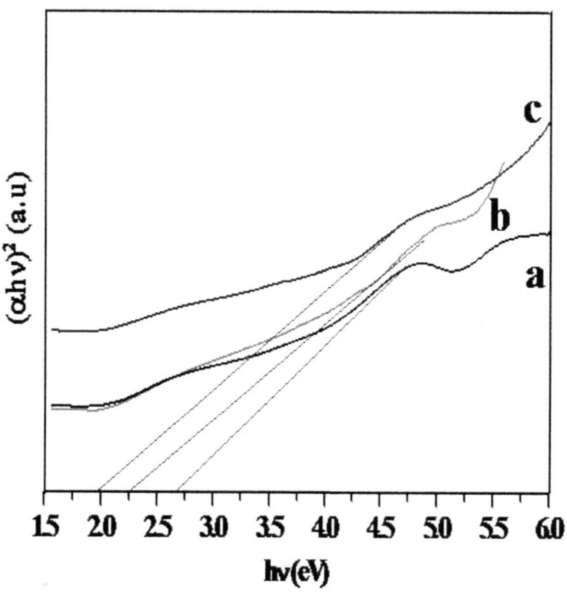

FIGURE 11.4 Typical Tauc's plot of (a) CTAB (0.1 mM)-CuS (b) SDBS (0.1 mM)-CuS (c) Triton X–100 (0.1 mM)-CuS.

11.3.4 EFFECT OF SURFACTANTS

To explore the influence of sulfur sources on the morphologies of copper sulfide by using different sulfur sources (viz., sodium sulfide, thioacetamide, and L-cysteine) instead of thiourea. Large quantities of small nanostructures were obtained (SEM images not shown). When considering the yield of the product, thiourea used to prepare CuS samples was best among the other sulfur sources (83%). From the reaction time-dependent experiments, it was observed that the yield is very high at the early stage, indicating that reaction is very rapid. After that CuS was synthesized at 130°C in the presences of surfactants viz., CTAB (0.1 mM) SDBS (0.1 mM) and Triton X–100 (0.1 mM) while maintaining the molar ratio of the precursors (1: 2) and reaction time (10 hours). As expected, the resulting products were found to be not agglomerated with porous natures. The self-aggregation phenomenon nearly disappeared as shown in the SEM/EDS and TEM figures.

11.3.5 SEM/EDS

In particular, the development of pores on the sample surface can be observed, which is consistent with TEM and good dispersion of smaller stacked flakes provides a larger number of sites for reactants than are provided by aggregated particles. The SEM images confirm that the average diameters of the porous nanostructure of CTAB-CuS, SDBS-CuS, and Triton X–100-CuS samples are about 14–17 nm, 20–24 nm and 25–30 nm respectively. The SEM/EDS images of the products obtained at 130°C (5(a-i)) are found to be spherical nanoparticles. The SEM images of these sphere-like structures show that they appear to be constituted porous nanospheres less than 30 nm in diameter. In addition, many mesopores are clearly observed in the nanostructures (Figure 11.5).

For Triton X–100 CuS sample, the precursor of CuS will nucleate on the Triton X–100 long chains and then form into porous nanostructures in parallel to each other due to the template effect of Triton X–100. The similar formation process of MnO2 nanowires was reported by Wenyao and co-workers. The formation process of CTAB-CuS and SDBS-CuS sample is similar to that of Triton X-CuS sample. But the difference is that Triton X–100 is a non-ionic surfactant and SDBS & CTAB are ionic surfactants. The hydrophilic terminal of CTAB will adsorb the anion (NO_3^- and OH^-) in the solution first, thus, the distance of hydrophobic terminals among CTAB molecules will be enlarged and the hydrophobic terminals keep close to each other, which results in the side by side growth of few structures. However, for SDBS, the hydrophobic terminal will absorb the Cu^{2+} ion directly and then Cu^{2+} ion will nucleate and grow along the chains of SDBS. Compared to Triton X molecules, the distance among SDBS molecules is large because of the electrostatic repulsion, and they will be nearly parallel due to the similar volume between hydrophilic and hydrophobic terminals, which is the formative reason for nearly monodisperse nanostructures. A similar process is described for SnO2 nanowires. Finally, the size of the nanostructures is decided by the length of surfactant chains (Triton X–100>SDBS >CTAB).

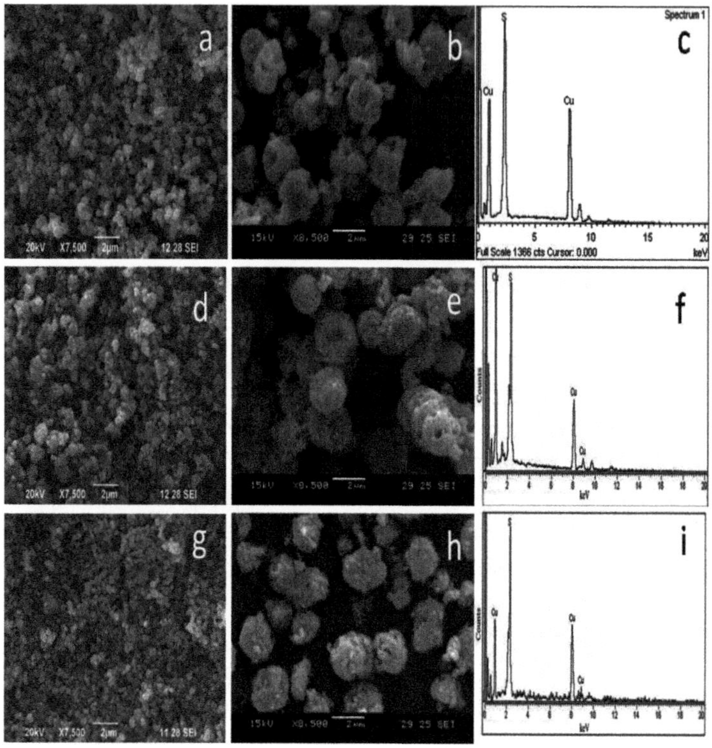

FIGURE 11.5 SEM images of **(a, (b)** CTAB (0.1 mM)-CuS **(d, (e)** SDBS (0.1 mM)-CuS **(g, h)** Triton X–100 (0.1 mM)-CuS and **(c), (f), (i)** are corresponding EDS spectrums.

The energy dispersive X-ray (EDX) spectrum of these samples shows that changing the template materials achieved almost similar CuS nanospheres with different diameters with hollows and pores. In Figure 11.5 (c, f, and i) the EDS spectrum from the prepared samples shows Cu and S without any impurity. The high S content is due to the large initial concentration of thiourea taken as a sulfur source for the synthesis. One possible explanation for this is that the formation of surface complexes through template material producing many active sites on the surfaces.

11.3.6 TEM WITH SAED

Figure 11.6 (a–c) shows the TEM microstructural details of porous CTAB templated CuS nanostructures. The morphology of CTAB-CuS consists of thinly stacked flakes of shapes with well-defined structures at the edges and the average diameter of around 30nm. The corresponding selected-area electron diffraction (SAED) pattern (Figure 11.5(d)) reveals that the crystalline structure can be indexed to the (101), (102), (103), (110), (108) and (116) planes of the hexagonal CTAB-CuS nanoflowers. These results confirm the XRD observation.

To gain further insight into the porous structure and pore size distribution of the CuS nanostructure, the nitrogen adsorption and desorption isotherms and pore size distributions of three CuS nanostructures were examined (not shown).

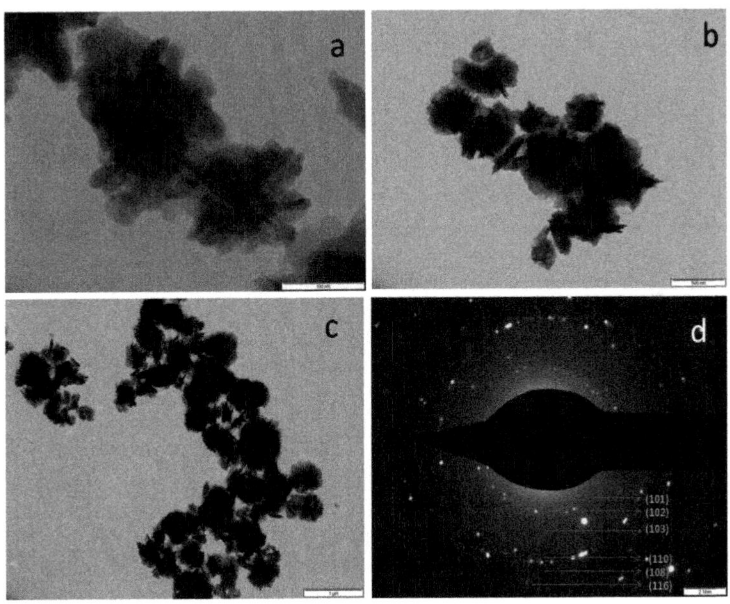

FIGURE 11.6 (a)-(c) TEM image of CTAB (0.1 mM)-CuS nanostructures **(d)** SAED pattern.

11.3.7 XPS ANALYSIS

To further investigate the CTAB-CuS products were analyzed by X-ray photoelectron spectroscopy (XPS) for the evaluation of their composition and purity. No peaks of any impurities are detected in the XPS spectra, indicating the high purity of the product. Figure 11.7a and b show the high-resolution XPS spectra of Cu 2p and S 2p respectively. Figure 11.6a shows the Cu 2p3/2 peak is found at 931.59 eV and Cu 2p1/2 peak is found at 951.37 eV which corresponds to CuS. Figure 11.7b presents the XPS spectrum for the S 2p state, the peak located at a binding energy of 162.53 eV corresponds to Cu – S atom [36]. The satellite structure is not detected, except that there is some asymmetry in the peak shape toward higher binding energies. As shown in Figure 11.7c the survey spectrum of the sample, this reveals that all the peaks have been marked. Evidently, the existence of all the peaks can be ascribed to the elements Cu 2p, S 2p and C 1s as well as O 1s elements from reference and oxygen in the sample is likely due to their exposure to the atmosphere. Hence from XPS analysis, it can confirm the presence of all desired elements used for the preparation of CuS nanostructures. In the near-surface range, the chemical composition of the product is Cu and S which is well agree with this experiment. From the XPS spectrum, there is no other elements are found it indicating the high purity of the sample.

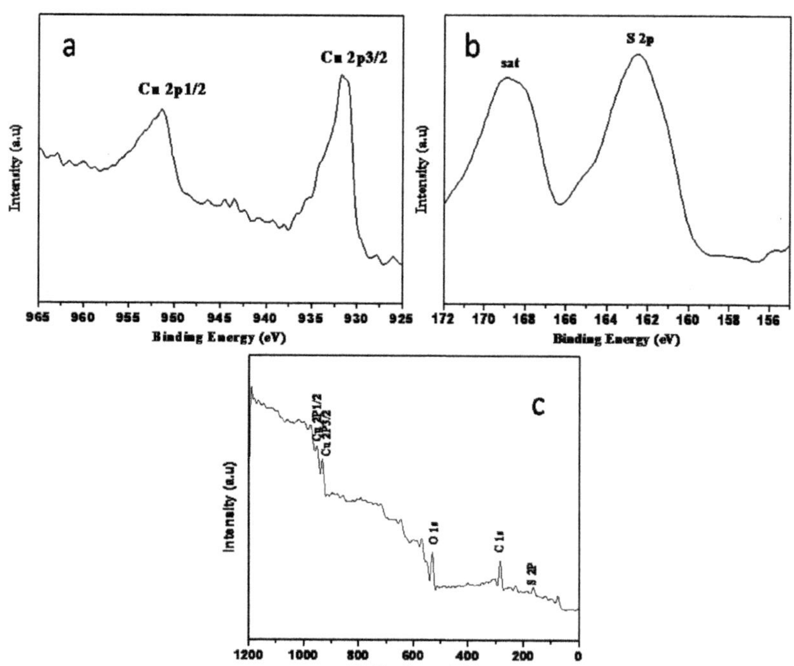

FIGURE 11.7 XPS spectra of CTAB stabilized CuS nanostructures **(a)** Cu 2p and **(b)** S 2p **(c)** full spectrum.

11.3.8 ELECTROCHEMICAL STUDIES

The capacitive behavior of the surfactants templated CuS electrodes were evaluated by cyclic voltammetry (CV) tests in a three-electrode configuration system with a 2M KOH aqueous electrolyte. Figure 11.8.(a–c) shows the cyclic voltammograms (CVs) of a CTAB-CuS, SDBS-CuS, and Triton X–100-CuS electrode with a potential range between 0.0 to 0.5 V (vs SCE) at different scan rates such as 5, 10, 20 and 50 mV/s.

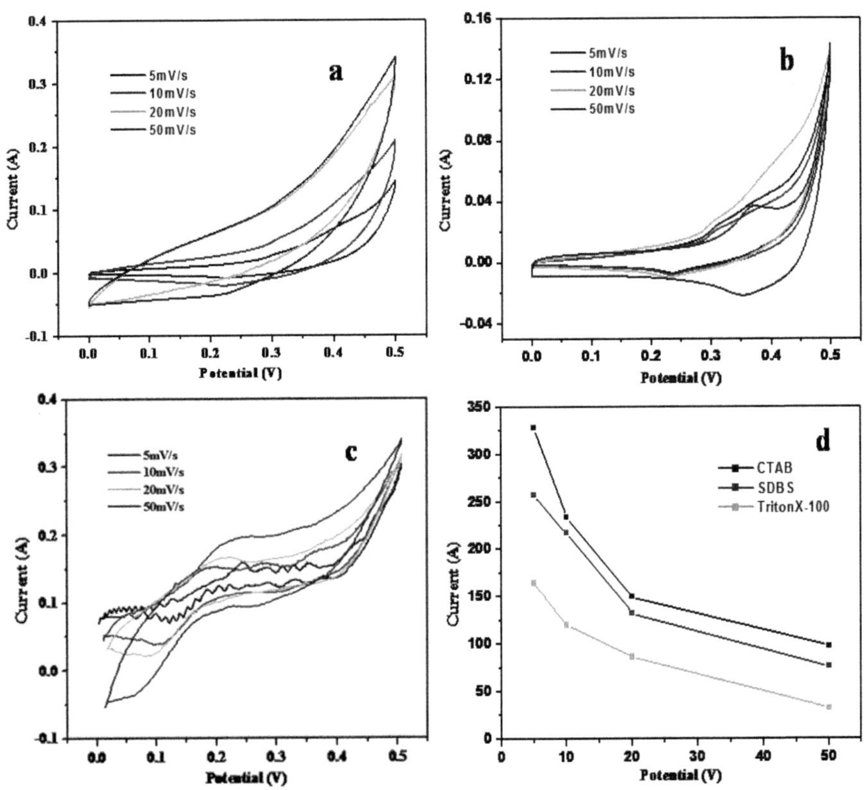

FIGURE 11.8 CV curves of **(a)** (0.1 mM) CTAB-CuS, **(b)** (0.1 mM) SDBS-CuS and **(c)** (0.1 mM) Triton X–100-CuS **(d)** specific capacitance calculated from the CV curves.

Figure 11.8a demonstrates the CV curves of modified CTAB-CuS electrode roughly rectangular shape superimposed with a broad faradaic curve in the range of 0.1 to 0.4 V [37]. This indicates the coexistence of electrical double layer and faradic pseudo capacitances for modified CuS electrode [38]. Compared with other electrodes, the area under the CV curve is greatly higher for the CTAB-CuS electrode has the largest CV area, indicating the highest levels of stored charge. This result can be attributed to the largest specific surface area of the CTAB-CuS sample. Figure 11.8b, SDBS-CuS electrode shows without cathodic and anodic peaks, signifying the ideal capacitive behavior ensuing from Coulombic interaction rather than the faradaic reactions. These curves are mostly based on the electrochemical reactions attributed to M–S with the alkaline electrolyte, where M represents Cu ions. Figure 11.8c shows the CV curves of Triton X–100-CuS electrode at high scan rate a broader oxidation and reduction curves are obtained compared with the lower scan rate.

At high scan rates, the shape of the CV curves is changed due to the limited diffusion time. Furthermore, the shape of the CV curve increases with increasing sweep rate from 5 to 50 mV/s, which indicates its good electrochemical reversibility. In order to explore the application of the CTAB-CuS electrode in supercapacitor electrode is successfully fabricated. The specific capacitance of the three electrodes has been quantified using,

$$Cs = \frac{Q}{\Delta V.m} \tag{11.5}$$

where Cs is the specific capacitance, Q is the anodic and cathodic charges on each scanning, m is the mass of the electroactive material and ΔV is the applied voltage window of the voltammetric curve (mVs^{-1}). The specific capacitances of three samples derived from the cyclic voltammetry are also compared.

The specific capacitance of Triton X-CuS, SDBS-CuS, and CTAB-CuS are 164.47, 257.13, 328.26 Fg^{-1} respectively. As expected, the specific capacitance of CTAB-CuS electrode is the highest, indicating that the CTAB-CuS electrode has a better discharge capacity than the other two samples, during the charge-discharge process, [to be done]. The result is in good agreement with the observation from CV curves (Table 11.2).

This should be associated with the synergistic effect of the advantageous structure and large specific surface area. Almost free-standing nanowire mesoporous structure on the surface is particularly beneficial for the diffusion of the electrolyte into the inner region of the electrode and the mass of the electrolyte within the electrode for fast redox reaction, resulting in the reduced diffusion resistance, high utilization percentage of the electroactive materials and enhanced diffusion kinetics. The large surface area and mesoporous structure can help the ions easily transport between materials. Also, it can greatly increase the electrode/electrolyte contact area and thus further enhance the electrochemical performance of the electrode.

TABLE 11.2 Specific Capacitance of Different Type of Surfactants Assisted CuS Electrode at Four Different Scan Rates

Samples	Scan rate (mV/s)			
	5	10	20	50
		Specific capacitance (F/g)		
CTAB-CuS	328.26	233.8	149.51	97.23
SDBS-CuS	257.13	197.05	112.52	76.27
TritonX–100-CuS	164.47	120.27	86.71	32.95

The CTAB-CuS electrode reveals a high specific capacitance of 328.26 Fg^{-1} at a scan rate of 5mV/s which is higher than the SDBS-CuS and TritonX–100-CuS electrodes. The specific capacitance of all electrode materials is decreasing with increasing scan rate, because of the longer diffusion length of reactive ions and higher resistance caused by the increment of the scan rates. This highest capacitance could be attributed to the well porous nanostructure of CTAB-CuS samples. The porous structure of nanomaterials can significantly develop the utilization and the specific capacitance of active electrode materials [39].

To understand the electrochemical behavior of the (0.1 mM) CTAB-CuS electrode was further investigated by EIS in the frequency range from 100 MHz to 100 kHz at open circuit potential. Nyquist plot of the modified CTAB-CuS electrode is shown in Figure 11.9, a single semi-circle obtained at the high-frequency region is attributed to the capacitor behavior and a straight line obtained at the low-frequency region suggesting ionic diffusion of electrolyte [40, 41]. From this electrochemical impedance analysis, the modified CTAB-CuS electrode reveals suitable electrode material for supercapacitor application.

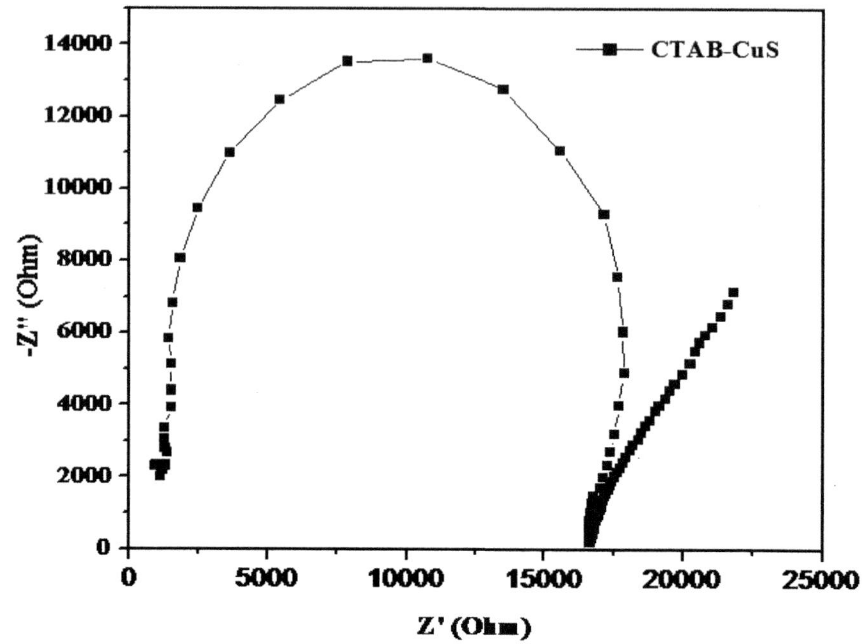

FIGURE 11.9 Nyquist plots of modified (0.1 mM) CTAB-CuS electrode.

11.4 CONCLUSION

In summary, three different CuS nanostructure were successfully prepared. CTAB, SDBS, and Triton X–100 were used as soft templates to control the self-assembly process of CuS nanostructures. The reactions were carried out in the reaction temperature of 130°C and at constant reaction time via a hydrothermal method. It was found out that a different stabilizer on the CuS affects the morphologies, size, and yields of the products, that is spherical CuS nanoparticles were obtained after 10hrs. Investigations showed that a longer reaction time increases the yield and the use of CTAB surfactant prevents the particles from forming a rod or plate-like structures and decrease the particle size due to steric hindrance after initial nucleation. The factors such as the adsorption process and the nature of the interaction of surfactant with the inorganic precursor, influence the nucleation of CuS and thereby affects the surface area, crystallite size, and organization of CuS nanostructures is further manifested in the specific capacitance measured on these materials. Among the three CuS electrodes has the highest specific surface area and optimal electrochemical properties because of the monodispersive mesoporous distribution. A different current density of charge-discharge measurement and cyclic stability for synthesized samples is currently under progress. Thus the as-obtained porous CTAB-CuS electrode reveals a high specific capacitance of 328.26 Fg^{-1} at a scan rate of $5mVs^{-1}$ is highly promising electrode material for energy storage applications (Supercapacitor).

ACKNOWLEDGMENT

We gratefully acknowledge the UGC, Government of India for the financial support for this work (Project file No: 43–533/2014(SR)).

KEYWORDS

- **anionic/cationic/nonionic surfactants**
- **cetyltrimethylammonium bromide**
- **cyclic voltammetry**
- **porous CuS**
- **sodium dodecyl benzene sulphonate**
- **specific capacitance**

REFERENCES

1. Winter, M., & Brodd, R., (2004). What are batteries, fuel cells, and supercapacitors, *Chem. Rev., 104*, 4245–4269.
2. Yang, L., Xianming, L., Weixiao, W., Jinbing, C., Hailong, Y., Chengchun, T., Jang-Kyo, K., & Yongsong, L., (2015). Hierarchical, porous CuS microspheres integrated with carbon nanotubes for high- performance supercapacitors. *Scientific Report, 5*, 1–11.
3. Ting, Z., Baoyu, X., Liang, Z., & Xiong, W., & David, L., (2012). Arrays of ultrafine CuS nanoneedles supported on a CNT backbone for application in supercapacitors, *J. Mater. Chem., 22*, 7851.

4. Lu, P., Xiao, J., Houzhao, W., Yunjun, R., Kui, X., Chi, C., Ling, M., Jianjun, J., & Nickel, S., (2015). Nanoparticles synthesized by the microwave-assisted method as promising supercapacitor electrodes: An experimental and computational study, *Electrochimica Acta. 182*, 361–367.
5. Arico, A. S., Bruce, P., Scrosati, B., Tarascon, J. M., & Schalkwijk, W. V., (2005). Nanostructured materials for advanced energy conversion and storage devices, *Nat. Mater., 4*, 366–377.
6. Simon, P., & Gogotsi, Y., (2008). Materials for electrochemical capacitors, *Nat. Mater., 7,* 845–854.
7. Liu, C., Li, F., Ma, L. P., & Cheng, H. M., (2010). Advanced materials for energy storage, *Adv. Mater., 22*, 28.
8. Wang, G., Zhang, L., & Zhang, J., (2012). A review of electrode materials for electrochemical supercapacitors, *Chem. Soc. Rev., 41*, 797–828.
9. Vol'fkovich, Y. M., & Serdyuk, T. M., (2002). Electrochemical Capacitors, Russ. *J. Electrochem., 38*, 935–959.
10. Deori, K., Ujjain, S. K., Sharma, R. K., & Deka, S., (2013). Morphology-controlled synthesis of nanoporous Co3O4 nanostructures and their charge storage characteristics in supercapacitors, *ACS Appl. Mater. Interfaces, 5*, 10665–10672.
11. Faraji, S., & Ani, F. N., (2014). Microwave-assisted synthesis of metal oxide/hydroxide composite electrodes for high power supercapacitors – A review, *J. Power Sources, 263*, 338–360.
12. Augustyn, V., Simon, P., & Dunn, B., (2014). Pseudocapacitive oxide materials for high-rate electrochemical energy storage, *Energy Environ. Sci., 7,* 1597–1614.
13. Antoniadou, M., Daskalaki, V. M., Balis, N., Kondarides, D. I., Kordulis, C., & Lianos, P., (2011). Photocatalysis and photoelec- trocatalysis using (CdS–ZnS)/TiO2 combined photocatalysts, *Appl. Catal., B., 107,* 188–196.
14. Huynh, W., Peng, X., & Alivisatos, A. P., (1999). CdSe Nanocrystal Rods/Poly(3-hexylthiophene) Composite Photovoltaic Devices, *Adv. Mater., 11*, 923–927.
15. Panthani, M. G., Akhavan, V., Goodfellow, B., Schmidtke, J. P., Dunn, L., Dodabalapur, A., Barbara, P. F., & Korgel, B. A., (2008). Synthesis of CuInS2, CuInSe2, and Cu(InxGa1?x)Se2 (CIGS) Nanocrystal "Inks" for Printable Photovoltaics, *J. Am. Chem. Soc., 130*, 16770–16777.
16. Steckel, J. S., Zimmoler, J. P., Coe-Sullivan, S., Stott, N. E., Bulovic, V., & Bawendi, M. G., (2004). Blue Luminescence from (CdS)ZnS Core-Shell Nanocrystals, *Angew. Chem., Int. Ed., 43*, 2154–2158.
17. Justo, Y., Goris, B., Kamal, J. S., Geiregat, P., Bals, S., & Hens, Z., (2012). Multiple dot-in-rod PbS/CdS heterostructures with high photoluminescence quantum yield in the near- infrared. *J. Am. Chem. Soc., 134*, 5484–5487.
18. Lewis, A. E., (2010). Review of metal sulfide precipitation. *Hydrometallurgy, 104*, 222–234.
19. Saranya, M., & Grace, A. N., (2012). Hydrothermal synthesis of CuS nanostructures with different morphology, *J. Nano Res., 18–19*, 43–51.
20. Saranya, M., Srishti, G., Iksha, S., Ramachandran, R., Santhosh, C., Harish, C., Vanchinathan, T., Bhanu, M., & Grace, A. N., (2013). Solvothermal preparation of ZnO/graphene nanocomposites and its photocatalytic properties, *Nanosci. Nanotechnol. Lett., 5*, 349–354.

21. Dhasadea, S. S., Patil, J. S., Kimc, J. H., Hand, S. H., Rathe, M. C., & Fularif, V. J., (2012). Synthesis of CuS nanorods grown at room temperature by electrodeposition method, *Mater. Chem. Phys., 137*, 353–358.
22. Wu, C., Shi, J. B., Cen, C. J., Cen, Y. C., Lin, Y. T., Wu, P. F., & Wei, S. Y., (2008). Synthesis and optical properties of CuS nanowires fabricated by electrodeposition with anodic alumina membrane, *Mater Lett., 62,* 1074–1077.
23. Liao, X. H., Chen, N. Y., Xu, S., Yang, S. B., & Zhu, J. J., (2003). A microwave assisted heating method for the preparation of copper sulfide nanorods. *J. Cryst. Growth, 252* 593–598.
24. Zhang, H. T., Wu, G., & Chen, X. H., (2006). Controlled synthesis and characterization of covellite (CuS) nanoflakes, *Mater. Chem. Phys., 98,* 298–303.
25. Li, B. X., Xie, Y., & Xue, Y., (2007). Controllable synthesis of CuS nanostructures from self-assembled precursors with biomolecule assistance., *J. Phys. Chem. C., 111,* 12181–12187.
26. Lai, C. H., Lu, M. Y., & Chen, L. J., (2012). Metal sulfide nanostructures: Synthesis, properties, and applications in energy conversion and storage, *J. Mater. Chem., 22,* 19–30.
27. Wu, Y., Wadia, C., Ma, W., Sadtler, B., & Alivisatos, A. P., (2008). Synthesis and photovoltaic application of copper (I) sulfide nanocrystals, *Nano Lett., 8*, 2551–2555.
28. Xia, X., Tu, J., Zhang, Y., Wang, X., Gu, C., Zhao, X., & Fan, H. J., (2012). High-quality metal oxide core/shell nanowire arrays on conductive substrates for electrochemical energy storage, *ACS Nano, 6*, 5531–5538.
29. Huang, Y., Hanning, X., Chen, S., & Wang, C., (2009). Preparation and characterization of CuS hollow spheres, *Ceram. Int., 35*, 905–907.
30. Gandhi, V., Ganesan, R., Abdulrahman, H. H., & Syedahamed, M. T., (2014). Effect of cobalt doping on structural, optical, and magnetic properties of ZnO nanoparticles synthesized by the coprecipitation method. *J. Phys. Chem. C., 118*, 9715–9725.
31. Sreelekha, N., Subramanyam, K., Amaranatha, D. R., Murali, G., Ramu, S., Rahul, K. V., & Vijayalakshmi, R. P., (2016). Structural, optical, magnetic, and photocatalytic properties of Co-doped CuS diluted magnetic semiconductor nanoparticles, *Applied Surface Science, 378*, 330–340.
32. Pei, L. Z., Wang, J. F., Tao, X. X., Wang, S. B., Dong, Y. P., Fan, C. G., & Zhang, Q. F., (2011). Synthesis of CuS and $Cu_{1.1}Fe_{1.1}S_2$ crystals and their electrochemical properties, *Mater. Charact., 9*, 354–359.
33. Wang, Y., Zhang, L., Jiu, H., Na, L., & Sun, Y., (2014). Depositing of CuS nanocrystals upon the graphene scaffold and their photocatalytic activities, *Appl. Surf. Sci., 303*, 54–60.
34. Gao, J., Li, Q., Zhao, H., Li, L., Liu, C., Gong, Q., & Qi, L., (2008). One-pot synthesis of uniform Cu2O and CuS hollow spheres and their optical limiting properties, *Chem. Mater., 20*, 6263–6269.
35. Pankove, J. I., (1971). *Optical Processes in Semiconductors*, Prentice- Hall, New Jersey.
36. Zhang, Y. C., Qiao, T., & Ya, X., (2004). A simple hydrothermal route to nanocrystalline CuS. *J. Cryst. Growth, 268*, 64–70.
37. Silambarasan, M., Ramesh, P. S., Geetha, D., & Venkatachalam, V., (2016). A report on 1D $MgCo_2O_4$ with enhanced structural, morphological, and electrochemical properties, *J. Mater Sci: Mater Electron, 28*, 6880–6888.

38. Fang, Y., Luo, B., Jia, Y., Li, X., Wang, B., Song, Q., Kang, F., & Zhi, L., (2012). Renewing functionalized graphene as electrodes for high-performance supercapacitors, *Adv. Mater., 24*, 6348–6355.
39. Silambarasan, M., Ramesh, P. S., & Geetha, D., (2016). Facile one-step synthesis, structural, optical, and electrochemical properties of $NiCo_2O_4$ nanostructures, *J. Mater Sci.: Mater Electron. 28*, 323–336.
40. Niu, Z. Q., Luan, P. S., Shao, Q., Dong, H. B., Li, J. Z., Chen, J., Zhao, D., Cai, L., Zhou, W. Y., Chen, X. D., & Xie, S. S., (2012). *Energy Environ. Sci., 5*, 8726.
41. Silambarasan, M., Padmanathan, N., Ramesh, P. S., & Geetha, D., (2016). Spinel $CuCo_2O_4$ nanoparticles: Facile one-step synthesis, optical, and electrochemical properties, *Mater. Res. Express., 3*(9), 1–11.

CHAPTER 12

AN EXPERIMENTAL STUDY ON THE INFLUENCE OF POLYMERIC NANO-TIO$_2$ IN CEMENT AND CONCRETE FOR ITS DISPERSION, STRUCTURAL CHARACTERIZATION, MECHANICAL PROPERTIES, AND ITS PERFORMANCE UNDER AGGRESSIVE ENVIRONMENT

MAINAK GHOSAL[1,2] and ARUN KR.CHAKRABORTY[2]

[1]*Adjunct Assistant Professor, JIS College of Engineering, Kalyani, Nadia, W. Bengal, India*

[2]*Research Scholar, Associate Professor, Department of Civil Engineering, Indian Institute of Engineering Science and Technology, Shibpur, Howrah 711103, India*

ABSTRACT

Titanium Dioxide has a very wide range of applications in the industry due to its self-cleaning properties. From paints and varnishes to sunscreen lotions to food colorings to cosmetics to construction its applications are increasing day-by-day. However, when used in cement mortar and concrete, though short-term gains are negligible, it enhances their long-term property to an appreciable extent. This paper aims to investigate the mechanical behavior of nano-TiO$_2$ in Ordinary Portland Cement (OPC) composites and also of a standard M–40 Grade concrete.

The optimized quantity of nano-TIO_2 both for short and long terms as found after addition to cement- mortar was 1% w.r.t cement weight and the same quantity addition was repeated for concrete also. This paper also aims to investigate the long-term effect of M–40 Grade concrete made with nano-TIO_2 when exposed under aggressive chemical attacks. The results corroborated the fact that through strength gains for cement mortar and M–40 Grade concrete made with 1% nano-TIO_2 is minor in the short term but their long-term durability property enhances multifold in natural atmosphere and also when exposed to chemical attacks. It should be noted that a second generation Polymer – Poly Carboxylate Ether was used to disperse the nano-TIO_2 as it was insoluble in water. The water/cement ratios (w/c) of all the mixes were kept constant at 0.4.

12.1 INTRODUCTION

Nano is a Greek word which means "dwarf." Nanotechnology applications is not a new technology, it finds a place in history also [1]. Nanomaterials, when added to cement & concrete, enhances its material properties in certain ways[6,23]. Furthermore, surface properties and structural characteristics of different construction materials are deteriorating during their service life due to exposure to sunlight and effect of altering and increasing polluted environment, i.e., exposure to chloride and sulfate attack. To deal with these problems, development of a new concrete material having self-cleaning, solar-powered remediation devices for the polluted environment and biocidal capacities have received significant attention in recent years [2–22, 25]. India's rising Auto Population and long Coastline (7517 km) has affected Concrete structures by DURABILITY problems due to SO_{4-}^{2} and Cl^{-}. Sulfate (SO_{4-}^{2}) attack arises from automobile exhausts while chloride (Cl^{-}) attack arises from seawater or saline areas as shown in Figure 12.1. TiO_2 has the unique potential to address this. This is due to the fact that:

1. When Titanium dioxide receives energy from light, electrons (-) are released from the surface. The electrons released are combined with oxygen in the air then creates "O^{2-}" (Superoxide anion) as shown in Figure 12.2 (Part 1).

2. The surface where the electrons were released is tinged with the electric charge of plus, and takes electrons from the moisture of the air and then return to its original state. On the other hand, the moisture that lost the electron becomes a hydroxyl radical (•OH) as shown in Figure 12.2 (part 2). [*Before reacting with H_2O, there are cases where electrons are taken and broken down directly from organic compounds.]
3. The "O^{2-}" and "•OH" that resulted in a process of transfer of this electron can produce strong oxidative decomposition, which decomposes organic compounds such as oil causing dirt and adhesion, bacteria, harmful chemical gas, viruses, and molds even into simple harmless substances that they are released into air as shown in Figure 12.2 (Part 3).

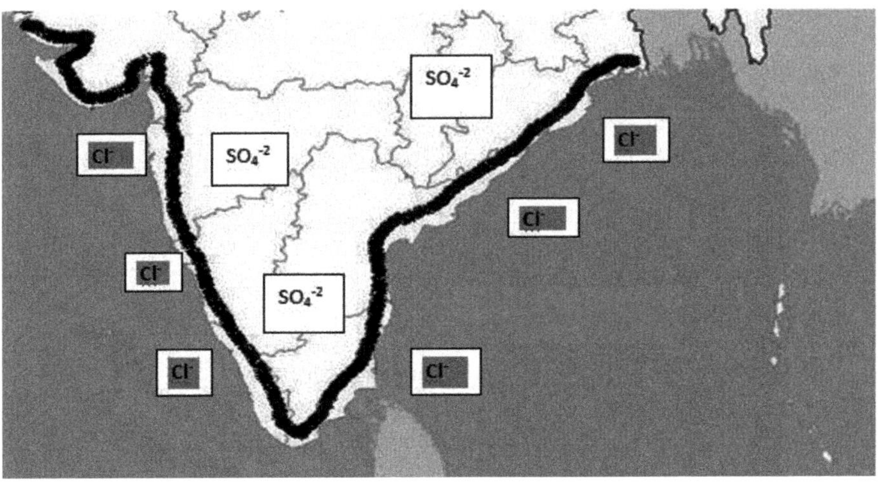

FIGURE 12.1 A schematic diagram showing preponderance of Cl^- ions at coast & SO_{4-}^{2} ions at inland.

In India, Titanium Dioxide is found in the beach sands of Kerala. Titanium dioxide occurs in nature as the well-known minerals known as rutile, anatase, and brookite. However, in, in India, only the Rutile form is used to manufacture Nano-Titanium Dioxide. In 1909, the German scientist Dr. Schomberg found traces of monazite in the sand flakes on the imported coir from Sankaramangalam (Kerala). In 1932, M/s F. X.

Periera and Sons (Travancore) Pvt. Ltd. established a fully-fledged plant for TiO_2 production. The Kerala Government took over the plant in 1956. It became a Limited Company in 1972 under the name "Kerala Minerals & Metals Ltd" [KMML]. The Mineral Separation Unit (MS Unit) of Kerala Minerals & Metals Ltd. (India's first and only manufacturer of Rutile Grade Titanium Oxide by chloride process), a State Government public limited company is engaged in the separation of ilmenite, Rutile, Leucoxene, Monazite, Silliminite, etc. from beach sand by Gravitational, Magnetic & High Tension Electrostatic Techniques for separation of minerals from the sand.

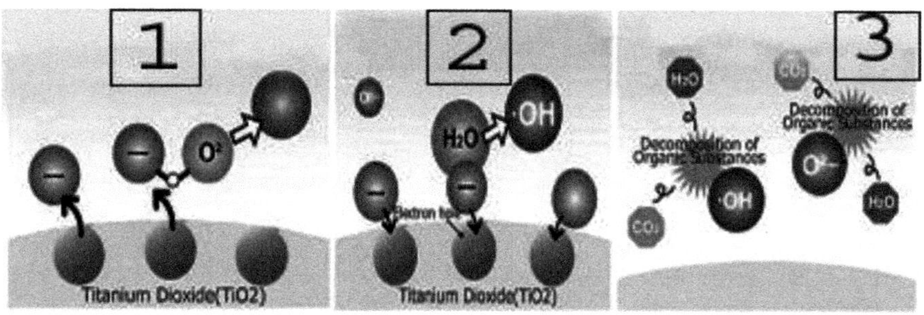

FIGURE 12.2 Line process showing the working of nano-TiO_2 under Sunlight (Photocatalytic).

The Jubilee Church of Rome shown in Figure 12.3 was constructed by using Nano-TIO_2 on its external plastered surface and is still performing well in terms of retarding the advances of chemicals on its silver-white exterior. Similarly, many waterproofing agents in the market use Nano-Titanium Dioxide as a principal compound. Titanium being an inert element is useful to ward off any external harmful chemical attacks. Now, the question arises is what is TiO_2? Titanium dioxide is a white powder (superfine particles) that has long been used as a white pigment. It is tasteless, odorless, and is harmless to humans. Titanium dioxide can be found in many products, ranging from paint to food and drugs to cosmetics. When applied to cement mortar and concrete one can easily make out the difference using his naked eye, between the normal concrete and with the concrete or mortar made with Nano-TIO_2 as shown in Figure 12.4.

FIGURE 12.3 Jubilee Church of Rome.

FIGURE 12.4 Difference between nano-titanium oxide mortar cubes and ordinary mortar cubes.

12.2 EXPERIMENTAL SECTION

12.2.1 MATERIALS AND CHEMICALS USED

1. Cement – Ordinary Portland Cement supplied by Ambuja Cements Ltd.
2. Reinforcement Bar – NA
3. Sone Aggregate – Pakur/Local variety.
4. Fine Aggregate – Natural (River) Sand.
5. Water – Drinking (Tap) Water.
6. Chemical Admixture – Superplasticizer type: Polycarboxylate Ether (PCE) supplied by Structural Waterproofing Co.
7. Nano Additives – *Nano-Titanium Oxide* supplied by M/s Kerala Minerals and Metals Ltd. as shown in Figure 12.5.

FIGURE 12.5 Nano-titanium oxide as supplied by M/s KMML.

12.2.2 TEST DATA

A) Cement = OPC 43 Grade
B) Sp.Gravity of Cement = 3.08 (as Lab. Experiment suggests)
C) Chemical Admixture = Superplastcizer (PolyCarboxylate Ether)
D) Sp. Gravity of
   i) Coarse Aggregate (for 20 mm = 2.831 and for 12.5 mm = 2.845)
      Now, Average Specific Gravity of Coarse Aggregates = 0.9 x 2.845 + 0.1 x 2.831 = 2.8436
   ii) Fine Aggregate (River Sand confirming to Zone II) = 2.688
E) Water Absorption
   i) Coarse Aggregate =3.09
   ii) Fine Aggregate=Nil.
F) Free Surface Moisture
   i) Coarse Aggregate = 1.716
   ii) Fine Aggregate = 0.3
G) Sieve analysis: As performed in the Laboratory.
   i) Coarse Aggregate:

IS. Sieve Sizes (mm)	Analysis of Coarse Aggr.		% of Different Fractions			Remarks
	Fraction I (12.5 mm passing)	**Fraction II (20 mm passing)**	**I 90%**	**II 10%**	**Combined 100% (I + II)**	**(Conforming to Table 2 of BIS: 383 for Graded Agg. of 20 mm nominal size)**
20	99.44	50.322	0.9 x 99.44	0.1 x 50.322	94.5	95–100
10	56.7	1.062	0.9 x 56.7	0.1 x 1.062	51	25–55
4.75	–	–	–	–	–	0 to 10
2.36	–	–	–	–	–	–

ii) Fine Aggregate:

IS. Sieve Sizes (mm)	Weight Retained (gms.)	%Weight Retained	Cum% Weight Retained	% Passing	Remarks (Conforming to Zone II of BIS:383 for Fine Aggregates.
4.75	–	–	–	100	90–100
2.36	67	6.77	6.77	93.23	75–100
1.18	101	10.20	16.97	83.03	55–90
600μ	277	27.98	44.95	55.05	35–59
300μ	367	37.07	82.02	17.98	0–30
150μ	161	16.26	98.28	1.72	0–10
75μ	17	1.72	100	–	
Pan	6	–	–	–	

Total Weight taken = 1000 gms.

The Specific properties of Nano Titanium Oxide(TiO_2) used here are as follows (Table 12.1):

TABLE 12.1 Specific Properties of Nano Titanium Oxide (TiO_2)

Nano Titanium Oxide%	97
Rutile content%	98
pH	7
Average particle size (TEM)	30–40 nm
Treatment	Nil
Moisture%	1.75–2
Bulk Density	0.31 gm/cc
Water Solubility	In-soluble

12.2.3 DISPERSION OF NANO-TIO_2

The n-TiO_2 powder as supplied by M/s KMML was found to be insoluble in distilled water. It was also found to be insoluble in conventional polymers like Melamine Formaldehyde which is used as an admixture in construction industries. So a modern 3rd generation polymer called Poly Carboxylate Ether (PCE) was used for dispersion after ultrasonication for about 30 minutes in an external ultrasonicator bath (250W Piezo-U-Sonic Ultrasonic Cleaner). The energy of ultrasonic waves is extremely high, which exfoliates the present agglomerated nanoparticles in the supersaturated solution, already dispersed by simply dissolving the n-TiO_2 powder in PCE. It was noticed that 2.5% n-TiO_2 solution was relatively more supersaturated than 1% n-TiO_2 in PCE.

12.2.4 STRUCTURAL CHARACTERIZATION OF NANO-TIO_2

The n-TiO_2 powder as obtained from M/s KMML was then characterized using X-ray diffraction (XRD) techniques to study the crystalline structure by using a powder X-ray diffractometer system with 2θ in the range of 20°C–80°C at room temperature. Samples were collected from Laboratory after crushing the cubes and were further powdered finely by grinding and sieved to remove coarse large granules, prior to XRD characterization.

The Figure 12.6 shows the XRD image of Nano-TiO_2.

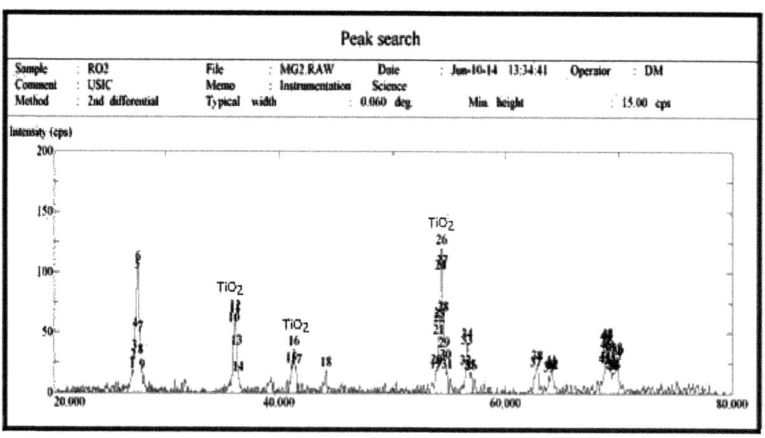

FIGURE 12.6 XRD image of nano titanium di-oxide used.

12.2.5 FABRICATION OF TITANIUM DI-OXIDE CEMENT MORTAR NANOCOMPOSITES AND ORDINARY CEMENT MORTAR COMPOSITES

Mortar Cubes of 70.7 mm x 70.7 mm x 70.7 mm size were casted with 1 part of cement + 3 parts of sand with water added as per the normal consistency formula of Indian standards, i.e., according to the standard formula P'= (P/4 +3)(1 part Cement + 3 parts Sand). Here P' = Quantity of water and P = Consistency of Cement used, i.e., amount of water used to make 300 gms cement paste to support a penetration of 5–7 mm in a standard Vicat mold with a Vicat needle. Before filling up fresh mortar pastes into the molds, the molds were cleaned, brushed, and oiled properly. The filled molds were subjected to external vibrations as shown in Figure 12.8 to ensure complete compaction of mortar pastes. For Titanium Di-Oxide Cement Mortar Nanocomposites, Nano Titanium Oxide is added in proportions ranging from 1.0% and 2.5% w.r.t cement wt. before compaction but after proper dissolutions in a suitable Super Plasticizer (Poly Carboxylate Ether) keeping the w/c ratio fixed at 0.4. The cubes were then ordinary cured underwater in a curing tank under standard conditions, at a constant temperature of 27 + 2°C and tested for compressive strength at 7 days, 28 days, 90 days, 180 days, and 365 days.

12.2.6 DETERMINATION OF MECHANICAL PROPERTIES OF TITANIUM DI-OXIDE CEMENT MORTAR NANOCOMPOSITES AND ORDINARY CEMENT MORTAR COMPOSITES

The Nanocomposites, as well as the ordinary mortar composites, were evaluated for their mechanical behavior by testing their compressive strength using a 200-ton capacity Compression Testing Machine (CTM), HEICO 49.55. The surface dry samples were tested at increasing curing times/ages of 1, 3, 7, 28, 90, 180, and 365 days as per BIS requirements, i.e., BIS: 269–1976. The cube samples were positioned at right angles to the casting position while testing. The axis of each specimen was carefully aligned with the center of thrust of plates. A gradual load was applied and increased to a constant loading rate of 5.2 N/mm^2 until the failure of the sample took place. An average of the values for three cubes was reported as the compressive strength, to minimize error (Figures 12.7–12.10).

FIGURE 12.7 Mixing of nanoconcrete by mixer machine.

FIGURE 12.8 Internal vibrator for nanoconcrete.

FIGURE 12.9 External vibrator for nanocomposites.

FIGURE 12.10 Curing chamber.

12.2.7 FABRICATION OF M–40 GRADE TITANIUM DIOXIDE NANO-CONCRETE AND M–40 GRADE CONTROL CONCRETE

Concrete cubes of 100 mm x 100 mm x 100 mm size with cement, fine aggregate [FA], coarse aggregate [CA] & water in proportions as per the mix design followed by Indian standards BIS:10262 (2009) for M–40 Grade concrete for 100 mm slump keeping the w/c = 0.4. The mix proportions were cement = 400 kg/m^3, CA = 1293.04 Kg/m^3 [CA1(90%) = 1163.74 kg/m^3; CA2(10%) = 129.3 kg/m^3], FA = 687.54 kg/m^3, water = 157 kg/m^3. The coarse aggregate used was of the saturated surface dry condition. Before filling up fresh concrete into the molds, the molds were cleaned, brushed, and oiled properly. The ingredients were mixed in a small-scale concrete mixer, self-powered with an electric motor, where Cement, sand, and other aggregates are loaded in a manually operated hopper and then poured in the mixing drum for final mixing and then can be unloaded by tilting the drum. The filled molds were subjected to internal vibrations as shown in Figure 12.7 to ensure complete compaction of the concrete. For Titanium Dioxide Nano-Concrete, 1% Nano Titanium Dioxide w.r.t cement wt. was added as per the optimized strength results at 28 days strength as obtained from 2.6. The cubes were then ordinarily cured underwater in a curing tank under standard conditions, at a constant temperature of 27+ 2°C and tested for compressive strength at 28 days and then immersed under $MgCl_2$ and $MgSO_4$ solutions for 90 days and 180 days, respectively as shown in Figure 12.11.

FIGURE 12.11 Concrete and nano concrete samples exposed to chloride and sulphate ions.

12.2.8 DETERMINATION OF MECHANICAL PROPERTIES OF M–40 GRADE TITANIUM DIOXIDE NANO-CONCRETE AND M–40 GRADE CONTROL CONCRETE

The M–40 Grade Titanium Dioxide Nano-Concrete as well as the ordinary M–40 Grade Concrete were evaluated for their mechanical behavior by testing their compressive strength using a 200-ton capacity Compression Testing Machine (CTM), HEICO 49.55. The surface dry samples were tested at increasing curing times/ages of 28, 90 and 180 days after taking out from water, air, $MgCl_2$, and $MgSO_4$ solutions, as per BIS requirements, i.e., BIS: 516–1959. A gradual load was applied and increased to a constant loading rate of 140Kg/cm^2 until the failure of the sample took place. An average of the values for three cubes was reported as the compressive strength, to minimize error.

12.2.9 MICRO-STRUCTURAL ANALYSIS OF M–40 GRADE TITANIUM DIOXIDE NANO-CONCRETE AND M–40 GRADE CONTROL CONCRETE

The microstructure of the fracture surfaces of M–40 Grade Titanium Dioxide Nano-Concrete and M–40 Grade Concrete was characterized using field emission scanning electron microscopy with Energy-dispersive X-ray spectroscopy (EDS, EDX, or XEDS), sometimes called energy dispersive X-ray analysis (EDXA) or energy dispersive X-ray micro-analysis (EDXMA), (SEM machine model no.S3400N; ion splatter E1010 Hitachi; EDAX software used was EMAX).All the samples were gold coated using a sputtering technique prior to microscopy analysis. The pore structure of the samples cured for 28 days was analyzed using intrusion porosimetry. The size of each sample chosen for analysis was kept close to 1 cm and samples were dried at 100°C in a vacuum oven for about 24 hours. To reduce error, three repetitions were carried out on each sample (Figures 12.12–12.17).

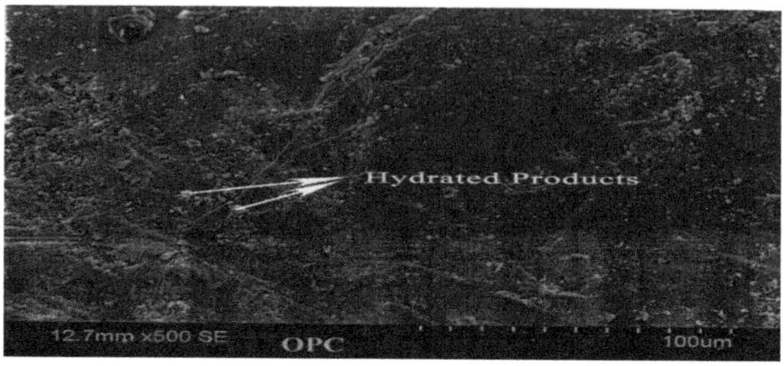

FIGURE 12.12 SEM image of 180 days microstructure of M– 40-grade control concrete.

FIGURE 12.13 SEM image of 180 days microstructure of M– 40-grade titanium dioxide nano-concrete.

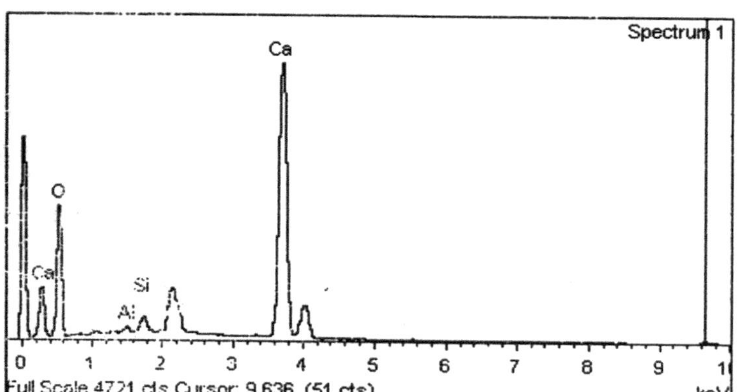

FIGURE 12.14 EDS spectrum of the M– 40-grade control concrete at 180 days.

Standard :
O SiO2 1-Jun-1999 12:00 AM
Al Al203 1-Jun-1999 12:00 AM
Si SiO2 1-Jun-1999 12:00 AM
Ca Wollastonite 1-Jun-1999 12:00 AM

Element	Weight %	Atomic %
O K	53.28	73.79
Al K	1.45	0.37
Si K	1.08	0.85
Ca K	45.19	24.99
Totals	100.00	

FIGURE 12.15 Elemental composition the M– 40-grade control concrete at 180 days.

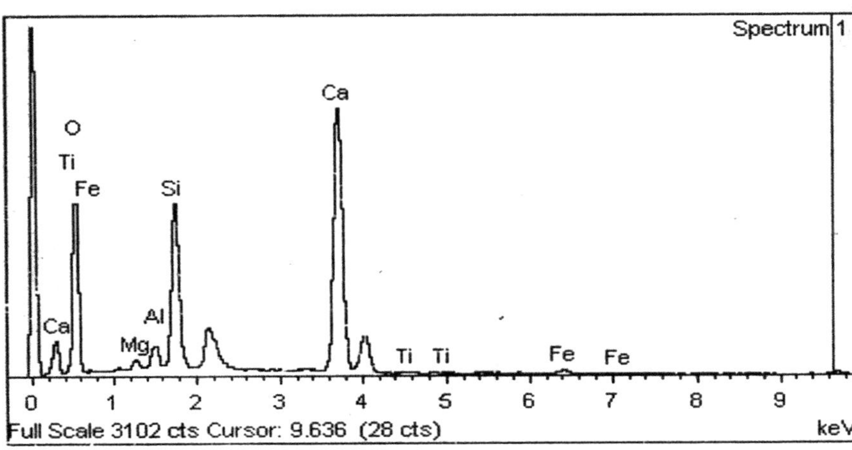

FIGURE 12.16 EDS spectrum of the M– 40-grade titanium dioxide nano-concrete at 180 days.

Standard :

O	SiO2	1-Jun-1999 12:00 AM
Mg	MgO	1-Jun-1999 12:00 AM
Al	Al203	1-Jun-1999 12:00 AM
Si	SiO2	1-Jun-1999 12:00 AM
Ca	Wollastonite	1-Jun-1999 12:00 AM
Ti	Ti	1-Jun-1999 12:00 AM
Fe	Fe	1-Jun-1999 12:00 AM

Element	Weight %	Atomic %
O K	51.16	70.42
Mg K	0.60	0.54
Al K	1.36	1.11
Si K	10.36	8.12
Ca K	34.84	19.14
Ti K	5.58	0.04
Fe K	1.61	0.63
Totals	100.00	

FIGURE 12.17 An experimental study on the influence of polymeric nano-TIO_2 in.

12.3 TEST RESULTS

The following Tables 12.2 and 12.3 illustrates the results obtained in mechanical tests (2.6) and (2.8).

12.4 DISCUSSION OF TEST RESULTS

The Test Results shows that:

i) Nano-TiO_2 additions never resulted in any appreciable change in physical characteristics of Cement – shape, size, weight, density, except color change. However, Nano-Additions in PCE for TiO_2 reduced the setting time (From 3 days to 1.5 days).
ii) There is 12.59% strength gain for cement mortar cubes made with Nano-TiO_2 when compared with ordinary cement mortar cubes at 28 days but for nanoconcrete the gain is 2.16% w.r.t M–40 concrete. However, the long-term strength gain for cement mortar cubes made with Nano-TiO_2 when compared with ordinary cement mortar cubes at 365 days was found to be 37% as shown in Figure 12.21, the reason for which is not very clear.

TABLE 12.2 Strength (N/mm^2) for Various Proportions/Ages of Nano-TiO_2 Added OPC Mortar (% Increase w.r.t Ordinary Control Cement Cubes

Sl No.	% Nano additions in Cement (OPC)	1 day strength (% increase)	3-day strength (% increase)	7-day strength (% increase)	28-day strength (% increase)	90-day strength (% increase)	180-day strength (% increase)	365-day strength (% increase)
1.	0	16.08	18.25	21.08	31.89	31.2	30.01	30.01
2.	1% TiO_2 (optimized)	Not Demoldable	27.81 (52.38%)	25.24 (19.73%)	36.71 (12.59%)	35.92 (15.13%)	33.42 (11.36%)	41.16 (37.15%)
3.	2.5% TiO_2	Not Demoldable	21.67 (18.74%)	20.34 (–3.51%)	34.97 (9.58%)	37.80 (21.15%)	40.95 (36.45%)	28.16 (–6.16%)

TABLE 12.3 Compressive Strength (in N/mm^2) of 100 mm Cubes at 28 Days at Different Exposure Conditions at w/c Ratio of 0.4

Type of Concrete	Exposure Conditions	Strength at 28 days	Weight at 28 days	Strength at 90 days (% increase in Strength)	Weight at 90 days	Strength at 180 days (% increase in Strength)	Weight at 180 days
Control Concrete	In Air	43.03	2.576Kg	49.71	2.621Kg	48. 34	2.589Kg
	In $MgCl_2$	–	–	48.51	2.606Kg	57.63	2.734Kg
	In $MgSO_4$	–	–	47.04	2.617Kg	47. 96	2.607Kg
M–40 Concrete (1.0% optimized Nano-TIO_2 addition)	In Air	43.96(2.16%)	2.737Kg	43.34[–12.81%]	2.6 93Kg	41.83[–13.47%]	2.685Kg
	In $MgCl_2$	–		36.89[–14.88%]	2.806Kg	62.48[49.36%]	2.709Kg
	In $MgSO_4$	–		35.68[–17.67%]	2.741Kg	68.67[64.16%]	2.637Kg

iii) For 90 days compressive strength at different exposure conditions it is found that the loss of strength is about 15–17% for all air, $MgCl_2$ and $MgSO_4$ exposures when compared to M–40 control concrete, as shown in Figures 12.22–12.24.
iv) For 180 days compressive strength at different exposure conditions it is found that there is an appreciable gain in strength of about 49% and 64% for $MgCl_2$ and $MgSO_4$ exposures when compared to M–40 concrete at the same exposures, as shown in Figures 12.20 and 12.21. This may be attributed to the fact that due to the chloride and sulfate attack by $MgCl_2$ and $MgSO_4$ on the M–40 concrete and formation of MgO-TiO_2 nano-caging in the TiO_2 added nano-concrete at latter ages (180 days), resulting in stable phase equilibrium of the $MgCl_2$/$MgSO_4$ system. Further studies need to be done in the nano-caging area.
v) There is no linearity involved in the weight gain/loss after exposure in $MgCl_2$/$MgSO_4$ when compared with ordinary controlled concrete.
vi) Microstructural SEM-EDS investigational studies as shown in Figures 12.15 and 12.16, at 180 days of M–40 Grade Titanium Dioxide Nano-Concrete and M–40 Grade Control Concrete, indicate the presence of hydrated products in control concrete which is the reason for its strength gain of 13.47%, while the presence portlandite ($Ca(OH)_2$) plates is in evidence in Nanoconcrete. Also, EDAX studies as shown in Figures 12.17–12.21 observe the presence of larger amounts of calcium silicate (hydrated products) in control concrete.

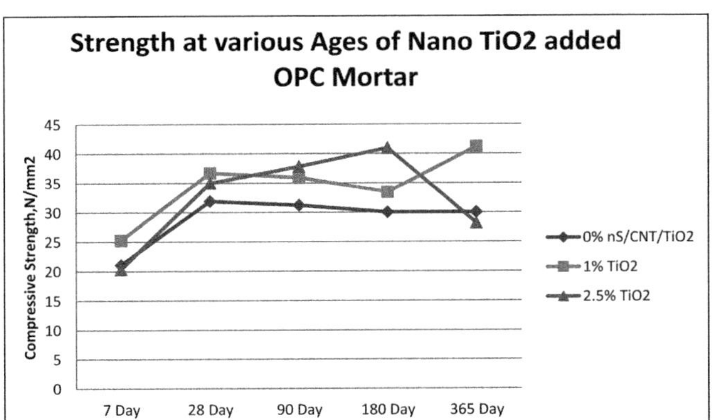

FIGURE 12.18 Chart showing strengths at various ages of different% of TiO_2 addition in OPC mortar.

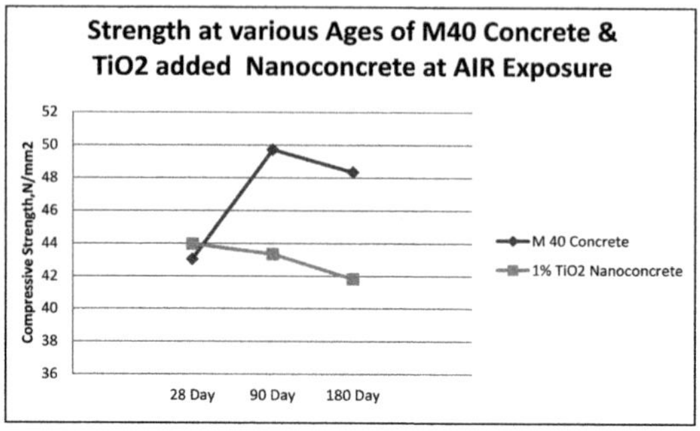

FIGURE 12.19 Chart showing strengths at various ages M40 concrete and TiO_2 added nanoconcrete at air exposure.

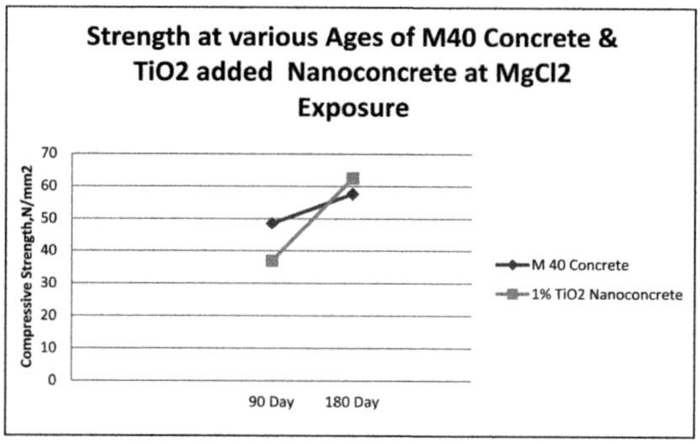

FIGURE 12.20 Chart showing strengths at various ages M40 concrete and TiO_2 added nanoconcrete at $MgCl_2$ exposure.

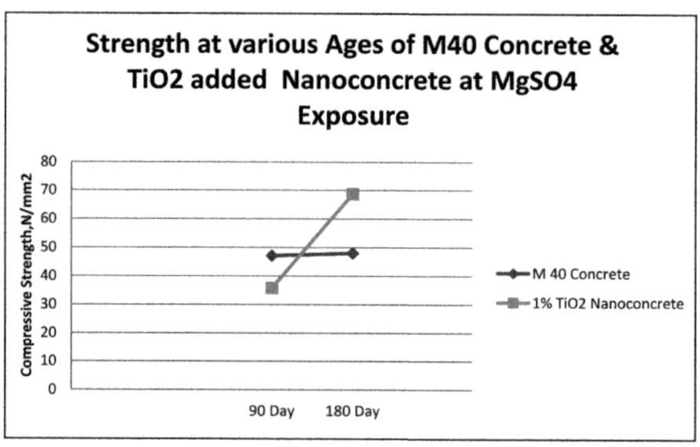

FIGURE 12.21 Chart showing strengths at various ages M40 concrete and TiO_2 added nanoconcrete at $MgSO_4$ exposure.

12.5 CONCLUSIONS

a) The results showed that the optimizations for Nano-Titanium Oxide, TiO_2 = 1.0% by weight of cement for gain in strength in OPC mortar up to 28 days which has already been supported by many researchers [24]. For long-term strength, the trend is not clear.
b) The optimum percentages based on cement mortar when used in concrete also produced good results up to 28 days w.r.t. controlled concrete.
c) Further research is needed on microstructural studies for characterization of nanomaterials in concrete.

KEYWORDS

- **cement**
- **chemical**
- **nanoconcrete**
- **nanomaterials**
- **nano-titanium oxide**

REFERENCES

1. Wikipedia.org. http://en.wikipedia.org/wiki/Lycurgus_Cup.
2. Paz, Y., (2010). Application of TiO_2 photocatalysis for air treatment patients' overview, *Applied Catalysis B: Environmental, 99*, 448–460.
3. Fujishima, A., Zhang, X., & Tryk, D., (2007). Heterogeneous photocatalysis: From water photolysis to applications in environmental cleanup, *International Journal of Hydrogen Energy, 32*, 2664–2672.
4. Rajagopal, G., Maruthamuthu, S., Mohanan, S., & Pa, N., (2006). Biocidal effects of photocatalytic semiconductor TiO_2, *Colloids*, and *Surfaces, B: Biointerfaces, 51*, 107–111.
5. Plassais, A., Ruot, B., Olive, F., Gulliot, L., & Bona, L., (2009). TiO_2-containing cement pastes and mortars: Measurements of the photocatalytic efficiency using a Rhodamine B based colorimetric test, *Solar Energy, 83*, 1794–1801.
6. Sanchez, F., & Sobolev, K., (2010). Nanotechnology in the concrete-a review. *Construction and Building Material, 24*(11), 2060–2071.
7. Irie, H., Tee, S. P., Shibata, T., & Hashimoto, K., (2005). Photo-induced wettability control on TiO_2 surface, *Electrochemical*, and *Solid State Letter, 8*, 23–25.

8. Djebbar, K., & Sehili, T., (1998). Kinetics of heterogeneous photocatalytic decomposition of 2,4-dichlorophenoxyacetic acid over TiO_2 and ZnO in aqueous solution, *Pesticide Science, 54*, 269–276.
9. Anpo, M., & Kamath, P. V., (2010). *Environmentally Benign Photocatalysts: Application of Titanium Oxide Based Materials* (1st ed.), Springer.
10. Fujishima, A., Hashimoto, K., & Watanabe, T., (1999). *TiO_2 Photocatalysis: Fundamentals and Applications*, Tokyo: BKC Inc, 1–28.
11. Folli, A., Pade, C., Hansen, T. B., Marco, T. D., & Macphee, D. E., (2012). TiO_2 photocatalysis in cementitious systems: Insights into self-cleaning and depolllution chemistry. *Cement and Concrete Research, 42*, 539–548.
12. Kim, S., Hwang, S. J., & Choi, W., (2005). Visible light active platinum ion doped TiO_2 photocatalyst, *Journal of Physical Chemistry B, 109*, 24260–24267.
13. Zhu, J., Ren, J., Huo, Y., Bian, Z., & Li, H., (2007). Nanocrystalline Fe/TiO_2 visible photocatalyst with a mesoporous structure prepared via a nonhydrolytic-gel route, *Journal of Physical Chemistry C, 111,* 18965–18969.
14. Ramirez, A., Demeestere, K., De Belie, N., Mantyla, T., & Levanen, E., (2010). Titanium dioxide coated cementitious materials for air purifying purposes: preparation, characterization, and toluene removal potential, *Building*, and *Environment, 45*, 832–838.
15. Sato, S., (1986). Photocatalytic activity of NO_x in the visible light region, *Chemical Physics Letters, 123*, 126–128.
16. Zaleska, A., (2008). Doped-TiO_2: A review, *Recent Patents on Engineering, 2*, 157–164.
17. Yousefi, A., Allahverdi, A., & Hejazi, P., (2013). Effective dispersion of nano-TiO_2 powder for enhancement of photocatalytic properties in cement mixes. *Construction and Building Material, 41*, 224–230.
18. Bellardita, M., Addamo, M., Di Paola, A., Marci, G., Palmisano, L., Caesar, L., & Borsa, P. M., (2010). Photocatalytic activity of TiO_2/SiO_2 systems, *Original Research Article Journal of Hazardous Materials, 174*(1–3), 707–713.
19. Wang, J., Lu, C., & Xiong, J., (2014). Self-cleaning and depolllution of fiber-reinforced cement materials modified by neutral TiO_2/SiO_2 hydrosol photoactive coatings, *Applied Surface Science, 298*, 19–25.
20. Senff, L., Tobaldi, D. M., Lemes-Rachadel, P., Labrincha, J. A., & Hotza, D., (2014). The influence of TiO_2 and ZnO powder mixtures on photocatalytic activity and rheological behavior of cement pastes, *Construction*, and *Building Material, 65*, 191–200.
21. Li, Z., Gao, B., Zhen, C. G., Mokaya, R., Sotiropoulos, S., & Puma, G. L., (2011). Carbon nanotubes/titanium dioxide (CNT/TiO_2) core-shell nanocomposites with tailored shell thickness, CNT content and photocatalytic/photoelectrocatalytic properties. *Applied Catalysis B: Environmental, 110*, 50–56.
22. Kim, S. J., Im, J. S., Kang, P. H., Kim, T., & Lee, Y. S., (2008). Photocatalytic activity o CNT-TiO_2Nano compost in degrading anionic and cationic dyes, *Carbon Letter 9,* 294–297.
23. Ghosal, M., & Chakraborty, A. K., (2015). A comparative study of nano embedments on different types of cement. *International Journal of Advances in Engineering & Technology (IJAET), 8*(2), 92–103.
24. Martins, T., et al., (2016). An experimental investigation on nano-TIO_2 and fly ash based high-performance concrete, *The Indian Concrete Journal, 90*(1), 23–31.
25. Mapa, M., et al., (2016). Development of nanomodified photocatalyst enabled cement composites, *The Indian Concrete Journal, 90*(1), 14–22.

PART III

Theoretical Protocols for Polymers and Clusters

CHAPTER 13

COOPERATIVE MOLECULAR AND MACROSCOPIC DYNAMICS OF AMORPHOUS POLYMERS: A NEW ANALYTICAL VIEW AND ITS POTENTIAL

JOSÉ JOAQUIM C. CRUZ PINTO[1] and JOSÉ REINAS S. ANDRÉ[2]

[1]*Retired Full Professor from the University of Aveiro/CICECO, Department of Chemistry, 3810–193 Aveiro, Portugal, E-mail: jj.cruz.pinto@ua.pt*

[2]*Coordination Professor, UTC-Engineering, and Technology, UDI-Research Unit for Inland Development, Guarda Polytechnic Institute, Technology, and Management School, Guarda, Portugal, E-mail: jandre@ipg.pt*

ABSTRACT

This contribution addresses the cooperative nature of the molecular and macroscopic dynamics of amorphous-phase polymers as the result of truly many-body processes. A new theory of clustering of specifiable primitive relaxors is proposed and applied to predict materials' responses as functions of temperature and experimental time scale, which appears able to quantitatively explain all features of the experimental behavior.

13.1 FOREWORD

The main focus will here be on non-linear stress relaxation, but the theory and method(s) are applicable to all types of physical response, as well as to most types of (polymer or non-polymer) materials, given the wide applicability of the concept of primitive relaxor and ensuing analysis. A wide range of experimental methods and techniques will be invaluable in the future extensive testing of the proposed theory on this and other types of response – thermal, creep, dynamic mechanical, dielectric, etc.

13.2 HOW MATERIALS BEHAVE?

The established features of most (polymer or non-polymer) materials' experimental behavior, as widely recognized in the literature, include [1–6]:

1. The possibility of very extended response times (often 20 or more decades of log time);
2. Non-linearity with respect to the excitation (e.g., strain or stress in mechanical response);
3. Non-affinity of local stresses/strains to the overall average ones;
4. Extreme sensitivity to temperature, T (Williams-Landel-Ferry-WLF or Vogel-Tammann-Fulcher-VTF behavior at low T, Arrhenius at high T);
5. Some influence of intermolecular packing [5];
6. Increased cooperativity at low temperatures of the motions of identifiable "primitive relaxors" [1] contributing to the main α-relaxation;
7. The onset of a glass transition upon T changes, depending on how fast T varies, or the frequency at which the material is tested;
8. The universal identification of two critical dynamic properties [2] – a crossover temperature, T_c, and frequency, v_c, below which the response becomes increasingly slow, many-body, and cooperative (i.e., involving growing clusters of those primitive relaxors), dynamically heterogeneous, and more strongly T-dependent (high average activation energy);
9. The striking universality of most of the above [2], even for some low molecular weight materials;
10. Characteristic average response times of the order of 10^2 s at the glass transition temperature, T_g, changing only a $\mp$ a few degrees for $\pm$ 1 decade expansion in the time scale, and which do not strictly diverge at lower T, except at 0 K.

Publications [2–21], [28–36], and [42–48] of Ref. [3] are a selective but reasonably representative sample of the very extensive work published mainly in the last twenty years, establishing, and analyzing all the above features.

The assumed primitive relaxors may be, e.g., rotating chain crankshafts or other structures in polymers, atom/atom or molecule/molecule pairs, or atom-or-molecule/hole pairs, in atomic or molecular materials, able to interchange their positions and contribute to the overall physical response. So, the concept of primitive relaxor [1] and ensuing formulation may prove applicable to the widest range of materials.

Ninth figure of Ref. [6] and Figures 13.1 and 13.2 of this chapter illustrate the specific and critical items (4) and (6)–(10). The top half of Figure 13.2 specifically illustrates feature (10) already using the predictions of the present theory for a PMMA – poly(methyl methacrylate) – as developed and discussed below, from 24-hour stress relaxation measurements at 4 and 5% tensile strains and temperatures (40 and 50°C) well below T_g [5, 6] after adequate weighing of the relaxors' relative contributions [3, 4]. The resulting calculated T_g (~T_c/12) [2] for the PMMA was ~106°C and, from similar measurements for a polycarbonate – PC –, an T_g of just over 140°C was obtained [6], readily suggesting the possible soundness of the proposed theory. As to features (1)–(3) and (5), they are even better known, and will of course also be explicitly considered in the present contribution.

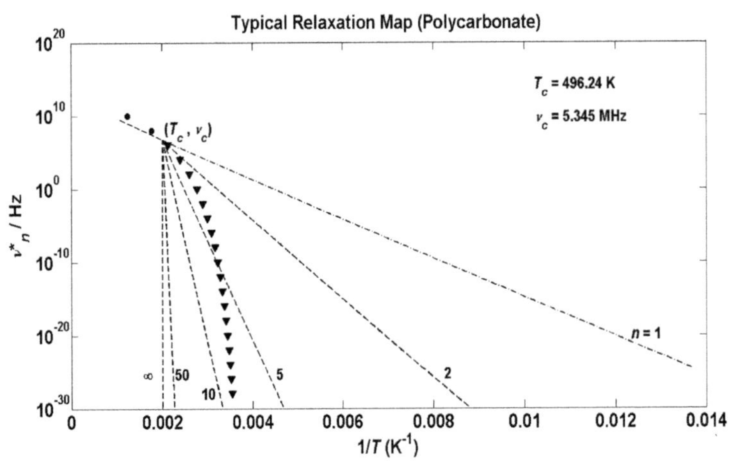

FIGURE 13.1 Typical Super-Arrhenius relaxation map (▼ – PC), showing the cooperative nature of the main α transition.

While plots of experimental data similar to the ▼-line of Figure 13.1 are relatively common in the literature for other materials, the explicit reference to and inclusion of the possible contributing elementary motions (by considering the various potential clusters of primitive relaxors via the lines for $n = 1, 2.$ is a new development and already one of the results of applying the present theory; it will be analyzed and discussed in the remainder of this text against experimental data for three different polymers. Further, as recently shown [6], all previously published data for polymer and non-polymer materials [2] (for which values have been measured or estimated by different authors for the crossover temperatures and frequencies and minimum activation energies, $E_{a,1}$) can be plotted on a single diagram, together with the present theory's predictions for a PMMA and a PC [3, 6], thus illustrating the universality of the behavior and confirming the critical role of the three most relevant parameters (themselves important dynamic physical properties), namely T_c, v_c and $E_{a,1}$.

Plots equivalent to those of Figure 13.2 are also widely available in the literature from data for a variety of materials (again with the exception of the lines referring to each of the assumed cluster sizes, n, in the top diagram), but we again deliberately wanted to illustrate features (4), (6) and (10) of the behavior by means of the proposed theory's own predictions for the same PMMA, to set the stage and highlight from the start that we may, in fact, be dealing with a promising and sound new theoretical approach, before presenting and discussing in the ensuing text the detailed foundations and further predictions of the theory.

13.3 SUMMARY OF THE FOUNDATIONS AND DEVELOPMENT OF THE THEORY

We may fruitfully draw a parallel between the clustering of primitive relaxors and a multi-molecular chemical reaction, though of course without actual chemical change, and formulate it according to the transition state theory (TST) – one of the well-established foundations of physical chemistry and chemical engineering, despite some imperfections. The analogy, though surely unexpected at first, should be obvious, and all we will be doing is to apply abductive reasoning (inference to the best explanation) [7] to features (1) – (10) of Section 1.2. Max Planck and Albert Einstein did something similar more than a century ago when they conjectured about, and successfully explained, the spectrum of thermal radiation

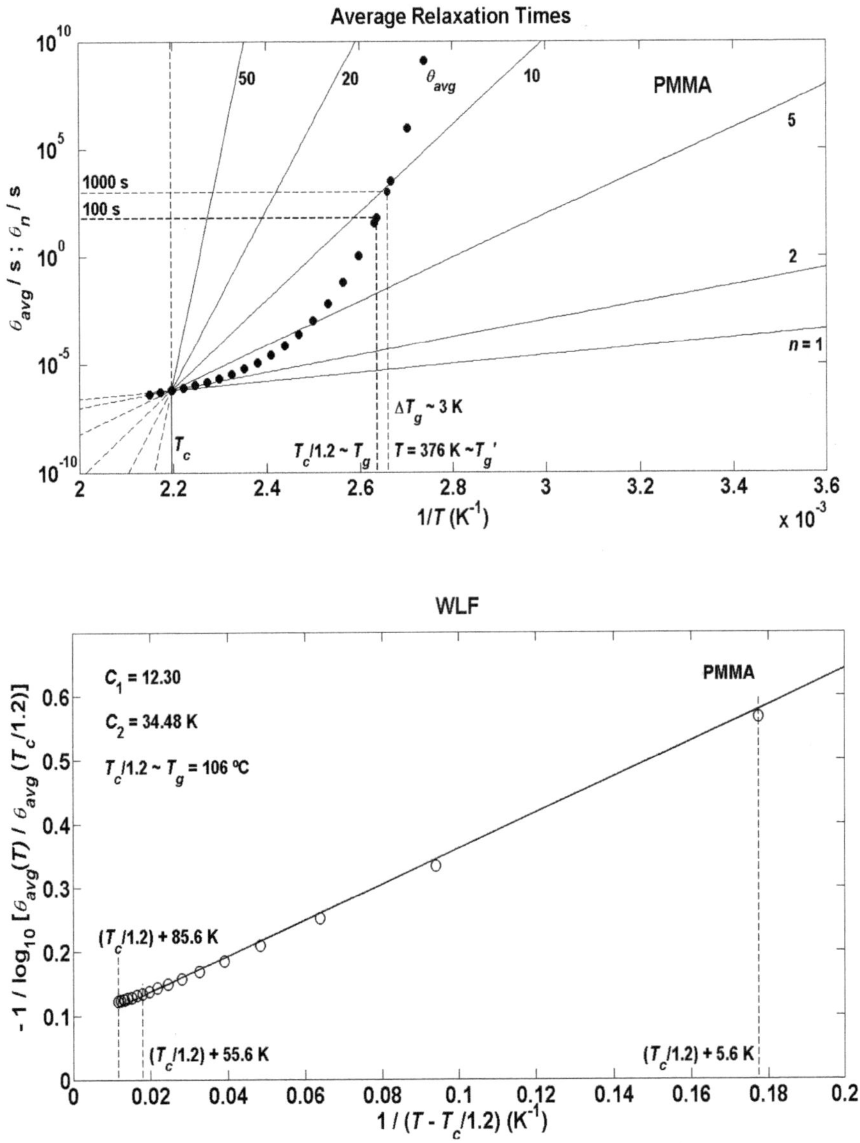

FIGURE 13.2 Top: PMMA's Average Relaxation Times, θ_{avg}, in the Linear Viscoelastic Domain (predicted from stress relaxation experiments [5, 6] *ca.* 60 K below T_g). Bottom: Illustration of PMMA's WLF behavior in the Same Linear Domain (near and above $T_g \sim T_c / 1.2$ [2], but below $T_g + 100$ K).

and the photoelectric effect (by the energy quantization and the photon hypotheses, respectively). Also, now, and every day, the same happens when physicians try to figure out what might be wrong with a patient, an IT-technician or programmer troubleshoots a system or a computer code, and detectives or criminal prosecutors try to make sense of a given crime scene. In our case, the "crime scene" is the set of features (1) – (10) specified above, which still remain largely unexplained [1–3], and the solution that is being proposed here is clearly less strange than the radiation quantization appeared to Planck himself. Of course, the abducted (assumed/inferred) explanation, "wild" as it may seem (like Planck's), must prove realistic and provide a thorough, logical, and superior interpretation of all features of the behavior relative to any possible alternatives, and be subject to complete validation or rejection.

In troubleshooting terms, what we should be asking is why none of the known condensed and soft matter theories and formulations satisfactorily explain and quantify each and every one of those ten features, such as to quantitatively (or at least approximately) predict the correct physical responses at any temperature and for any timescale, and also identify and yield the numerical values of relevant physical properties of the materials. Actually, K. L. Ngai's insight [1] led him to rightly suggest where the difficulty may lie: we need to tackle the problem as the "many-body" one that it truly is – head-on, and from the start.

This chapter will not re-edit or replace the detailed statistical mechanical development of the theory's fundamentals as recently presented [3], where TST was applied and expanded to explicitly account for and quantify the relative numbers of the various types of clustered activated states [3] that unquestionably precede (like in every chemical reaction [8]), most of the microscopic transitions responsible for the behavior of materials. In this way, the transition frequencies [3, 4] of clusters of any size, n, at thermal equilibrium, $\nu_n^*(T)$ (somewhat analogous to Maxwell's molecular velocities in an equilibrated gas), could be obtained [3] and, simultaneously, found a clear physical and mathematical definition for the important crossover properties [2] T_c, ν_c and $E_{a,1}$ as

$$\nu_n^*(T) = \nu_1^\# \left\{ \left(\frac{z_{1,r}^\#}{z_1} \right) \cdot \frac{\exp\left[-\left(E_{a,1} - h\nu_1^\# / 2 \right) / \left(k_B T \right) \right]}{\sinh\left(\dfrac{h\nu_1^\#}{2k_B T} \right)} \right\}^n , \tag{13.1}$$

where h and k_B are Planck's and Boltzmann's constants, and the crossover temperature and frequency turn out such that $T_c \ni \nu_n^*(T_c) = \nu_1^{\#} = \nu_c$, and where the ratio

$$\left(z_{1,r}^{\#} / z_1\right) = \sinh\left(\frac{h\nu_c}{2k_B T_c}\right) \cdot \exp\left[\left(E_{a,1} - h\nu_c / 2\right) / \left(k_B T_c\right)\right] \qquad (13.1a)$$

characterizes how sterically strained (e.g., by electrostatic repulsions) are the relaxors' activated states [3]. At the crossover temperature, as defined here, all the above frequencies become identical, and an immediate very significant observation is that all the above individual frequencies become exactly zero only at 0 K, with no divergence whatsoever at any $T > 0$ K, a result whose physical reasonableness and relevance should be highlighted.

13.4 FORMULATION OF STRESS RELAXATION

The formulation proceeds, for stress relaxation [5, 6], to calculating the frequencies of all direct (stress-reducing) and reverse (stress-increasing) cluster transitions, considering that the above $E_{a,1}$ for each primitive relaxor is changed according to the stored elastic energies before, at, and after activation. This is so because, as the transition (activated) states are reached from an initially un-relaxed state, as well as when each relaxor completes its direct (stress-reducing) transition to the final relaxed state, the local molecular skeleton strains decrease, thus generally releasing some energy that will help in the thermal activation itself; the reverse, of course, will occur in each reverse (stress-increasing) transition. As for the clustering case (Section 1.3), this section will not re-edit or replace, but just summarize, the detailed analysis previously published [5, 6] with explicit reference to their Figures. 13.1 and 13.2, but now denoting the states there named g_i and t_i (with i = 1 or 2) by u_i (un-relaxed) and r_i (relaxed), respectively.

Let us consider at this stage the least complex form of a two-state model [5, 6] (represented by the lower part of those references' Figure 13.1) in the absence of viscous flow, *i.e.,* assuming a lightly cross-linked material.

By (1) formulating the combined effect on stress (σ) by the direct minus reverse transitions of the relaxors of each given size n as:

$$-\frac{d\sigma}{dt}=\omega(\sigma_0-\sigma_\infty)\left\{\underbrace{f_{u_1}}_{f_{u_1,0}\left[1-a(\sigma_0-\sigma)\right]}e^{-E'_{ur}/(k_BT)}-\underbrace{f_{r_1}}_{f_{u_1,0}a(\sigma_0-\sigma)}e^{-E'_{ru}/(k_BT)}\right\} \tag{13.2}$$

where the populations of the corresponding un-relaxed and relaxed states, f_{u_1} and f_{r_1} per unit resisting area (cross-sectional in tensile, or tangential in shear, testing), vary of course linearly with the level of stress relaxation from the initial value, σ_0,

(2) expressing each of the above effective activation energies (E'_{ur} and E'_{ru}) as modifications of $E_{a,1}$ (or $nE_{a,1}$ in the case of clusters of primitive relaxors) according to the preceding paragraph and Eqs. (2) and (3) of Ref. [6], and

(3) calculating the parameter *a*n above from the condition at t infinite (where $d\sigma/dt=0$), the above dynamic equation reduces to the one for a standard non-linear solid, which has an analytical solution of the form $E_r(t)=E_\infty+(E_0-E_\infty)e^{-t/\theta}$, with relaxation times θ (or θ_n for clusters) given by [6]

$$\theta_n=\frac{1}{\omega_n}\exp\left(\frac{n\cdot E_{a,1}}{k_BT}\right)\exp\left[-\frac{n\cdot v_{01}}{8k_BT}\frac{(E_0-E_\infty)(3E_0+E_\infty)}{E_0}\varepsilon_0^{\,2}\right] \tag{13.3}$$

where v_{01} is the total (occupied + free) volume of each primitive relaxor. The angular frequencies ω_n (in rad/s, not in Hz as the ν_n^* of Eq. 13.1) must, of course, be those obtained by the clustering theory outlined in the preceding section, i.e.,

$$\omega_n(T)=2\pi\nu_c\left[\left(\frac{z_{1,r}^{\#}}{z_1}\right)\cdot\frac{e^{h\nu_c/(2k_BT)}}{2\sinh\left(\dfrac{h\nu_c}{2k_BT}\right)}\right]^n,\ \text{or} \tag{13.4}$$

$$\omega_1(T)=2\pi\nu_c\frac{\sinh\left(\dfrac{h\nu_c}{2k_BT_c}\right)}{\sinh\left(\dfrac{h\nu_c}{2k_BT}\right)}\exp\left[\frac{h\nu_c}{2k_B}\left(\frac{1}{T}-\frac{1}{T_c}\right)\right]\exp\left(\frac{E_{a,1}}{k_BT_c}\right)\ \text{and} \tag{13.4a}$$

$$\omega_n(T)=2\pi\nu_c\left[\frac{\omega_1(T)}{2\pi\nu_c}\right]^n. \tag{13.4b}$$

In the formulation leading to Eqs. (13.2) and (13.3), a simplification was made whereby, at any given time, all primitive relaxors within each cluster are either un-relaxed or relaxed, but a modification proves possible and is being developed, in which a more accurate binomial distribution is considered among those primitive relaxors. The result is that the final relaxation times of Eq. (13.3) finally come out divided by the relaxor size, n, without radically changing the physics of the behavior. This modified formulation will be dealt with in detail in subsequent publications.

With the above simplified two-state model representation, the formulation yields a time-dependent distribution of local stresses (which are either σ_0, or σ_∞), while the full four-state model predicts a total of four different local stress values [6]. Feature (3) of the behavior in Section 1.2 is therefore rightly contemplated.

The above equations imply that, in the linear viscoelastic domain, where $E_{a,1}$ is not significantly altered by the applied strain ($\varepsilon_0 \sim 0$) and, given that $\sinh[h\nu_c/(2k_BT)] \approx h\nu_c/(2k_BT)$ to within less than 0.5% for temperatures down to 10^{-4} K (for PC) or even 10^{-6} K (for PMMA) [3, 6], the various characteristic relaxation times, θ_n, of Eq. (13.3) may be written as

$$\theta_n \approx \frac{1}{2\pi\nu_c}\left\{\left(\frac{T_c}{T}\right)\exp\left[\frac{E_{a,1}-h\nu_c/2}{k_BT_c}\left(\frac{T_c}{T}-1\right)\right]\right\}^n \tag{13.5}$$

Relevant observations are that:

1) in any of the viscoelastic domains, the above individual relaxation times turn out obeying the same basic relationship with that of the primitive relaxors, θ_1, that was proposed by Ngai [1] in his coupling model (CM) for the average relaxation time, *i.e.* $\theta_n(T)=\left[\omega_c\theta_1(T)\right]^n/\omega_c$, where ω_c is the crossover angular frequency $2\pi\nu_c$ and,
2) as in the linear limit, one will have $\theta_n(T,\varepsilon_0=0)=\left[2\pi\nu_n^*(T)\right]^{-1}$ (where the frequencies $\nu_n^*(T)$ are, by their nature, those simply driven by thermal fluctuations), the theory also ensures full compliance with the fluctuation-dissipation theorem.

The above θ_n, in whichever (linear or non-linear) form, show that normalization is possible and straightforward to each material's T_c(via a reduced temperature T/T_c, and a reduced minimum activation energy is given by the temperature-independent factor in the argument of the exponential of Eq. (13.5), as well as to v_c (via $\theta_n v_c$).

Some oversimplification of this version of the two-state model may be relaxed [6], as the full four-state model of Figures (13.1) and (13.2) of the latter reference may be and is, in fact, being applied to this problem in ongoing work, to reach more accurate results and conclusions. However, the predicted physical behavior does not radically change and, in any case, the applicability of the new clustering theory, as outlined in Section 1.3, is in no way affected.

13.5 RESULTS OF THE COMBINED FORMULATIONS OF CLUSTERING AND STRESS RELAXATION

Similar to the plots of Figure 13.1, the complete relaxation map for PMMA has also been calculated, yielding a crossover temperature and frequency in close agreement with published data [2], which will be reported elsewhere. Despite being based on TST and explicitly assuming constant volume at this stage (via a constant υ_{01} in Eq. 13.3, rather than *e.g.*, a linearly expanding one with temperature), the predicted WLF behavior demonstrated in the bottom half of Figure 13.2 should be highlighted; a variable υ_{01} might provide future adjustments to the WLF parameters. Figure 13.3 illustrates that the same applies to the related prediction of VTF behavior, with two separate VTF relationships being necessary to correlate the data at both high and low temperatures, as experimentally observed [2]. This seriously questions (and actually contradicts) old interpretations of this type of behavior as the exclusive result of volume changes with temperature. How free volume changes might in future work be more fully taken into account, in addition to the υ_{01} expansion, has already been suggested [5].

Another very significant result is that the present clustering theory allows formulating the material's Laughlin-Uhlmann-Angell's fragility index [9–12], m (cf. Figure 13.4), thereby providing a long-sought model of tunable fragility [3,13], specifically as

$$m = \left[d \log_{10} \left(\eta \; or \; \theta_{avg} \right) / \; d\left(T_g \,/\, T \right) \right]_{T=T_g} \approx$$

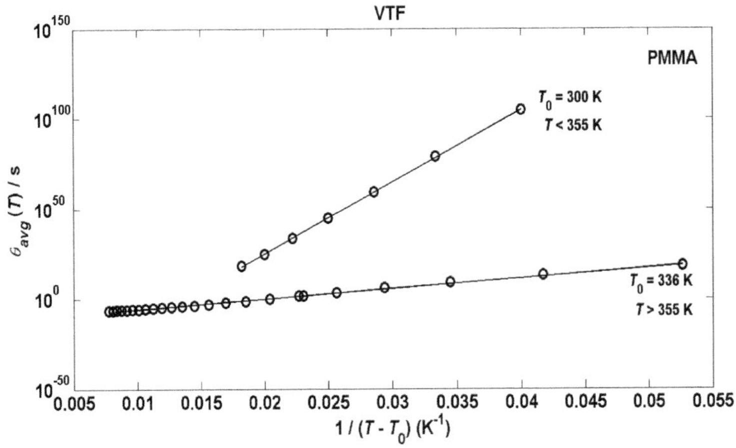

FIGURE 13.3 Illustration of VTF behavior.

$$\approx 0.521\left(0.833+\frac{E_{a,1}-h\nu_c/2}{k_B T_c}\right)\left(\sum_n n\,F'_n\right)_{T=T_g}+\left[\sum_n \frac{dF'_n}{d\left(T_g/T\right)}\log_{10}\theta_i\right]_{T=T_g}, \quad (13.6)$$

the tuning parameters being T_c, v_c and $E_{a,1}$, where η denotes a viscosity, the F'_n stand for the relative weights of the various cluster sizes, as separately evaluated [3] and briefly outlined within the following section, and θ_{avg} is a generalized geometric mean [3] of the θ_n, defined as $\theta_{avg}=\exp\left(\sum_n F'_n \ln\theta_n\right)=\prod_n \theta_n^{F'_n}$. A more general and accurate new definition of fragility (less dependent on the time scale via the constraint $T=T_g$) has also been proposed [3], by imposing $T=T_c$ instead, as

$$m^*=\left[d\ln\left(\theta_{avg}\right)/d\left(T_c/T\right)\right]_{T=T_c}=\left(1+\frac{E_{a,1}-h\nu_c/2}{k_B T_c}\right)\left(\sum_n n\,F'_n\right)_{T=T_c} \quad (13.7)$$

A timely observation is that, with the above definition of θ_{avg}, Ngai's coupling model relationship turns out fully obeyed, *i.e.*, $\theta_{avg}(T)=\left[\omega_c\theta_1(T)\right]^{n_{avg}(T)}/\omega_c$, where $n_{avg}(T)=\sum_{n\geq 1} nF'_n(T)$ is the average cooperativity, or cluster size, in the material's response, seemingly yielding a first physical interpretation of the "quintessential" parameter of Ngai's model – the extension parameter β of Kohlrausch's exponential – as the reciprocal of the average cooperativity.

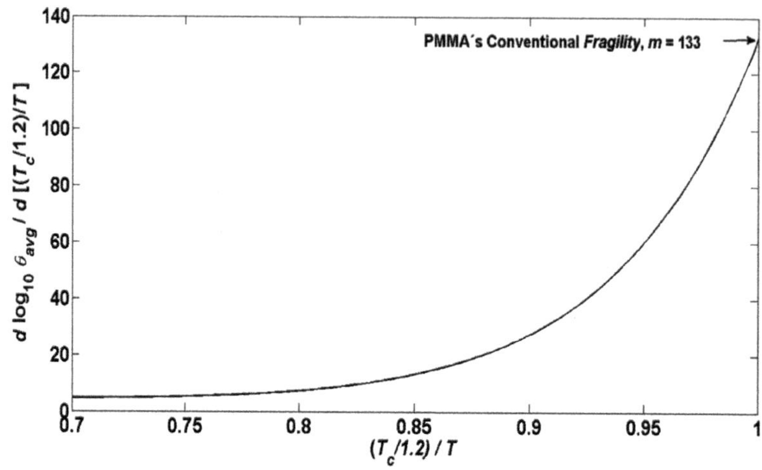

FIGURE 13.4 Calculated PMMA's Laughlin-Uhlmann-Angell's fragility index [9–12].

Figures 13.5 and 13.6 below show the experimental [5, 6] and the newly predicted relaxation moduli, $E_r(t)$, for amorphous PMMA and PC (via an improved treatment of the initial strain ramp), though still adopting an earlier approximate evaluation of the statistical weights of the various clusters (corresponding to the F_n'), by means of truncated log-normal relaxation spectra [5, 6] (cf. also the next section). The quality of the agreement with the experimental data is nevertheless better than 1%, and the resulting crossover properties obtained for PMMA closely agree with the known experimental values [2], while those for PC (cf. Figure 13.1) are to our knowledge original estimates, from which the right T_g value (~140 °C) could again be obtained.

Very significant is also the excellent agreement achieved with recent new experimental data for a semi-crystalline polymer – PET-poly(ethylene terephthalate) –, documented in Figure 13.7, together with (we think) the first-ever estimates of its crossover properties and the right approximate value of T_g. This assumed that stress relaxation is confined to the amorphous phase, as expected at the low temperatures (well below T_g) and not excessive strains of the experiments. In this context, the capability of the model to yield the right order of magnitude of the relaxed modulus, E_∞, of such semi-crystalline material should be highlighted. This was achieved by considering that only a fraction of the primitive relaxors (those not within any crystalline structures) may contribute to the relaxation, instead

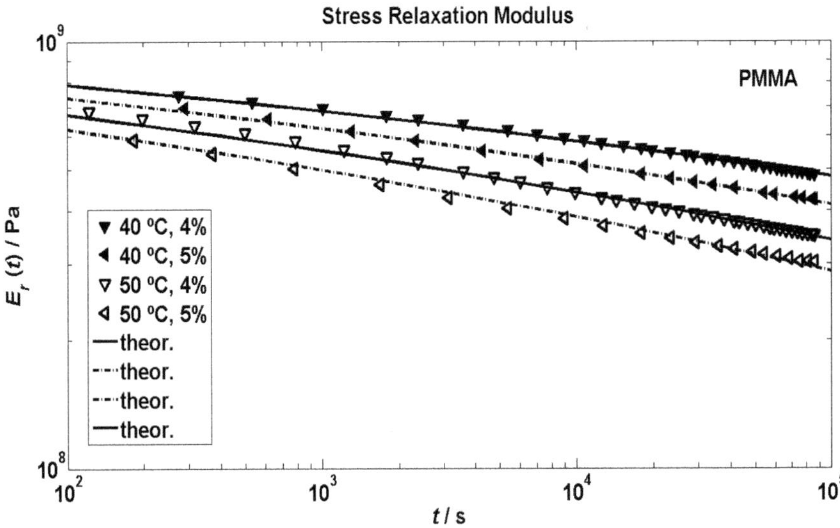

FIGURE 13.5 Predicted (lines) and experimental [5, 6] (symbols) relaxation moduli for PMMA (T_c = 455.2 K → T_g ~ 106 °C; v_c = 2.48 × 10^5 Hz; $E_{a,1}$ = 35.65 kJ/mol; υ_{01} = 1.04 × 10–^{3} m^3/mol; E_0 = 1.26 GPa; E_∞ = 1.00 MPa).

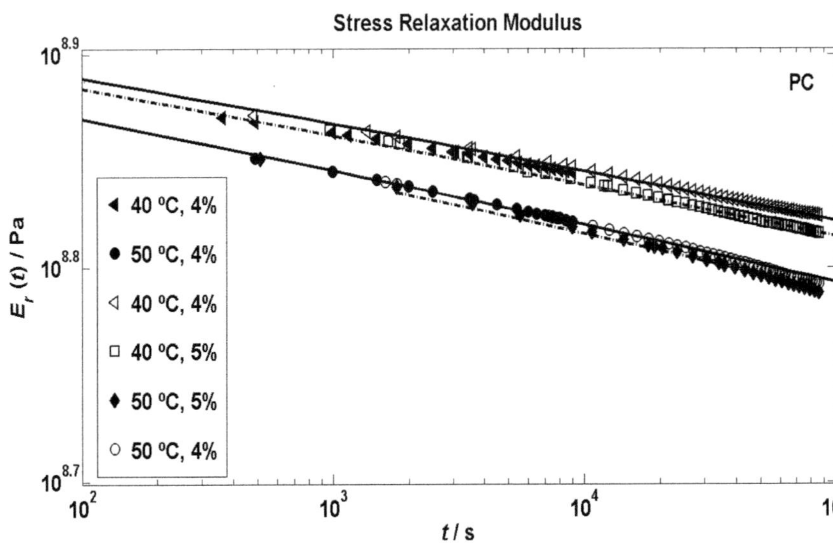

FIGURE 13.6 Predicted (lines) and experimental [5, 6] (symbols) relaxation moduli for PC (T_c = 496.2 K → T_g ~ 140 °C; v_c = 5.34 ×10^6 Hz; $E_{a,1}$ = 49.99 kJ/mol; υ_{01} = 1.16 × 10–^{3} m^3/mol; E_0 = 1.00 GPa; E_∞ = 0.56 MPa).

of taking $f_{u_{1,0}} = 1$ in Eq. 13.2 above in the two-state model. Evidence so far out of recent calculations suggests that E_∞ and $f_{u_{1,0}}$ might be related by $E_\infty / E_0 = 1 - f_{u_{1,0}}$, meaning that $f_{u_{1,0}}$ gives, as it should, a direct measure of the so-called relaxation strength, $(E_0 - E_\infty)/E_0$. Subsequent work may explicitly relate E_∞ to crystallinity itself.

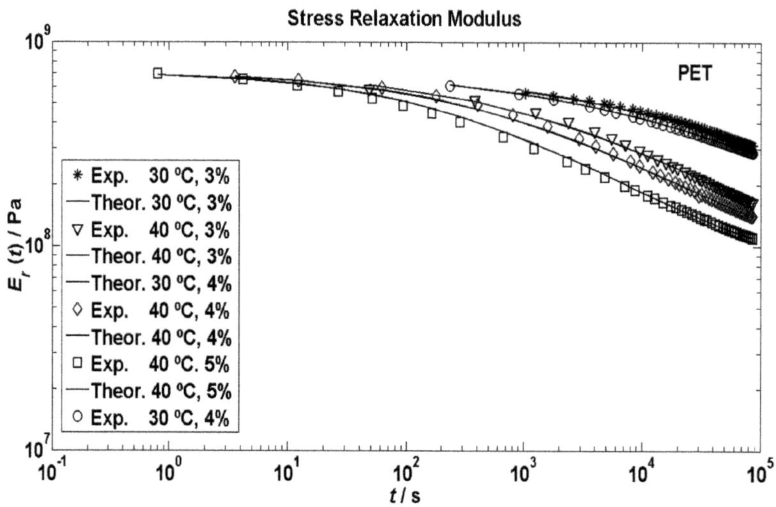

FIGURE 13.7 Predicted (lines) and experimental (symbols) relaxation moduli for semi-crystalline PET ($T_c = 412.7$ K → $T_g \sim 71$ °C; $v_c = 7.88 \times 10^3$ Hz; $E_{a,1}$ 79.40 kJ/mol; υ_{01} 3.00×10^{-3} m³/mol; $E_0 = 0.70$ GPa; E_∞ 0.77×10^8 Pa).

The theory also:

1) quantifies the average and maximum cooperativities or cluster sizes (Figure 13.8);
2) yields an approximate (not exact) time-temperature superposition of the entire response curves (Figures 13.9 and 13.10), in addition to the more exact superposition of the average response times already shown in Figures 13.2 (bottom) and 13.3, and
3) predicts that the discontinuous relaxation spectra (due to the integer n), though infinitely sharp at T_c, widen at decreasing temperatures, as suggested by experimental data, clearly pointing to rheological complexity being the rule rather than the exception

(cf. Figure 13.11 obtained for PC, the F'_n being the ordinates of the plots, whose sums equal exactly 1).

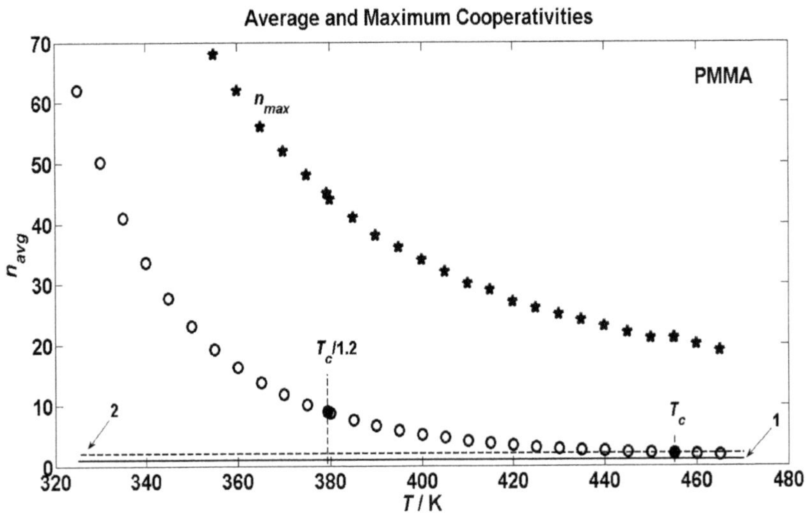

FIGURE 13.8 Predicted average and maximum cooperativities (cluster sizes) for PMMA.

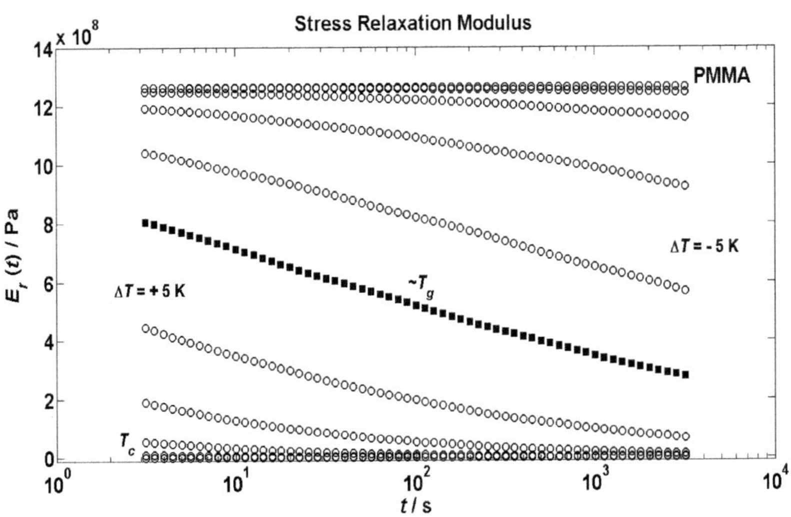

FIGURE 13.9 Predicted Stress relaxation moduli for PMMA at easily accessible time scales by experiments above and below T_g.

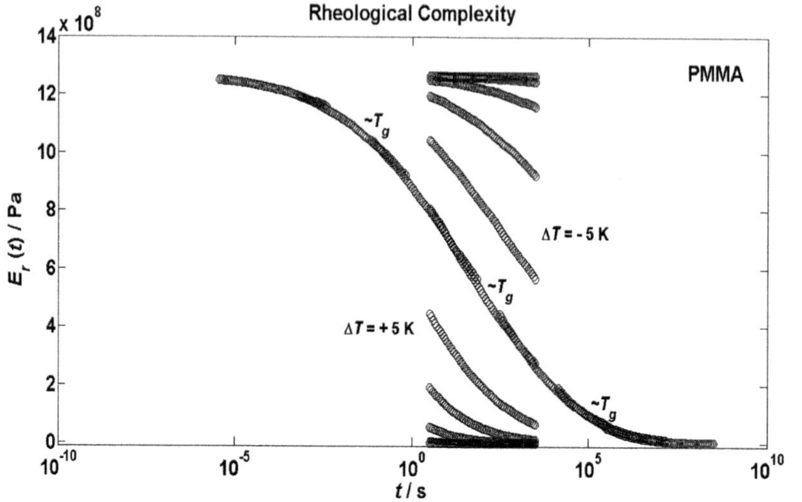

FIGURE 13.10 Approximate superposition by horizontal displacements of the curves of Figure 13.9.

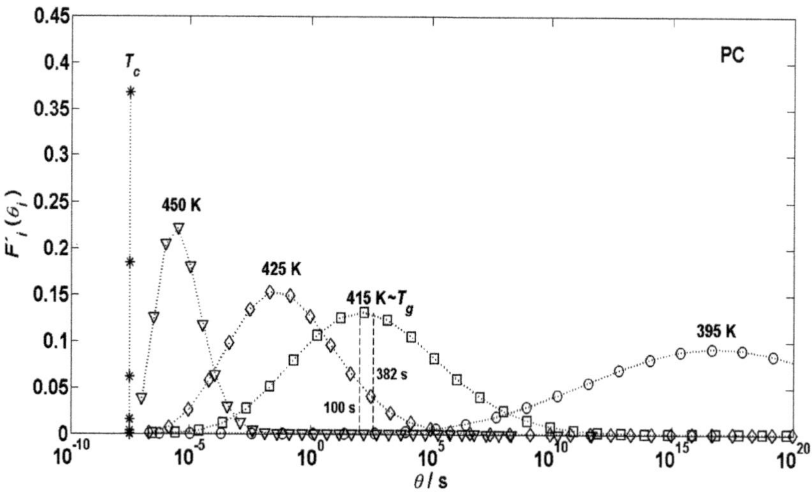

FIGURE 13.11 Predicted discontinuous spectra and rheological complexity for PC (cluster sizes, n, are counted from the leftmost symbol at each temperature).

The latter result is further analyzed and justified in the next section, where this most difficult aspect of the theory – the actual relative weights of the various clusters –, though currently still being updated, is outlined and shown to eventually achieve a better representation of the actual spectra than the approximate truncated log-normal one used so far.

13.6 THE RELAXATION SPECTRA

As a long time ago Feltham [14] had already shown that a conventional log-normal distribution would approximately represent most materials' relaxation spectra, $H(\theta)$, our first calculations to predict the relaxation moduli (cf. Figures 13.5–13.7) used that same approach, with the only modification of truncating the distribution at and below a minimum relaxation time, θ_1, corresponding to that of the primitive relaxors ($n = 1$), according to

$$\frac{H(\theta)}{E_0 - E_\infty} = \frac{b_0 / \sqrt{\pi}}{\left[1 + erf(b_0)\right] \ln\left(\theta_{avg} / \theta_1\right)} \exp\left\{-\left[b_0 \ln\left(\theta / \theta_{avg}\right) / \ln\left(\theta_{avg} / \theta_1\right)\right]^2\right\}. \quad (13.8)$$

This is consistent with a numerically adjustable standard deviation of the order of $\ln\left(\theta_{avg} / \theta_1\right) / \left(b_0 \sqrt{2}\right)$ and an integral in $\ln\theta$ of the above expression exactly equal to 1. This, combined with the classical Alfrey's approximation, whereby $e^{-t/\theta} \sim 0$ for $\theta < t$ and ~ 1 for $\theta \geq t$, yielded as the final approximate modulus expression [5, 6] used in the calculations,

$$E_r(t) \approx E_\infty + \left(E_0 - E_\infty\right) \frac{1 + erf\left[b \ln\left(\theta_{avg} / t\right)\right]}{1 + erf\left(b_0\right)}, \quad (13.9)$$

with $b = b_0 / \ln(\theta_{avg} / \theta_1)$ and $E_r(t \leq \theta_1) = E_0$, where b_0 was numerically adjusted to the experimental data.

Figure 13.12 shows the PMMA's discontinuous cluster relaxation spectra (symbols) that would correspond to the truncated log-normal ones directly extracted from the experimental data (lines), while Figure 13.13 illustrates the effect of Alfrey's approximation on the final responses of an idealized standard linear solid and of a normal superposition of such solids (top), as well as on the predicted combined behavior of the whole range of clusters (bottom) is characterized by the discontinuous H_n and values of Figure 13.12. Although the effect does not seem very large, it was nevertheless recognized that directly predicting more accurate cluster weight distributions, F'_n, needs to be taken as one of the next important objectives of the theory, to avoid the illustrated small shift of the calculated curves to slightly longer times, relative to the exact ones. The latter was obtained by accurate numerical integration of the response integral (top) or by the present clustering theory (bottom). It is the discontinuous character of the spectra that gives rise to the undulations seen in the modulus curves of the plots of Figures 13.13 (bottom) and 13.14, (discussed further below) around the relaxation times of each family of clusters.

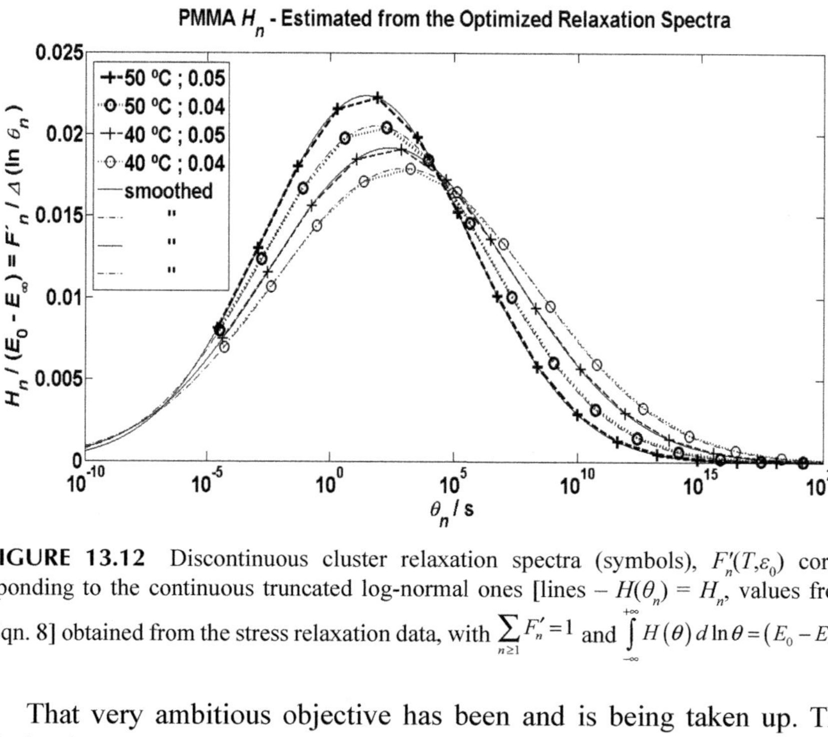

FIGURE 13.12 Discontinuous cluster relaxation spectra (symbols), $F'_n(T,\varepsilon_0)$ corresponding to the continuous truncated log-normal ones [lines – $H(\theta_n) = H_n$, values from Eqn. 8] obtained from the stress relaxation data, with $\sum_{n\geq 1} F'_n = 1$ and $\int_{-\infty}^{+\infty} H(\theta)\, d\ln\theta = (E_0 - E_\infty)$.

That very ambitious objective has been and is being taken up. The derivations are complex, requiring detailed treatment in a separate publication, but their evolution and current status are summarized in the final paragraphs of this section. The procedure started with conventional and conceptually simple combinatorial arguments, because clustering was first assumed to be a random association of primitive relaxors [3, 5, 6], though physically expected to involve mostly or even exclusively adjacent, and in some degree "entangled," primitive relaxors (whatever the entanglement mechanism that might be operating). The ensuing derivation led to a first estimate of the cluster relative weights as proportional to $n/n! = 1/(n-1)$, which already includes the required proportionality of the response to the cluster size, n; this assumes that all cluster sizes would be effectively accessible at all times to the primitive relaxors making up the material.

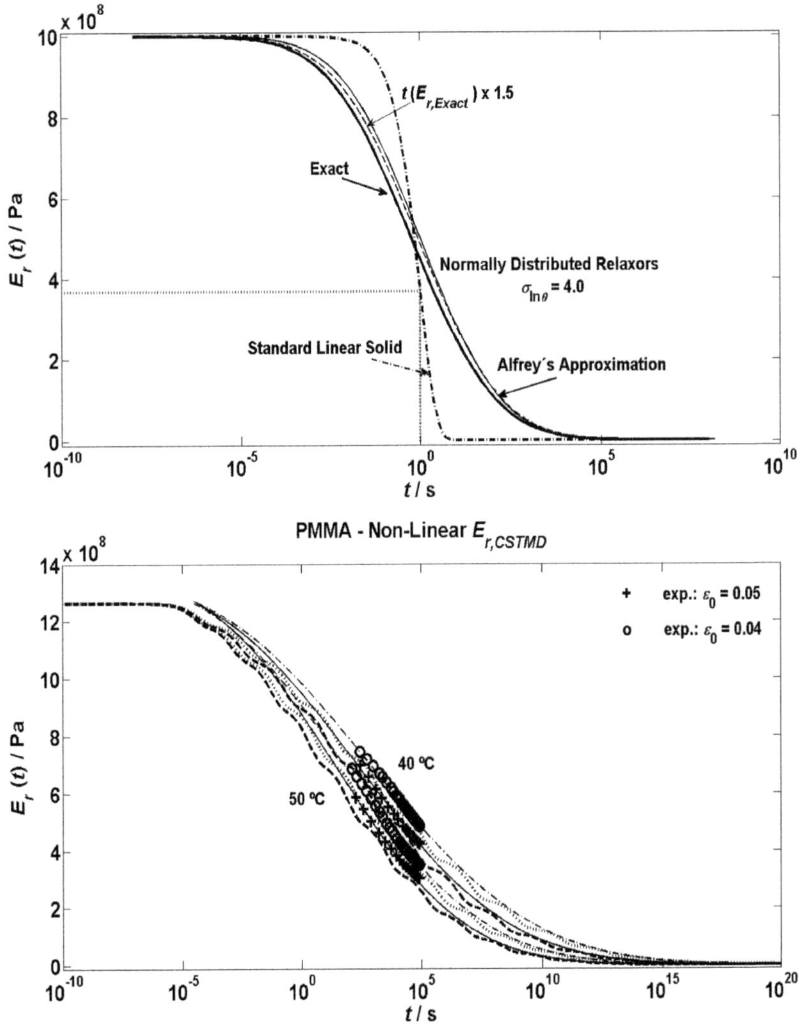

FIGURE 13.13 Effect of Alfrey's approximation – Top: Response of a standard linear solid with $\theta = 1$ s and of a normal superposition of a range of such solids with $\theta_{avg}=1$ s; Bottom: This clustering theory's predictions for PMMA at 40 and 50°C, and 4 and 5% strains (undulated curves), assuming the population of clusters of Figure 1.12, next to the smooth approximate curves for the truncated log-normal distributions [5, 6].

However, that is not so, as of course the various cluster sizes compete with each other for the same set of primitive relaxors, leading to intermittency, in the sense that a given fast moving primitive relaxor (independently, or within a small cluster) may, at some different time, be moving slower, as part of a larger cluster. It is this widely accepted feature of the behavior that defines the so-called dynamic heterogeneity. This behavior requires that the above cluster statistical weights must, in addition, be weighed proportionately to some persistence time, which in turn should in principle be proportional to the specific response time, θ_n, of each family of clusters. This yielded the F'_n values used in most of the calculations so far (namely those that led to Figures 13.1–13.4 and 13.8–13.11) as

$$F'_n(T) = \frac{1}{(n-1)!} \frac{\theta_n(T)}{\sum_{m \geq 1} \{\theta_m(T)/(m-1)!\}}, \quad \sum_{n \geq 1} F'_n = 1. \tag{13.10}$$

These cluster weights (cf. Figure 13.11 for PC) rightly portray features (4) and (6) of Section 1.2, and strongly support rheological complexity.

But, when these F'_n were used to predict the stress relaxation behavior of the same polymers at the very low temperatures of the actual experiments (without assuming the initial truncated log-normal spectra), the modulus curves came shifted to excessively long times, meaning that clusters were being oversized by the theory at such low temperatures. This prompted the new and still ongoing analysis of the above weighing problem, whereby the clustering process is being re-formulated as a truly "chain-like addition," with some probability p_c (likely controlled by interactions or "entanglements" of whatever nature, $1 - p_c$ is the associated probability of no further interaction), of additional primitive relaxors to the outer perimeter of any growing cluster within or crossing the resisting area, in the case of stress relaxation. The first results of this still-evolving formulation are encouraging, as illustrated in Figure 13.14, which also summarizes the formulation itself (inset expressions) and yields average cluster sizes between 5 and 6 (if one counts the E_r undulations from the left), as also approximately obtained for the PMMA experiments with the continuous distributions [5, 6]. However, they also suggest that the above clustering probabilities, p_c, might be slightly increasing with both temperature and strain. This, we think, might be physically explained, but such detailed physical and quantitative justification must be looked at and left for future developments. Therefore, the use of these new F'_n seems at present, as the continuous distributions, dependent on two adjustable parameters – p_c and a coordination number, n_c (having a maximum value of 6 in stress relaxation, assuming a compact hexagonal arrangement of the primitive relaxors in the resisting area). Worth noting is that, despite the 5–6 average cooperativity values, more than sixty primitive relaxors make up the largest clusters that determine the behavior for very long time scales at the low temperatures of the experiments, such as to ensure that $\sum_{n\geq 1} F'_n = 1$ to within 10^{-2} × machine precision in all calculations.

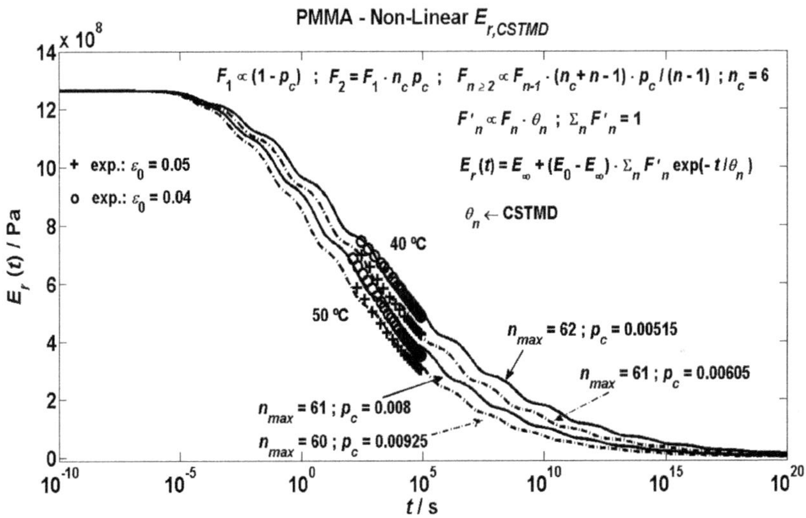

FIGURE 13.14 Latest predictions (lines) vs. experiments [5, 6] (symbols) for PMMA's stress relaxation moduli by the updated cluster distributions (inset expressions), at 40 and 50°C and 4 and 5% Strains.

It should also be stressed that the infinite possible variations of the exact microscopic surroundings of the various primitive relaxors and clusters within any real material justify that the above discontinuous spectra (which directly result from the theory) may, in the end, smooth out to the continuous ones revealed by the experiments. Smoothing of the above pure theoretical spectra is a straightforward numerical operation, by replacing the uniform clusters within each specific F'_n fraction by a set of normally distributed ones around each of the θ_n. Preliminary calculations show that the final results are very insensitive to the assumed standard deviation of such distributions.

13.7 INSIGHTS AND CONCLUSIONS

The developed cooperative or clustering theory of materials dynamics (CSTMD – Cooperative Segmental Theory of Materials Dynamics or, perhaps better, CTMD – Clustering Theory of Materials Dynamics) appears to reasonably describe all physical features (1) – (10) of the experimental behavior, as well as a wide range of specific observations and conclusions published by many previous authors [15–19] (cf. publications' specific titles), with positive contributions on:

- the calculation of important dynamic physical properties – T_c, v_c, E_a,1 and $\upsilon_{0,1}$ – in addition to the instantaneous, E_0, and relaxed, E_∞, moduli, from stress relaxation measurements during 24 hours (also possible from creep, dynamic mechanical, etc. and thermal measurements, to be explored in future work);
- easy and fast extrapolation to different temperatures, long timescales, and well into the non-linear viscoelastic domain;
- good agreement with the experimental data;
- the validation and prediction of varying non-affine levels of local stress and strain within the material, relative to the average macroscopic ones;
- achieving a detailed cooperative (many-body) microscopic description of materials dynamics, with a clear association of molecular mobility, response dynamics and glass transition;
- predicting a temperature-dependent (increasing) cooperativity and dynamic heterogeneity (cluster average size and cluster size dispersion) as temperature decreases, with infinite spectra sharpness at T_c;
- providing a universal model of dynamics and of tunable fragility for amorphous materials, allowing normalization relative to each materials' T_c and v_c;
- the emergence of temperature as more important than free volume to condensed matter dynamics;
- the right type and (limited) range of time/frequency-temperature equivalence or superposition, but with materials showing thermorheological complexity, due to non-uniform activation energies;
- the observation of no divergence of response times at low temperatures [1], except at 0 K, and so a thermodynamic glass transition may arguably not exist;[1] The only limits to the timescales are those imposed by the numbers' representation in the numerical

processor's memory (16 decimal digit mantissa and exponents between –323 and +308 in 32 bit systems).

- allowing an analytical (and easily coded) formulation of the dynamics of materials which leads to equally (very) fast computations, irrespective of temperature and time-scales – << 1 s to predict complete sets of response curves for a wide range of temperatures, even well below T_g, for each single set of the above dynamic physical properties – inaccessible to present and possibly future molecular dynamics simulations.

This theory (CSTMD or CTMD), in addition to reasonably describe the main features of materials' responses, makes a seemingly wider range of specific falsifiable predictions than Ngai's coupling model [1] (CM) and Götze's mode coupling theory [20] (MCT). Therefore, it calls for a detailed open-minded analysis and falsification (or validation), with the help of the widest possible range of experimental techniques.

KEYWORDS

- **amorphous condensed matter**
- **clustering**
- **cooperativity**
- **glass transition**
- **non-linear stress relaxation**
- **poly(carbonate)**
- **poly(ethylene terephthalate)**
- **poly(methyl methacrylate)**

REFERENCES

1. Ngai, K. L., (2011). *Relaxations and Diffusion in Complex Systems*, Springer: New York Dordrecht Heidelberg London.
2. Donth, E. J., (2001). *The Glass Transition: Relaxation Dynamics in Liquids and Disordered Materials*, Springer-Verlag: Berlin Heidelberg New York.

3. Cruz, P. J. J. C., (2016). A cooperative theory of amorphous materials dynamics, *Free Access Online Supporting Material for Reference, 6*, p. 38.
4. Cruz, P. J. J. C., (2008). Cooperativity and materials' dynamics – New insights and quantitative predictions, *A. I. P. Conf. Proc., 982*, 452–457. doi: 10.1063/1.2897836.
5. André, J. R. S., & Cruz, P. J. J. C., (2014). Modeling nonlinear stress relaxation. *Polym. Eng. Sci., 54*, 404–416. doi: 101002/pen.23581.
6. Cruz, P. J. J. C., & André, J. R. S., (2016). Towards the accurate modeling of amorphous nonlinear materials – Polymer stress relaxation (I). *Polym. Eng. Sci., 56*, 348–360. doi: 101002/pen.24260.
7. Walton, D., (2014). *Abductive Reasoning*, The University of Alabama Press: Tuscaloosa.
8. Polanyi, J. C., & Zewail, A. H., (1995). Direct observation of the transition state. *Acc. Chem. Res., 28*, 119–132. doi: 10.1021/ar00051a005.
9. Laughlin, W. T., & Uhlmann, D. R., (1972). Viscous flow in simple organic liquids. *J. Phys. Chem., 76*, 2317–2325. doi: 10.1021/j100660a023.
10. Angell, C. A., (1985). Strong and fragile liquids. In: Ngai, K. L., & Wright, G. B., (eds.), *Relaxation in Complex Systems*. Naval Research Laboratory (U. S.).
11. Angell, C. A., (1991). Relaxation in liquids, polymers, and plastic crystals – Strong/fragile patterns and problems. *J. Non-Cryst. Solids.*, (131–133), 13–31. doi: 1016/022–3093(91)90266–9.
12. Angell, C. A., (1995). Formation of glasses on liquids and biopolymers. *Science, 267*, 1924–1935. doi: 10.1126/science.267.5206.1924.
13. Tarjus, G., (2007). *Invited Plenary Lecture* (Unpublished), 5th International Workshop on Complex Systems, Sendai, Japan.
14. Feltham, P., (1955). On the representation of rheological results with special reference to creep and relaxation. *Br. J. Appl. Phys., 6*, 26. doi: 10.1088/0508–3443/6/1/311.
15. Simon, S. L., & McKenna, G. B., (2009). Experimental evidence against the existence of an ideal glass transition. *J. Non-Cryst. Solids., 355*, 672–675. doi: 10.1016/j.jnoncrysol.2008.11.027.
16. Hecksher, T., Nielsen, A. I., Olsen, N. B., & Dyre, J. C., (2008). Little evidence for dynamic divergences in ultraviscous molecular liquids. *Nat. Phys., 4*, 737–741. doi: 10.10138/nphys1033.
17. Kivelson, S. A., & Tarjus, G., (2008). In search of a theory of supercooled liquids. *Nat. Mat., 7*, 831–833. doi: 10.10138/nmat2304.
18. Ngai, K. L., (2007). Why the glass transition problem remains unsolved ?. *J. Non-Cryst. Solids., 353*, 709–718. doi: 10.1016/j.jnoncrysol.2006.12.033.
19. Prevosto, D., Capaccioli, S., Lucchesi, M., Rola, P. A., & Ngai, K. L., (2009). Does the entropy and volume dependence of the structural α-relaxation originate from the Johari-Goldstein β-relaxation ?. *J. Non-Cryst. Solids., 355*, 705–711. doi: 10.1016/j.jnoncrysol.2008.09.043.
20. Götze, W., (2009). *Complex Dynamics of Glass-Forming Liquids – A Mode-Coupling Theory*, Oxford University Press: Oxford New York.

CHAPTER 14

VIBRATIONAL MODES OF THE VAN DER WAALS-LONDON SILICON CLUSTERS SI_2 AND SI_3 IN THE GAS-PHASE

SREĆKO BOTRIĆ[1] and IVAN ZULIM[2]

[1]*Faculty of Electrical Engineering, Mechanical Engineering and Naval Architecture, Department of Mathematics and Physics, University of Split, Ruđera Boškovića 32, HR–21000 Split, Croatia, E-mail: srecko.botric@fesb.hr*

[2]*Faculty of Electrical Engineering, Mechanical Engineering and Naval Architecture, Department of Electronics and Computing, University of Split, Ruđera Boškovića 32, HR–21000 Split, Croatia, E-mail: zulim@fesb.hr*

ABSTRACT

We indicate the possibility for the Van der Waals-London clusters Si_2 and Si_3 to exist as the vibrational structures. By applying the Lagrange's formalism of classical mechanics to small oscillations of weakly bound clusters Si_2 and Si_3, we found that their harmonic frequencies are in the infrared spectral range (IR). We also obtained for the He_2 dimer the harmonic frequency 32.4 cm^{-1} which is in accordance with the density functional theory (DFT) estimations yielding 33.1 cm^{-1}.

14.1 INTRODUCTION

In this article, we show that the vibrational frequencies of small Si clusters, as derived from Van der Waals-London interaction are consistent with available experimental findings. This finding has a wider consequence that indicates the possibility that the first stage in the formation of those clusters is by that interaction. This consequence is also supported by the experimental data that indicate the possibility of a continuous transformation of Van der Waals-London bond (weak bonding) to covalent bond in various molecules and clusters [1]. This is also consistent with the viewpoint of thermodynamics: a metastable thermodynamic state defined by a given number of Van der Waals-London bonds (interactions) goes over to the stable state defined by a given number of covalent bonds, the transformation being an exothermic process.

Our analysis here stems from the hypothesis that the process of formation of neutral silicon clusters can be understood as a sequence of continuous changes from the Van der Waals-London bonding to the covalent bonding. However, while the question of how do weakly bonded silicon clusters transform into covalently bonded ones is beyond the scope of this article, we will try to show that the Van der Waals-London clusters can be the first stage structures of silicon clusters. We further suggest that while those are metastable they are still capable of having vibrational structures, at least in the case of very small clusters such as Si_2 and Si_3. As is well known the experimentally determined values of the binding energies of the neutral covalent silicon clusters Si_2 and Si_3 are 3.21 eV and 7.6 eV, respectively. These binding energies were also calculated by applying the density functional theory (DFT) and the theoretically obtained values 3.62 eV and 7.82 eV [2, 3], are in good agreement with experiment. Correspondingly, the harmonic frequency of the most stable covalent silicon cluster Si_2 is 480 cm^{-1}, while the harmonic frequencies of the most stable covalent silicon cluster Si_3 are: 340 cm^{-1}, 340 cm^{-1} and 508 cm^{-1} [3]. As it will be shown these frequencies are close to the values we obtained for the most stable Van der Waals-London Si_2: 396.077 cm^{-1} and for the most stable Van der Waals-London cluster Si_3: 343.005 cm^{-1} and 485.196 cm^{-1}.

Our calculation of the harmonic frequencies of Van der Waals-London clusters Si_2 and Si_3 will be performed by applying Lagrange's formalism of classical mechanics to small oscillations of clusters in which the interaction energy between two neutral silicon atoms is given by commonly used Lennard-Jones potential.

The structure of this paper is as follows. In Section 14.2, we use the expression derived in Appendix to obtain two approximate expressions that are valid in a very small region around the minimum of Lennard-Jones potential. Those two expressions are applicable to small oscillations due to the Van der Waals-London force in the clusters composed of two and three identical neutral atoms. The Section 14.3 consists of the two subsections. The Subsection 14.3.1 is devoted to calculating the harmonic frequency of weakly bound silicon cluster Si_2 and He_2 dimer as well. In the Subsection 14.3.2, the harmonic frequencies of weakly bound silicon cluster Si_3 are calculated.

14.2 POTENTIAL ENERGY OF WEAKLY BOUND VAN DER WAALS-LONDON CLUSTERS COMPOSED OF TWO AND THREE IDENTICAL NEUTRAL ATOMS

The Van der Waals-London interaction may be viewed as an interaction between the two simultaneously fluctuating dipoles: the charge fluctuation in one atom- the instantaneous dipole 1-induces charge fluctuation in the other atom- the instantaneous dipole 2. The interaction between those two fluctuating dipoles leads to a long-range correlation energy.

This noncovalent interaction is of pure quantum-mechanical origin; it is a weak and long-range interaction [4].

For a cluster composed of a pair of neutral atoms that are bound by the Van der Waals-London interaction the potential energy is approximately determined by Lennard-Jones potential

$$E_p(r) = E\left[\left(\frac{r_m}{r}\right)^{12} - 2\left(\frac{r_m}{r}\right)^{6}\right], \tag{14.1}$$

where r denotes the distance between centers of the atoms (represented by two spheres of equal radii) and r_m is the distance at which potential energy reaches its minimum. This corresponds to the equilibrium positions around which the oscillations of neutral atoms occur. Let the equilibrium position of the atom 1 be at A(0,0,0), the origin of Cartesian system, and let the equilibrium position of atom 2 be at B(α,β,γ). Hence:

$$r_m = \left[\alpha^2 + \beta^2 + \gamma^2\right]^{\frac{1}{2}}. \tag{14.2}$$

In other words, r_m is the most probable distance between the two atoms that are bound by the Van der Waals-London forces such that the positions

A and B define the most stable configuration of the cluster. This yields that the instantaneous distance $r(t)$ between the centers of the atoms is given by

$$r(t) = \left[(\alpha + x_2 - x_1)^2 + (\beta + y_2 - y_1)^2 + (\gamma + z_2 - z_1)^2 \right]^{\frac{1}{2}}, \tag{14.3}$$

where the instantaneous values of displacements of atom 1 around A are given by the Cartesian coordinates $x_1(t)$, $y_1(t)$, $z_1(t)$ and the instantaneous displacements of atom 2 around B are determined by the Cartesian coordinates $x_2(t)$, $y_2(t)$, $z_2(t)$. The potential energy given by (1) depends on the instantaneous displacements of both atoms

$$E_p = E\left[\left(\frac{r_m}{r(t)}\right)^{12} - 2\left(\frac{r_m}{r(t)}\right)^{6}\right]. \tag{14.4}$$

Let us introduce the three variables u, v, and w representing the relative displacements:

$$u = x_2(t) - x_1(t),\ v = y_2(t) - y_1(t),\ w = z_2(t) - z_1(t), \tag{14.5}$$

such that:

$$r(t) = r(u,v,w) = \left[(\alpha + u)^2 + (\beta + v)^2 + (\gamma + w)^2 \right]^{\frac{1}{2}}. \tag{14.6}$$

The energy given by (4) becomes then the function $Y(u,v,w)$ of the three variables u,v,w:

$$\Upsilon(u,v,w) = E\left[\left(\frac{r_m}{r(u,v,w)}\right)^{12} - 2\left(\frac{r_m}{r(u,v,w)}\right)^{6}\right]. \tag{14.7}$$

Let us expand now the function $Y(u, v, w)$ in a Taylor series around the equilibrium position $u = 0$, $v = 0$, $w = 0$, where $Y(u, v, w)$ reaches its minimum. Assuming that the relative displacements about the equilibrium, u, v and w are small, we neglect all the terms higher than the second one. In this approximation the potential energy (7) is given, as shown in the Appendix, by the nonnegative quadratic form:

$$\Upsilon(u,v,w) = -E + \frac{m_{at}\omega_0^2}{4r_m^2}\left[\alpha^2u^2 + \beta^2v^2 + \gamma^2w^2 + 2\alpha\beta uv + 2\alpha\gamma uw + 2\beta\gamma vw\right], \tag{14.8}$$

where the characteristic angular frequency ω_0 is given by:

$$\omega_0 = \frac{12}{r_m}\sqrt{\frac{E}{m_{at}}}, \tag{14.9}$$

and the m_{at} is the atomic mass of the vibrating atom:

For a cluster composed of atom 1 and atom 2 with their equilibrium positions at A(0,0,0) and B(r_m,0,0), respectively, i.e., for $\alpha_{12} = r_m$, $\beta_{12} = 0$, $\gamma_{12} = 0$, the expansion (8) is reduced to:

$$\Upsilon_{12}(u,v,w) = -E + \frac{m_{at}\omega_0^2}{4}u^2 \tag{14.10}$$

where $u = x_2(t) - x_1(t)$.

For a cluster composed of atoms 1, 2 and 3 (three spherical particles that have equal radii) with their equilibrium positions at (the vertices of equilateral triangle) A(0,0,0), B(r_m,0,0) and $C\ (\frac{1}{2}r_m\ \frac{\sqrt{3}}{2}r_m, 0)$, respectively, there are three pairs of interacting atoms. The corresponding Y-functions of Eq. (14.8) are determined then by the following parameters:

For the atom 1 and atom 2 pair: $\alpha_{12} = r_m$, $\beta_{12} = 0$, $\gamma_{12} = 0$; $u = x_2 - x_1$, $v = y_2 - y_1$, $w = z_2 - z_1$, which yields that $\Upsilon_{12}(u,v,w) = -E + \frac{m_{at}\omega_0^2}{4}u^2$. For atom 1 and atom 3 pair: $\alpha_{13} = \frac{1}{2}r_m$, $\beta_{13} = \frac{\sqrt{3}}{2}r_m, \gamma_{13} = 0$;

$u = x_3 - x_1$, $v = y_3 - y_1$, $w = z_3 - z_1$, and

$$\Upsilon_{13}(u,v,w) = -E + \frac{m_{at}\omega_0^2}{4}\left[\frac{1}{4}u^2 + \frac{3}{4}v^2 + \frac{\sqrt{3}}{2}uv\right].$$

Similarly, for atom 2 and atom 3 pair: $\alpha_{23} = -\frac{1}{2}r_m$, $\beta_{23} = \frac{\sqrt{3}}{2}r_m$, $\gamma_{23} = 0$;

$u = x_3 - x_2$, $v = y_3 - y_2$, $w = z_3 - z_2$, and

$$\Upsilon_{23}(u,v,w) = -E + \frac{m_{at}\omega_0^2}{4}\left[\frac{1}{4}u^2 + \frac{3}{4}v^2 - \frac{\sqrt{3}}{2}uv\right]$$

The potential energy of the cluster composed of two identical atoms is simply given by:

$$E_{p2} = -E + \frac{m_{at}\omega_0^2}{4}(x_2 - x_1)^2 \tag{14.11}$$

while the potential energy of the cluster composed of three identical atoms is determined by:

$$E_{p3} = Y_{12}(u, v, w) + Y_{13}(u, v, w) + = Y_{23}(u, v, w) =$$

$$= -3E + \frac{m_{at}\omega_0^2}{4}(x_2 - x_1)^2 +$$

$$+\frac{m_{at}\omega_0^2}{4}\left[\frac{1}{4}(x_3 - x_1)^2 + \frac{3}{4}(y_3 - y_1)^2 + \frac{\sqrt{3}}{2}(x_3 - x_1)(y_3 - y_1)\right] +$$

$$+\frac{m_{at}\omega_0^2}{4}\left[\frac{1}{4}(x_3 - x_2)^2 + \frac{3}{4}(y_3 - y_2)^2 - \frac{\sqrt{3}}{2}(x_3 - x_2)(y_3 - y_2)\right] \quad (14.12)$$

14.3 THE FREQUENCY OF SMALL OSCILLATIONS IN A SMALL CLUSTER

We will carry out here the analysis of small oscillations in the clusters Si_2 and Si_3 by applying Lagrange's formalism of classical mechanics as commonly done for polyatomic molecules [5].

14.3.1 A TWO-ATOM CLUSTER

According to Eq. (14.11) a cluster composed of two identical atoms is a system of two interacting harmonic oscillators. The Lagrangian of this system is:

$$L = \frac{1}{2}m_{at}\left(\dot{x}_1^2 + \dot{x}_2^2 + \dot{y}_1^2 + \dot{y}_2^2 + \dot{z}_1^2 + \dot{z}_2^2\right) + E - \frac{m_{at}\omega_0^2}{4}(x_2 - x_1)^2 \quad (14.13)$$

Since $\frac{\partial L}{\partial y_1} = \frac{\partial L}{\partial y_2} = \frac{\partial L}{\partial z_1} = \frac{\partial L}{\partial z_2} = 0$, only oscillations along the x-axis exist while the remaining five degrees of freedom are due to translation (three degrees) and rotation (two degrees).

The small oscillations in the system are described then by the Euler-Lagrange equations:

$$m_{at}\ddot{x}_1 - \frac{m_{at}\omega_0^2}{2}(x_2 - x_1) = 0 \text{ and } m_{at}\ddot{x}_2 + \frac{m_{at}\omega_0^2}{2}(x_2 - x_1) = 0, \quad (14.14)$$

where ω_0 is given by Eq. (14.9).

Noting that from Eq. (14.13) one can deduce the equality, $\frac{\partial L}{\partial x_1} + \frac{\partial L}{\partial x_2} = 0$, we have from Euler-Lagrange equations that $\frac{d}{dt}(\frac{\partial L}{\partial \dot{x}_1} + \frac{\partial L}{\partial \dot{x}_2}) = 0$ and thus that

$\dot{x}_1 + \dot{x}_2$ = const. Taking zero initial conditions for both displacements and velocities one obtains that $x_2 = -x_1$. Using this conclusion we have in Eqs. (14) that the displacements x_1,x_2 oscillate with the angular frequency ω_0 as defined by Eq. (14.9). When Eq. (14.1) is applied to a cluster composed of two neutral silicon atoms the binding energy E is determined by the expression $E = \frac{3}{8}\frac{\alpha_e^2}{r_m^6} I$ [6], where for the silicon atom $r_m = 2.1\times10^{-10}\, m$ [1], the ionization energy I = 8.1517 eV = 1.304×10^{-18} J, $\alpha_e = 3.733\times10^{-30}\, m^3$, and one obtains that E = 0.4967 eV = 7.947×10^{-20} J.

Upon insertion of $r_m = 2.1 \times 10^{-10}\, m$, $m_{at} = 4.662 \times 10^{-26}$ kg and $E = 7.947 \times 10^{-20}$ J into Eq. (14.9) we get that for a cluster composed of two neutral silicon atoms:

$$\omega_0 = 2\pi v_0 = 7.4607 \times 10^{13}\, s^{-1}$$
$$v_0 = 1.1874 \times 10^{13}\ \text{Hz} \tag{15}$$

Taking the value 2.9979×10^8 ms^{-1} for the velocity of light the harmonic frequency calculated from Eq. (14.15) is 396.077 cm^{-1} that belongs to the infrared band (IR). This value is close to the value 480 cm^{-1} mentioned above that was obtained a long ago by applying DFT method to strongly bound neutral silicon cluster Si_2 [3].

It is also of interest to apply Eq. (14.9) to the He_2 dimer. The binding energy of He_2 is

$E = 15.17 \times 10^{-23}$ J, $r_m = 2.97\times10^{-10}\, m$ [7] and the atomic mass is $m_{at} = 6.64 \times 10^{-27}$ kg.

This yields that $v_0 = 9.72 \times 10^{11}$ Hz i.e., that the harmonic frequency equals 32.4 cm^{-1}, where the value 2.9979×10^8 ms^{-1} is taken for the velocity of light. This is in accordance with an estimation 33.1 cm^{-1} based on single and double coupled clusters method with perturbative triple corrections CCSD(T) as quoted in Ref.[7].

14.3.2 A THREE-ATOM CLUSTER

Following the above Eq. (14.12), the Lagrangian representing the small oscillations in a cluster composed of three identical atoms is given by:

$$L=\frac{1}{2}m\left(\dot{x}_1^2+\dot{x}_2^2+\dot{x}_3^2+\dot{y}_1^2+\dot{y}_2^2+\dot{y}_3^2+\dot{z}_1^2+\dot{z}_2^2+\dot{z}_3^2\right)+3E-\frac{m\omega_0^2}{4}(x_2-x_1)^2-$$
$$-\frac{m\omega_0^2}{16}\begin{bmatrix}(x_3-x_1)^2+(x_3-x_2)^2+3(y_3-y_1)^2+3(y_3-y_2)^2\\+2\sqrt{3}(x_3-x_1)(y_3-y_1)-2\sqrt{3}(x_3-x_2)(y_3-y_2)\end{bmatrix} \quad (14.16)$$

Since $\frac{\partial L}{\partial z_1}=\frac{\partial L}{\partial z_2}=\frac{\partial L}{\partial z_3}=0$, the small oscillations of such a cluster exist only in the x-y plane while the remaining six degrees of freedom is due to translation (three degrees) and rotation (three degrees).

The corresponding Euler-Lagrange equations describing the small oscillations, in terms of three interacting harmonic oscillators, in the x-y plane for the three atoms cluster concerned are given then by:

$$\ddot{x}_1-\frac{\omega_0^2}{2}(x_2-x_1)-\frac{\omega_0^2}{8}(x_3-x_1)-\frac{\sqrt{3}\,\omega_0^2}{8}(y_3-y_1)=0$$

$$\ddot{x}_2+\frac{\omega_0^2}{2}(x_2-x_1)-\frac{\omega_0^2}{8}(x_3-x_2)+\frac{\sqrt{3}\,\omega_0^2}{8}(y_3-y_2)=0$$

$$\ddot{x}_3+\frac{\omega_0^2}{8}(2x_3-x_1-x_2)+\frac{\sqrt{3}\,\omega_0^2}{8}(y_2-y_1)=0$$

$$\ddot{y}_1-\frac{3\omega_0^2}{8}(y_3-y_1)-\frac{\sqrt{3}\,\omega_0^2}{8}(x_3-x_1)=0$$

$$\ddot{y}_2-\frac{3\omega_0^2}{8}(y_3-y_2)+\frac{\sqrt{3}\,\omega_0^2}{8}(x_3-x_2)=0$$

$$\ddot{y}_3+\frac{3\omega_0^2}{8}(2y_3-y_1-y_2)+\frac{\sqrt{3}\,\omega_0^2}{8}(x_2-x_1)=0 \quad (14.17)$$

Now, since, $\frac{\partial L}{\partial x_1}+\frac{\partial L}{\partial x_2}+\frac{\partial L}{\partial x_3}=0$ and $\frac{\partial L}{\partial y_1}+\frac{\partial L}{\partial y_2}+\frac{\partial L}{\partial y_3}=0$ we have from Euler-Lagrange equations that:

$\frac{d}{dt}(\frac{\partial L}{\partial \dot{x}_1}+\frac{\partial L}{\partial \dot{x}_2}+\frac{\partial L}{\partial \dot{x}_3})=0$; $\frac{d}{dt}(\frac{\partial L}{\partial \dot{y}_1}+\frac{\partial L}{\partial \dot{y}_2}+\frac{\partial L}{\partial \dot{y}_3})=0$, which yields that:

$\dot{x}_1+\dot{x}_2+\dot{x}_3=$ const. and $\dot{y}_1+\dot{y}_2+\dot{y}_3=$ const. Taking zero initial conditions for both displacements and velocities one has that:

$x_3=-x_1-x_2$, $y_3=-y_1-y_3$. Using those relations, Eqs. (17) reduce to:

$$-\omega^2 x_1 - \frac{\omega_0^2}{2}(x_2 - x_1) + \frac{\omega_0^2}{8}(2x_1 + x_2) + \frac{\sqrt{3}\,\omega_0^2}{8}(2y_1 + y_2) = 0$$

$$-\omega^2 x_2 + \frac{\omega_0^2}{2}(x_2 - x_1) + \frac{\omega_0^2}{8}(x_1 + 2x_2) - \frac{\sqrt{3}\,\omega_0^2}{8}(y_1 + 2y_2) = 0$$

$$-\omega^2 y_1 + \frac{3\omega_0^2}{8}(2y_1 + y_2) + \frac{\sqrt{3}\,\omega_0^2}{8}(2x_1 + x_2) = 0$$

$$-\omega^2 y_2 + \frac{3\omega_0^2}{8}(y_1 + 2y_2) - \frac{\sqrt{3}\,\omega_0^2}{8}(x_1 + 2x_2) = 0 \qquad \text{(14.18}$$

Here, the second derivatives with respect to time $\ddot{x}_1$, $\ddot{y}_1$, $\ddot{x}_2$, $\ddot{y}_2$, in Eqs. (17) were replaced by the products $(-\omega^{2x}{}_1)$, $(-\omega^2 y_1)$, $(-\omega^{2x}{}_2)$, $(-\omega^2 y_2)$, respectively.

In order to solve the above system of linear homogeneous equations we consider the zeros of the determinant of the system (18):

$$\begin{vmatrix} x & (-a\sqrt{3}) & 2a & a \\ (-a\sqrt{3}) & x & (-a) & (-2a) \\ 2a & a & x & a\sqrt{3} \\ (-a) & (-2a) & a\sqrt{3} & x \end{vmatrix} = x^2(x^2 - 12a^2) = 0 \qquad (14.19)$$

where $x = \frac{3}{4}\omega_0^2 - \omega^2$ and $a = \frac{\sqrt{3}}{8}\omega_0^2$. This leads to:

$$\omega^2\left(\omega^2 - \frac{3}{4}\omega_0^2\right)^2\left(\omega^2 - \frac{3}{2}\omega_0^2\right) = 0 \qquad (14.20)$$

Expressed in terms of ω_0 the roots of Eq. (14.20) are:

$$\omega_1 = 0, \quad \omega_2 = \frac{\sqrt{3}}{2}\omega_0 = 0.866\,\omega_0 \ \text{and}\ \omega_3 = \sqrt{\frac{3}{2}}\,\omega_0 = 1.225\,\omega_0 \qquad (14.21)$$

where, $\omega_0 = 2\pi v_0 = 7.4607 \times 10^{13}\ s^{-1}$ ($v_0 = 1.1874 \times 10^{13}$ Hz). The numerical solutions are then:

$\omega_2 = 0.866\,\omega_0 = 6.46097 \times 10^{13}\ s^{-1}$ and $\omega_3 = 1.225\omega_0 = 9.13936 \times 10^{13}\ s^{-1}$.

Correspondingly, again, taking $2.9979 \times 10^8\ ms^{-1}$ for the velocity of light, the harmonic frequencies are 343.005 cm^{-1}, 343.005 cm^{-1} and 485.196 cm^{-1}.

These values have to be compared with harmonic frequencies 340 cm^{-1} and 508 cm^{-1} that were obtained by applying DFT method to the strongly bonded neutral silicon cluster Si_3[3], and to the vibration frequencies of the Si_3 cluster in the state $^1A_1(C_{2v})$ that are given experimentally (approximately) as 525 cm^{-1} and 550 cm^{-1} [8].

14.4 CONCLUSION

We found that the Van der Waals-London silicon clusters can account for the vibrational structures of the very small Si clusters Si_2 and Si_3. This was done by showing that the harmonic frequencies of these two weakly bonded silicon clusters are close to those of their covalent counterparts. Since these frequencies are in the infrared spectral range it is reasonable to expect their experimental verification.

ACKNOWLEDGMENT

It is a pleasure to thank Professor Isaac Balberg for a critical reading of the manuscript.

Appendix: The derivation of Eq. (14.8)

$$E_p(r) = E\left[\left(\frac{r_m}{r}\right)^{12} - 2\left(\frac{r_m}{r}\right)^{6}\right] \tag{A.1}$$

$$r = \left[(\alpha+u)^2 + (\beta+v)^2 + (\gamma+w)^2\right]^{\frac{1}{2}} \quad r_m = \left[\alpha^2 + \beta^2 + \gamma^2\right]^{\frac{1}{2}} \tag{A.2}$$

$$E_p(r) = E\left[\left(\frac{r_m}{r}\right)^{12} - 2\left(\frac{r_m}{r}\right)^{6}\right] = \mathrm{Y}(u,v,w)$$

$$\frac{\partial r}{\partial u} = \frac{\alpha+u}{r} \quad \frac{\partial r}{\partial v} = \frac{\beta+v}{r} \quad \frac{\partial r}{\partial w} = \frac{\gamma+w}{r}$$

$$\frac{dE_p}{dr}\frac{\partial r}{\partial u} = \frac{\partial Y}{\partial u} \quad \frac{dE_p}{dr}\frac{\partial r}{\partial v} = \frac{\partial Y}{\partial v} \quad \frac{dE_p}{dr}\frac{\partial r}{\partial w} = \frac{\partial Y}{\partial w}$$

$$\frac{\partial \mathrm{Y}}{\partial u} = 12E\left(\frac{r_m^6}{r^8} - \frac{r_m^{12}}{r^{14}}\right)(\alpha+u) = \mathrm{Y}_u$$

$$\frac{\partial Y}{\partial v} = 12E\left(\frac{r_m^6}{r^8} - \frac{r_m^{12}}{r^{14}}\right)(\beta + v) = Y_v$$

$$\frac{\partial Y}{\partial w} = 12E\left(\frac{r_m^6}{r^8} - \frac{r_m^{12}}{r^{14}}\right)(\gamma + w) = Y_w$$

$$\frac{\partial^2 Y}{\partial u^2} = 12E\left(\frac{14r_m^{12}}{r^{16}} - \frac{8r_m^6}{r^{10}}\right)(\alpha + u)^2 + 12E\left(\frac{r_m^6}{r^8} - \frac{r_m^{12}}{r^{14}}\right) = Y_{uu}$$

$$\frac{\partial^2 Y}{\partial v^2} = 12E\left(\frac{14r_m^{12}}{r^{16}} - \frac{8r_m^6}{r^{10}}\right)(\beta + v)^2 + 12E\left(\frac{r_m^6}{r^8} - \frac{r_m^{12}}{r^{14}}\right) = Y_{vv}$$

$$\frac{\partial^2 Y}{\partial w^2} = 12E\left(\frac{14r_m^{12}}{r^{16}} - \frac{8r_m^6}{r^{10}}\right)(\gamma + w)^2 + 12E\left(\frac{r_m^6}{r^8} - \frac{r_m^{12}}{r^{14}}\right) = Y_{ww}$$

$$\frac{\partial^2 Y}{\partial v \partial u} = 12E\left(\frac{14r_m^{12}}{r^{16}} - \frac{8r_m^6}{r^{10}}\right)(\alpha + u)(\beta + v) = Y_{uv}$$

$$\frac{\partial^2 Y}{\partial w \partial u} = 12E\left(\frac{14r_m^{12}}{r^{16}} - \frac{8r_m^6}{r^{10}}\right)(\alpha + u)(\gamma + w) = Y_{uw}$$

$$\frac{\partial^2 Y}{\partial w \partial v} = 12E\left(\frac{14r_m^{12}}{r^{16}} - \frac{8r_m^6}{r^{10}}\right)(\beta + v)(\gamma + w) = Y_{vw}$$

The function $E_p(r) = Y(u, v, w)$ can be represented by a Taylor series around the equilibrium position O defined by $r = r_m$; it corresponds to the values: u = 0, v = 0, and w = 0, where Y(u,v,w) reaches its minimum. If the quantities u, v and w, representing the relative displacements around the minimum of Y(u,v,w) are small, all the terms in the Taylor expansion of the order higher than a second can be neglected. So in this approximation one obtains:

$$Y(u, v, w) = Y(0) + Y_u(0)u + Y_v(0)v + Y_w(0)w \;+$$

$$+\frac{1}{2}\left[Y_{uu}(0)u^2 + Y_{vv}(0)v^2 + Y_{ww}(0)w^2 + 2Y_{uv}(0)uv + 2Y_{uw}(0)uw + 2Y_{vw}(0)vw\right]$$

where:

$$Y(0) = -E = E_p(r_m) \quad Y_u(0) = Y_v(0) = Y_w(0) = 0$$

$$Y_{uu}(0) = \frac{72}{r_m^4}\alpha^2 E \quad Y_{vv}(0) = \frac{72}{r_m^4}\beta^2 E \quad Y_{ww}(0) = \frac{72}{r_m^4}\gamma^2 E$$

$$\mathrm{Y}_{uv}(0)=\frac{72}{r_m^4}\alpha\beta E \quad \mathrm{Y}_{uw}(0)=\frac{72}{r_m^4}\alpha\gamma E \quad \mathrm{Y}_{vw}(0)=\frac{72}{r_m^4}\beta\gamma E \tag{A.3}$$

Thus, in the analysis of small oscillations due to the Van der Waals-London force, the Lennard-Jones potential $E_p = Y(u,v,w)$ is determined by the approximate expression that is valid in a very small region around the energy minimum of the interaction between a pair of neutral atoms:

$$\mathrm{Y}(u,v,w)=-E+\frac{m_{at}\omega_0^2}{4r_m^2}\left[\alpha^2u^2+\beta^2v^2+\gamma^2w^2+2\alpha\beta uv+2\alpha\gamma uw+2\beta\gamma vw\right], \tag{A.4}$$

where the characteristic angular frequency ω_0 is given by:

$$\omega_0=\frac{12}{r_m}\sqrt{\frac{E}{m_{at}}}. \tag{A.5}$$

KEYWORDS

- **harmonic frequencies in IR**
- **Lagrange's formalism of classical mechanics**
- **small oscillations**
- **weakly bound silicon cluster**

REFERENCES

1. Batsanov, S. S., (2001). Van der Waals radii of elements, *Inorganic Materials, 37*(9), 871–885.
2. Aghavachari, K. R., & Ogovinsky, V. L., (1985). Structure and bonding in small silicon clusters, *Phys. Rev. Lett., 55*(26), 2853–2856.
3. Ournier, R. F., Sinnot, S. B., & DePristo, A. E., (1992). Density functional study of the bonding in small silicon clusters, *J. Chem. Phys., 97*(6), 4149–4161.
4. Dappe, Y. J., Ortega, J., & Flores, F., (2010). Weak Chemical interaction and Van der Waals Forces: A combined density functional and intermolecular perturbation theory-application to graphite and graphitic systems, *Lect. Notes Phys., 795,* 45–79.
5. Feynman, R. P., & Hibbs, A. R., (2005). Emended. In: Daniel, F. S., (ed.), *Quantum Mechanics and Path Integrals* (Emended edn., pp. 203–206). Dover Publications, Inc. Mineola, New York, Chapter 8.

6. Chang, R., (2005). *Physical Chemistry for the Biosciences* (1 edn., pp. 497–498). University Science Books: Sausalito, California, Chapter 13.
7. Van Mourik, T., & Gdanitz, R. J., (2002). A critical note on density functional theory studies on rare-gas dimers, *J. Chem. Phys., 116*(22), 9620–9623.
8. Garcia-Serrano, J., Pal, U., Koshizaki, N., & Sasaki, T., (2001). Formation and vibrational structure of Si nano-clusters in ZnO matrix, *Rev. Mex. Fis., 47*(1), 26–29.

PART IV

Special Topics in Polymer Processing and Polymer Coating

CHAPTER 15

INITIAL FLOW ANALYSIS IN CRUCIBLE-NOZZLE DURING PLANAR FLOW MELT SPINNING PROCESS

SOWJANYA MADIREDDI

CVR College of Engineering, Hyderabad, Telangana 501510,
E-mail: madireddisowjanya@gmail.com

ABSTRACT

Planar flow melt spinning is a one-step casting process to produce amorphous ribbons. The process involves ejection of molten metal from the crucible through a slit nozzle on to a rotating cooling wheel by gas pressure. Initial non-uniform flow through the nozzle blocks the flow passage, leading to casting failure. Fully developed flow is required to obtain a uniform flow rate at the exit of the nozzle slit. This requires the accurate design of the lower portion of the crucible-nozzle. Manufacturing of the crucible with various designs and experimentation with the modified nozzle consumes time and cost. Hence a numerical model is developed to investigate the flow patterns in the crucible-nozzle for few designs. The volume of fluid technique with fluid-structure interaction has been employed along with conservation equations and temperature dependent viscosity equation. Molten metal flow in a straight or curved nozzle wall with straight slit edge, straight nozzle wall with chamfered slit edges and with no-slit passage have been simulated and compared. Nozzle with straight wall and no slit passage shows better flow patterns and fully developed flow is attained within 4 milliseconds. Temperature is more uniform in the flow passage and melt pressure is higher than the atmospheric pressure, recommends the nozzle with no slit passage. The model can be used for any material-nozzle design combinations employed in the planar flow melt spinning process to predict the flow through the slit passage. The model can also be used for simulation of flow in an extruder die in polymer processing to study the effect of sudden contraction of the reservoir with a large diameter to exit passage of smaller diameter like tube or orifice.

15.1 INTRODUCTION

Planar flow melt spinning (PFMS) is a rapid solidification process to produces amorphous ribbons for transformer core applications. A defect-free ribbon used as a core reduces the losses in the distribution transformer by 75%. The process involves melting of the alloy taken in a crucible using induction heating and ejecting the molten alloy through a nozzle fitted at the bottom of the crucible through a constricted slit passage onto a rotating copper wheel. A small pool of melt forms as a puddle on the cooling wheel. A thin sheet of the melt is dragged out of the puddle in the form of a ribbon. The puddle acts as a reservoir which continuously supplies melt for the production of ribbons. The manufacturing process is complex due to the involvement of higher ejection pressures, very high operating speeds and rapid solidification to obtain an amorphous structure in the end product (bypassing the crystallization phase). The first stage of the process is required uniform melting of the alloy and the second critical stage needs ejection of the molten metal through a constricted nozzle. Next stage requires a defect-free ribbon formation, i.e., amorphous, continuous, and polished ribbon. Figure 15.1(a) shows the schematic of the planar flow melt spinning process. Figure 15.1(b) shows a crucible with the nozzle and a slit at the bottom. Investigators mainly focused on the rapid solidification process in the melt pool above the cooling wheel. However, uniform flow through the crucible-nozzle is essential for successful casting. Flow through the nozzle may sometimes be non-uniform resulting in casting failure. The failure can be free fall of melt or solidification of melt in the flow passage leading to blockage. Very few investigators studied the pressure profiles [1] and the type of ribbons obtained [2] during the casting of ribbons using various nozzles (Figure 15.1(c–g)). The shape of the nozzle was observed to affect the downstream pressures thereby affecting the flow pattern. Surface conditions of the aluminum foils were also studied for various nozzle shapes [3]. The front flow or the downstream flow conditions in the melt puddle was observed to be affected by varying the flat face of the nozzle. Praisner et al. [4] experimentally investigated the process behavior using a ceramic crucible with a flat bottom and a plug to stop the free fall of the melt. Crucibles made of quartz [5] or graphite [6] with flat nozzle bottoms (Figure 15.1 (d)) have been widely employed to obtain wider ribbons. Present author [7] during the experimental investigation of the process, employed quartz crucibles with a nozzle having a flat bottom. Initial cast trials were unsuccessful due to free fall of the melt and choking of the nozzle due to non-uniform flow through the nozzle-slit. Figure 15.2(a) shows one of the ribbons obtained from the experiments due to non-uniform flow through the nozzle-slit. Of the initial 20 experiments conducted, only 5 were successful in ejecting the melt through the crucible. Figure 15.2(b) shows ribbons obtained at various process conditions using nozzles with rectangular slit passage of the same dimension. Each ribbon obtained is of different width. This motivated to study, compare, and analyze the flow through various nozzles to obtain a uniform flow at the exit and avoid flow blockage. Manufacture of the crucible with various nozzle designs and testing is expensive and time-consuming. Analysis of the flow through the nozzle during the experiments using high-speed camera is difficult due to the radiation from the high-temperature melt. Hence, a numerical model is developed to predict the approximate flow behavior. The present investigation is carried out for a better flow design of the nozzle to obtain an un-choked flow through the slit passage, uniform, and fully developed flow at the slit exit.

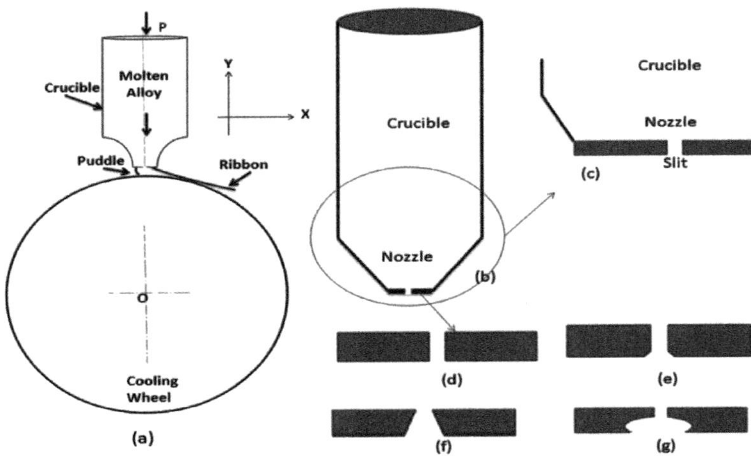

FIGURE 15.1 Schematic of (a) Planar flow melt spinning process (b) Crucible-Nozzle (c) enlarged view of the computational domain and (d) to (g) different shapes of nozzle-slit [1, 2].

FIGURE 15.2 (a) Ribbon obtained due to non-uniform flow (b) Ribbons obtained at various process conditions.

15.2 NUMERICAL MODEL

Computational fluid dynamics is used for the simulation and analysis of the problem. The procedure for CFD simulation involves:

a) selection of computational domain;
b) defining the physics;
c) mesh generation;
d) selection of solver method;
e) post-processing and analysis of the results.

Selection of computational domain involves (i) geometry selection (ii) defining the parameters for the geometry and (iii) defining the domain shape and size. Defining the physics of the problem involves (i) selection of heat transfer model if temperature effects are to be considered (ii) defining whether the flow is compressible or incompressible (iii) defining whether the flow is laminar or turbulent (iv) defining the flow as viscous or non-viscous (v) defining the boundary conditions applicable and (vi) defining the initial conditions.

Mesh generation involves (i) discretization of the domain into elemental volumes (ii) conversion of the mesh to structured or unstructured. Selection of solver method involves (i) solving for steady or unsteady condition (ii) deciding the number of iterations required and the time step for each iteration (iii) defining the convergence limits and (iv) defining the numerical scheme to be employed for the simulation. If the solution is not converged the time step is to be altered or the convergence limit is to be reset. Once the simulations are performed reports are to be generated in the form of XY plot and the results are to be validated with that of experimental if exists. If the experimental results are not available, mesh size is to be altered and the process is to be repeated. One is two critical values are to be tested. Mesh is to be refined until the values of the critical results are unaltered. The flow velocities or heat flux calculations can be checked with that of theoretical calculations for the validation of the model. Once the simulations are performed on the selected mesh, simulations results in the form of contours, vectors, and streamlines are to be analyzed. The domain in the melt spinning process is a combination of the crucible, the gap between crucible and wheel and rotating copper wheel shown in Figure 15.1(a). In the analysis of the melt spinning process, it is difficult to consider the complete domain for the simulations as the computational time is more and analysis is complex. Hence, the process is divided into the following domains:

i. The gap between the nozzle bottom wall and the rotating copper wheel is considered as one domain (Figure 15.3(a)). A 2D numerical model is developed (Figure 15.3b) for this domain to simulate the puddle formation. The model qualitatively compared the puddle formation at various process conditions with that of the experimentally obtained high-speed images [8]. The 2D model is further used to study the flow dynamics using path lines and pressure profiles [9] and to estimate the stable puddle formation [10] at various process conditions. A stable puddle is observed to be attained by 10 milliseconds after the melt exit from the crucible for all the process conditions except for pressures higher than 13.79 kPa. The time of stability for higher pressures is 15 milliseconds. The model is employed for the first time to numerically estimate the ribbon thickness obtained at various process conditions [10]. Thickness and surface topography are also compared with that of experimental results. Numerical results are observed to be well in comparison with the experimental values.

ii. The above domain is extended in the third dimension to simulate the flow of melt on the wheel across its width. The flow pattern gives the surface feature of the product. Hence, a 3D numerical model (Figure 15.4a) is developed to simulate [11, 12] surface topography of the ribbon. The model simulated the surface topography like dimple, streak, and polished which are obtained experimentally at the same process conditions. The reasons for various surface features like wavy, streak, dimple patterns etc are studied. Hence, the 3D model can be used to predict the topography of the end product (ribbon) at a given condition prior to the experimentation.

iii. Another 2D model is developed (Figure 15.5) to analyze the heat transfer in the cooling wheel [13] including the cooling wheel domain to the previous 2D domain. This model predicts the conditions at which a continuous/broken/no ribbon formation occurs. The model also predicts whether crystalline or amorphous ribbon is obtained for a given process condition.

However, in all these models the melt-ejecting from the nozzle-slit is assumed as uniform and fully developed (Figure 15.3(b)). However, during few experiments, it has been observed that the melt flow at the exit is not uniform leading to various casting problems.

iv. Hence, the crucible is taken as another domain and the present numerical model is developed to understand the flow behavior in the existing nozzle design and suggest a new design for a uniform and fully developed flow. Figure 15.6 shows the computational domain selected for the analysis and the mesh generated for the geometry shown in Figure 15.1(c) with a rectangular nozzle slit and inclined crucible walls.

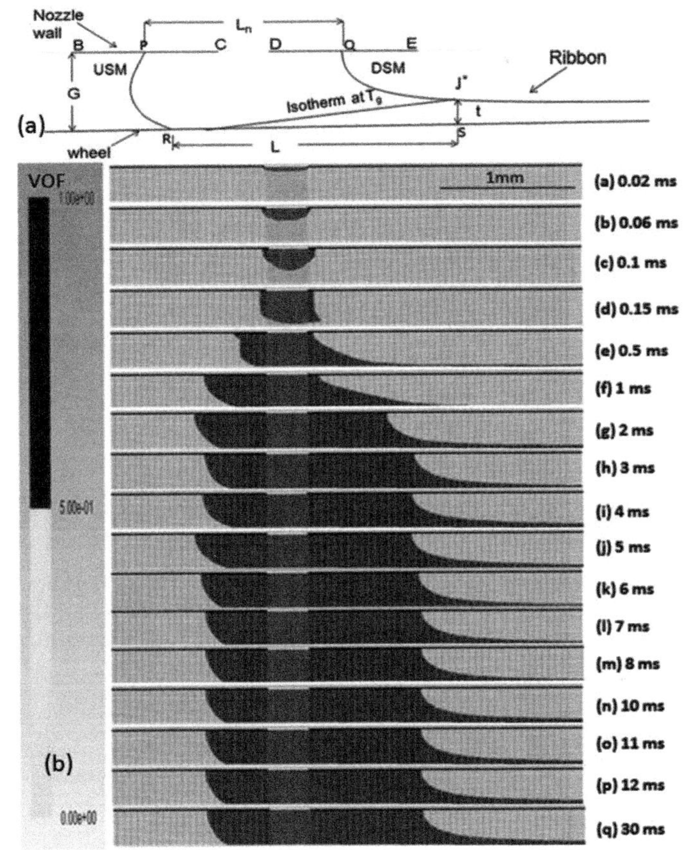

FIGURE 15.3 (a) Schematic of Nozzle-wheel gap domain (b) Puddle formation.

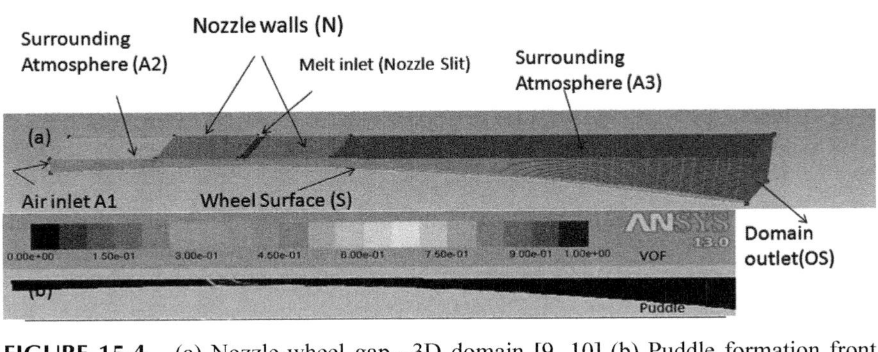

FIGURE 15.4 (a) Nozzle-wheel gap –3D domain [9, 10] (b) Puddle formation front view.

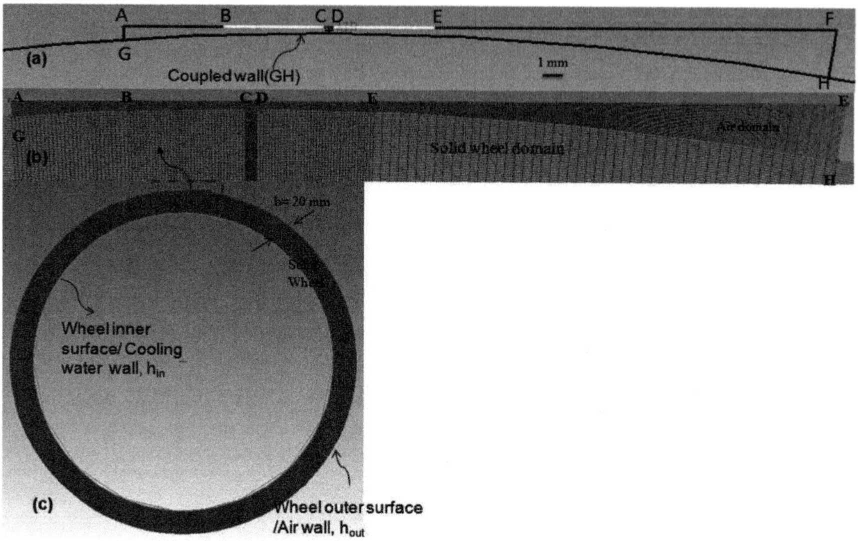

FIGURE 15.5 (a) Nozzle-wheel gap and wheel domain (b) Mesh generation in the gap and at the interface at wheel surface and (c) Mesh generation in the Cooling wheel [13].

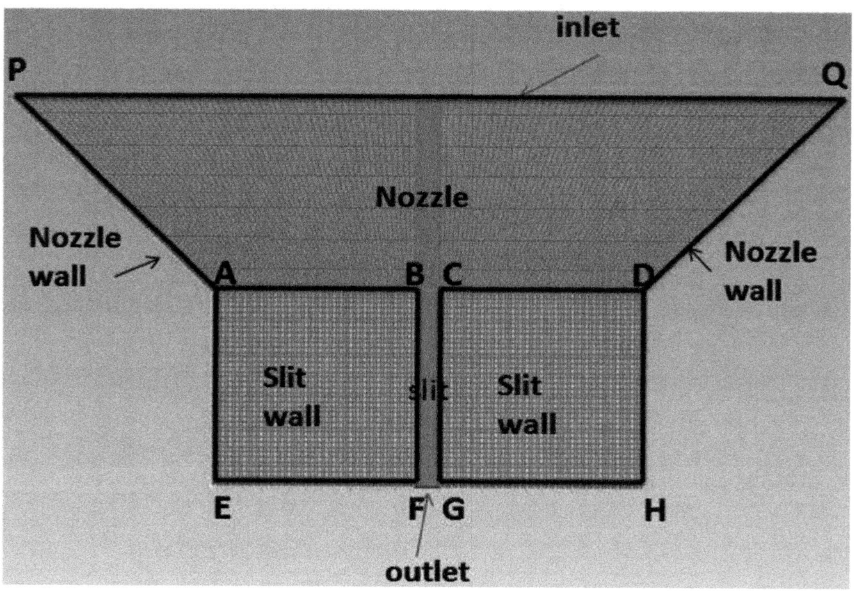

FIGURE 15.6 Geometry/Computational domain for Crucible-Nozzle bottom with Quad-Mesh generated.

AP and DQ are the inner sides of the (straight) nozzle walls. ABFEA and CDHGC are slit walls defined as solid domains with thickness 5 mm on each side of the slit passage BFGCB. This is to identify any heat loss to the walls during the process. Nozzle (PQDCBAP) and slit (BCGFB) domains are defined as air domains. Melt enters the nozzle domain by the application of pressure at inlet PQ and passes through the constricted slit passage through an interface BC (0.6 mm). Conservation equations of mass, momentum, and energy are solved with the explicit first order. The volume of fluid technique (VOF) is employed to show the flow of melt through the air domain. Geometric reconstruction scheme is employed to capture the free flow of melt in the slit passage.

The equations used in the present model are:

Continuity equation:

$$\frac{\partial \rho}{\partial t}+\nabla.(\rho\bar{v})=0 \tag{15.1}$$

Momentum equation:

$$\frac{\partial(\rho\bar{v})}{\partial t}+\nabla.(\rho\bar{v}\,\bar{v})=-\nabla p+\nabla.\left[\mu\left(\nabla\bar{v}+\nabla\bar{v}^{\mathrm{T}}\right)\right]+\rho\bar{g}+f_{\sigma} \tag{15.2}$$

where

$$f_{\sigma}=\sigma\frac{\rho k\nabla F}{\frac{1}{2}(\rho_{a}+\rho_{m})} \text{ and } k=-\nabla\frac{\nabla F}{|\nabla F|}$$

Energy equation:

$$\frac{\partial\left(\rho \mathrm{C_p T}\right)}{\partial \mathrm{t}}+\nabla.\left(\bar{v}\left(\rho \mathrm{C_p T}\right)\right)=\nabla.(\mathrm{K}.\nabla T) \tag{15.3}$$

Volume of Flow (VOF) equation:

$$\frac{\partial F}{\partial t}+\nabla.(F\bar{v})=0 \tag{15.4}$$

Material properties are calculated using the following equations:

$$\left.\begin{aligned}
\rho &= \rho_{m}F + \rho_{a}(1-F)\\
\mu &= \mu_{m}F + \mu_{a}(1-F)\\
C_{p} &= C_{p,m}F + C_{p,a}(1-F)\\
k &= k_{m}F + k_{a}(1-F)
\end{aligned}\right\} \tag{15.5}$$

where the suffix *'a'* denotes air and *'m'* denotes melt.

F is the volume fraction of the melt in the cell, which may vary from 0 to 1.

Assumptions:

- No heat flux at the walls PA and DQ.
- All the material properties (Fe-Si-B) except viscosity [8] are independent of temperature.
- The fluid (melt) is viscous and laminar.
- The crucible is made of quartz. Initial conditions; Nozzle and slit domains are initially filled with air.
- All (solid and air) the domains are initially at 300 K. Boundary conditions; condition at PQ is pressure inlet and condition at FG is pressure outlet.
- PA & DQ are stationary walls at no heat flux condition. BC is a domain interface.

Based on the assumptions, the following boundary and initial conditions are imposed on each of the model boundaries.

Boundary conditions:

- At boundary PQ – pressure inlet
- At BC – interface
- At AP and DQ – no heat flux
- At FG – pressure outlet
- At AB, CD, BF, and CG fluid-structure interaction (FSI)

Initial Conditions:

- Area PQDCBAP and area BCGFB initially filled with air and at 300 K.
- Area ABFEA and area CDHGC are solid domains initially at 300 K.

The properties of the materials used in the simulations are given in Tables 15.1 and 15.2.

TABLE 15.1 Properties of Molten Alloy $Fe_{78}Si_9B_{13}$

Designation	Parameters	Values	Remarks
ρ	Density	7180 kgm^{-3}	Chen et al., [14]
Cp	Specific heat	544 $Jkg^{-1}K^{-1}$	Chen et al., [14]
k	Thermal conductivity	8.99 W/mK	Chen et al., [14]
μ	Viscosity	$\mu(T) = 0.1(\exp(-3.6528 + 734.1/(T-674)))$ Pa.s	Liu et al., [15]
T_g	Glass Transition Temperature	873 K	Liu et al., [15]
σ	Surface tension	1.2 N/m	Chiriac et al., [16]
M.P	Melting Point	1180°C	Chiriac et al., [16]

TABLE 15.2 Properties of the Air and Quartz Crucible

Designation	Parameters	Values for Air (Ansys CFD material database)	Values for Quartz
ρ	Density	1.225 kgm^{-3}	2200 kgm^{-3}
Cp	Specific heat	1006.43 $Jkg^{-1}K^{-1}$	740 $Jkg^{-1}K^{-1}$
k	Thermal conductivity	0.0242 W/mK	1.38 W/mK
μ	Viscosity	1.7894e–05 Pa.s	-

15.3 RESULTS AND DISCUSSION

15.3.1 INITIAL FLOW SIMULATION THROUGH THE CRUCIBLE

The numerical model in the present study is developed to simulate the flow of molten metal through the bottom portion of the crucible, i.e., nozzle. This area is critical as the flow effects the puddle formation which in turn effects the amorphous ribbon formation. The melt (Fe-Si-B alloy) at temperature T = 1473 K and pressure P = 13.78 Pa enters the nozzle domain (ABCDQPA) at PQ. The simulations are performed at a time step of 10e–6 and are converged in less than 5 iterations. As the crucible is made of quartz, no significant increase in temperature is observed initially in the solid domain due to its poor thermal conductivity. Figure 15.7 shows the initial melt flow through the nozzle with straight inclined walls after 1 millisecond to 8 milliseconds of flow initiation at domain inlet PQ (Figure 15.6). VOF = 1 represents the region filled with melt and VOF = 0 represents the air. The inverted meniscus observed is due to the high surface tension (1.2 N/m) of the melt. Melt enters the slit passage (0.6

mm wide) at BC (Figure 15.6), by 2.6 milliseconds and is observed to flow through as a converged stream. The simulations are performed on 3 mesh sizes. This time of travel by the melt is used to select the present mesh. Time taken for the melt from PQ to the slit inlet BC is compared with the theoretical value using the simple equation for flow velocity

$$Velocity(V) = \frac{Distance(s)}{Time(t)} \tag{15.6}$$

The melt slowly fills up the passage BCGFB and exits the nozzle as a fully developed flow by 8 milliseconds. The non-uniform flow till 8 milliseconds may result in choking of the slit passage. The sudden contraction leads to pressure differences at BC.

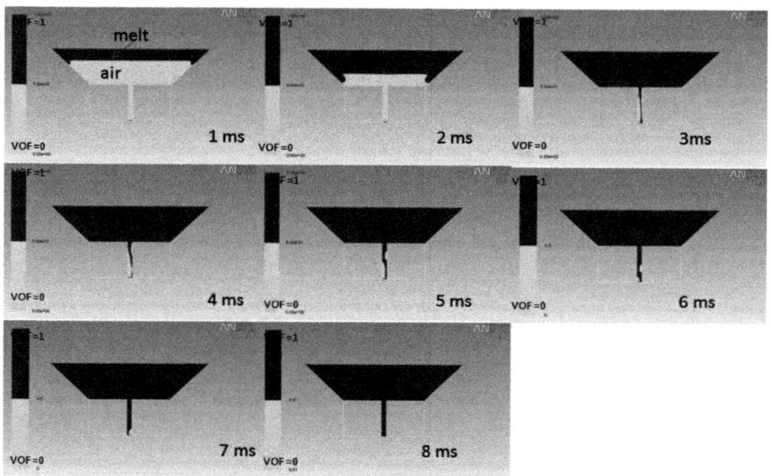

FIGURE 15.7 Flow simulations through the crucible from 1 millisecond to 8 milliseconds.

Figure 15.8 presents the path lines of the melt and air after 2,3,6 and 8 milliseconds of the initiation of flow at the domain inlet (PQ). As the flow enters the slit between 2 to 3 milliseconds there exist a pressure difference due to constricted passage resulting in re-circulation of the melt at the slit entry. Figure 15.8(b) and 8(c) clearly show these re-circulations and by 8 milliseconds the flow becomes uniform shown in Figure 15.8(d). We can infer from these results that the flow becomes uniform once the melt completely fills the slit passage. However, the melt may solidify and blocks the slit passage if the flow time in the slit passage is long.

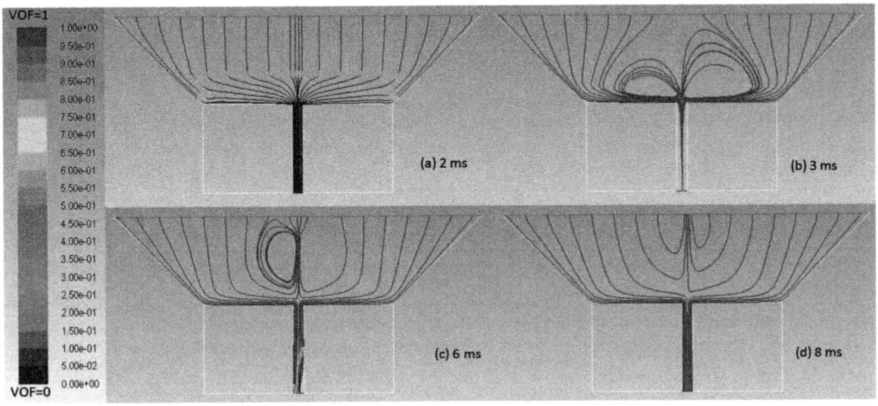

FIGURE 15.8 (See color insert.) Pathlines in the crucible after (a) 2 ms (b) 3 ms (c) 6 ms and (d) 8 ms.

15.3.2 COMPARISON OF FLOW THROUGH THE NOZZLE WITH STRAIGHT/CURVED WALL WITH VERTICAL SLIT

The present investigation is to reduce the time gap between the initial flow at the slit entry BC and fully developed flow at the exit FG. As, it is impractical to reduce the wall (AE or DH) thickness less than 5 mm, the effect of change of shape of the walls AP & DQ is studied. Figure 15.9(b) shows the flow when the straight nozzle walls are made as curved. The inverted meniscus is observed to turn into a straight line. This is due to the effect of wall curvature. Geometric re-construction scheme used in the simulation re-constructs the air-melt interface in each cell for every time step and iteration. The flow condition at 3 ms for a straight edge is occurred by 4 ms with the curved edge. This is because of the increase in volume due to the curved edges. No other significant change is observed in the flow pattern. As the time to obtain a fully developed flow is increased, a nozzle with straight edge walls is preferred. However, the path lines for curved edges (Figure 15.10) show that the entry into the slit is more uniform with a curved edge (Figure 15.10(b)) when compared to that with the straight edge (Figure 15.8(b)). This helps to reduce the slit entry blockage during melt ejection at lower pressures. But, it is interesting to see that the flow pattern at 6 ms and 8 ms is similar for both straight (Figure 15.8(c) and 8(d)) and curved edge (Figure 15.10(c) and 10(d)). As the crucible nozzle with curved edges takes more flow time, new models are developed with chamfering the slit edges using the nozzle with a straight edge. Slit edge is the flat bottom, inside the nozzle at the entrance to the slit passage shown as AB and CD in Figure 15.6.

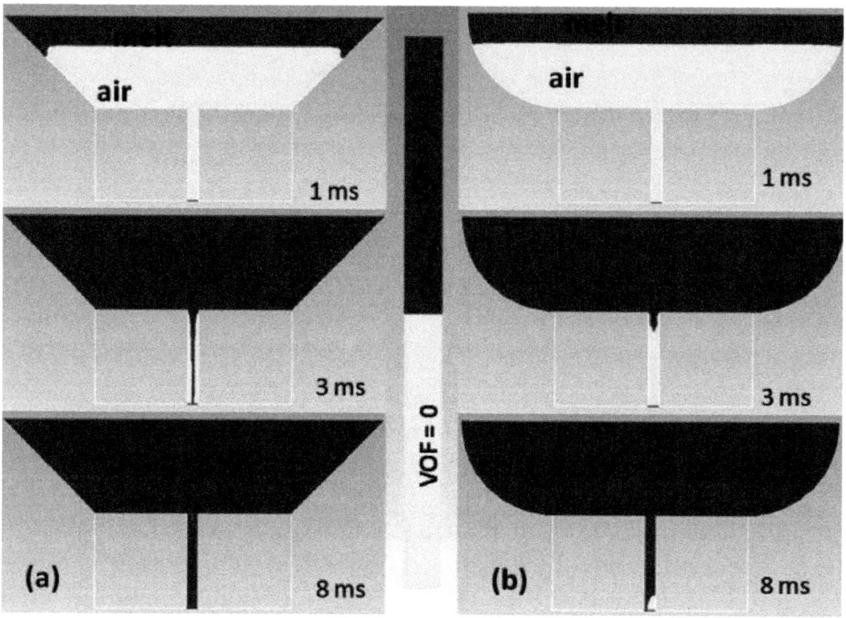

FIGURE 15.9 The flow of melt through nozzle and slit from 1 ms to 8 ms (milliseconds) (a) Straight wall (b) Curved wall.

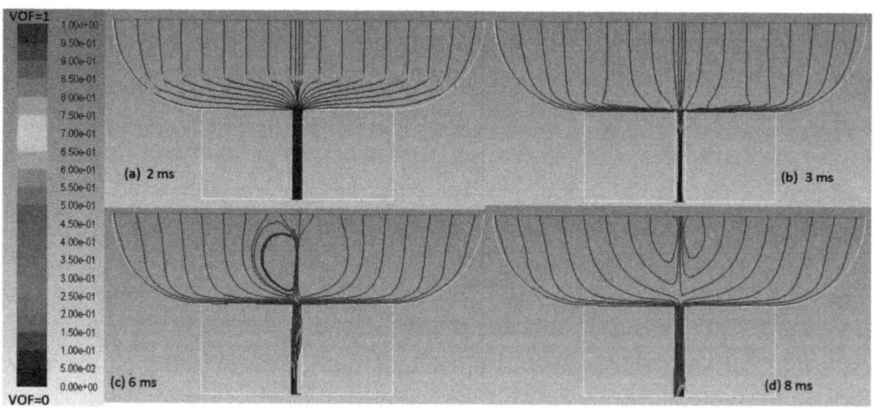

FIGURE 15.10 **(See color insert.)** Pathlines in the crucible after (a) 2 ms (b) 3 ms (c) 6 ms and (d) 8 ms.

15.3.3 EFFECT OF CHAMFERING OF THE SLIT WALL

The new model is developed by modifying the slit edge walls AB & CD chamfered to 1 mm with straight nozzle walls. The flow pattern is recorded for every millisecond and compared. Figure 15.11 shows the flow of melt in a nozzle with straight walls and 1 mm chamfering at 3, 7, 8, 9, 10, and 11 milliseconds. Fully developed flow is attained by 5 ms which is less than that of in a straight/ rectangular slit passage. But it is interesting to observe that air initially present in the nozzle forms a bubble in the melt and enters the nozzle slit at 8 ms and reaches to the slit exit by 10 ms (Figure 15.11(b) to 11(e)). The bubble formation is due to the entrapment of air between the inverted meniscus and chamfered slit edge. By 11 milliseconds the air bubble is pushed out of the slit exit. The melt which enters the slit and travels through the passage between 5 ms to 11 ms may solidify and block the slit due to the difference in temperature in air, melt, and slit wall. Hence, to avoid bubble formation simulations are repeated by chamfering the slit edge to 1.5 mm, 2 mm, 2.5 mm and 5 mm. For a chamfering radius of 5 mm, there exists no slit passage as the slit wall thickness is only 5 mm.

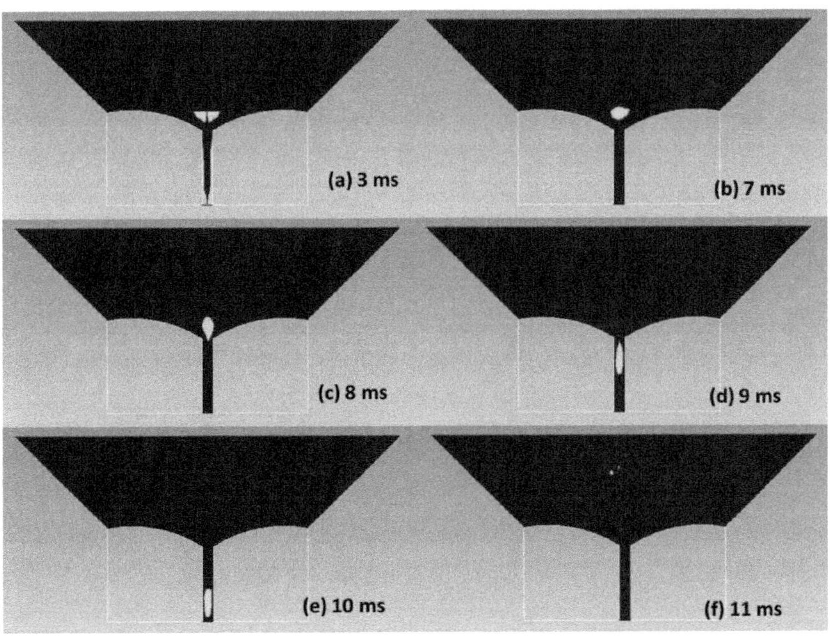

FIGURE 15.11 Flow through Crucible-Nozzle with 1 mm chamfering.

Figure 15.12 (a)–(c) show the flow at 3 ms and 12(d)–(f) show the flow near to fully developed at the exit for chamfering values of 1 mM, 2.5 mm and 5 mm (no-slit), respectively. It is interesting the observation that the melt slides along the side walls in the case of 5 mm chamfering or no slit passage condition. The melt along the axis line of the nozzle takes more time for the flow down to the slit exit. This is because of the inverted meniscus formation. However, the time taken for the development of fully developed flow is reduced from 8 ms to 5 ms with an increase in slit wall chamfering, from zero (straight) to 2.5 mm. With further increase in chamfering to 5 mm, the time to form a fully developed flow at the exit has drastically reduced to 3.4 milliseconds. But an air bubble is observed in the melt up to 20 milliseconds. The size of the air bubble is increased with increase in chamfering radius. Even though it is negligible, this can be avoided if the casting is performed in a vacuum. However, to avoid the air bubble in the melt flow path for casting in the air, a further change is made in the design of the nozzle with a combination of the curved crucible-nozzle wall with no-slit (5 mm chamfering) condition.

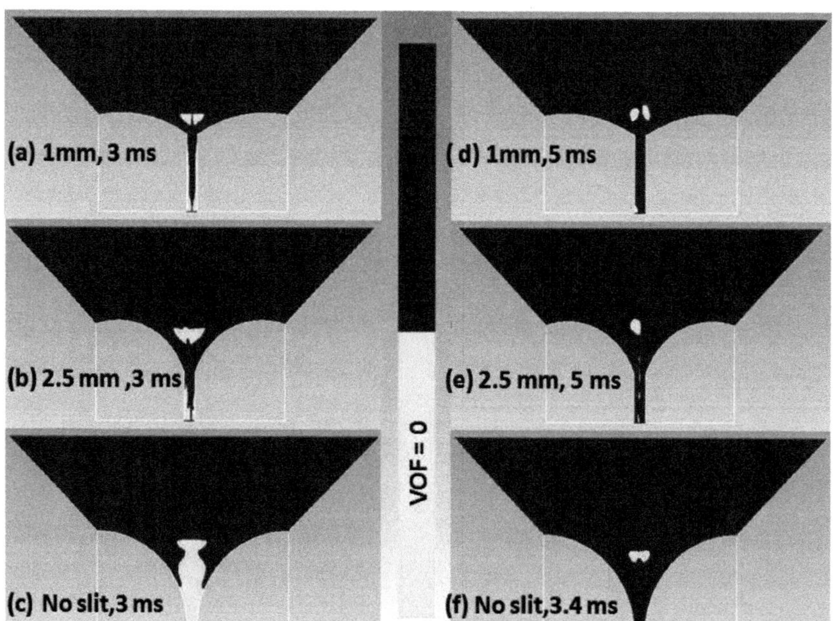

FIGURE 15.12 Melt flow for the different chamfering radius of slit wall and at different times of flow.

15.3.4 COMBINATION OF NO SLIT PASSAGE AND CURVED NOZZLE WALL

Figure 15.13 shows the initial flow pattern in a nozzle with curved walls and no-slit passage condition. At 3 ms, the melt reaches the bottom of the crucible. Due to the inverted meniscus and geometry of the slit edge wall, the flow becomes unstable (Figure 15.13(a)) and haphazard. By 4 ms the melt slides along the slit walls. The melt continues to flow along the walls and by the end of 6 ms, the flow exits the nozzle as fully developed. The air is completely pushed out by the melt flow and hence no air bubble remains in the melt. This design of the nozzle is complex to manufacture. To avoid the initial zig-zag flow at 3 ms, a nozzle with straight walls and 5 mm chamfering is suggested.

Figure 15.14 shows the path lines at 3, 4, 5 and 6 ms for a nozzle with straight walls and 5 mm chamfering. The corresponding VOF figure at 3 milliseconds is shown in Figure 15.12(c) and (f). There is no haphazard flow at 3 ms and melt is guided by the slit walls. By 3.4 milliseconds the flow is fully developed at the slit exit. This design is most suitable and is less complex than the curved nozzle wall with 5 mm chamfering. The flow time in the nozzle is reduced significantly from 11 milliseconds to 3.4 milliseconds.

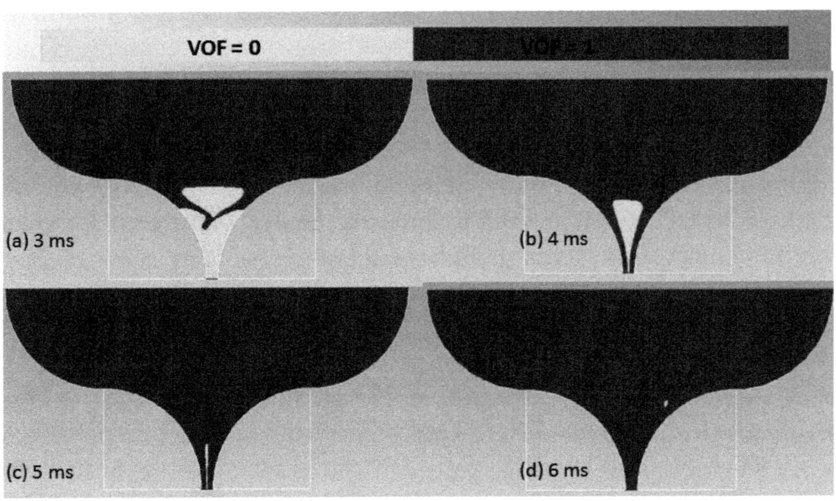

FIGURE 15.13 Melt flow in curved nozzle wall with no-slit passage with curved nozzle walls.

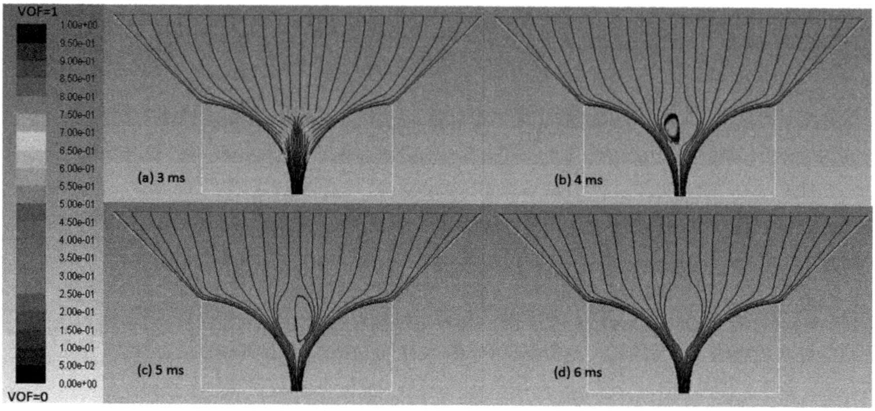

FIGURE 15.14 Pathlines at (a) 3 ms (b) 4 ms (c) 5 ms and (d) 6 ms for 5 mm chamfering radius (no slit passage) and straight nozzle walls.

15.3.5 TEMPERATURE VARIATION IN THE SLIT WALL AND PASSAGE

Melt flow through the slit must be with uniform velocity during the process. However, any heat absorption by the crucible material leads to the non-uniform supply of melt into the puddle and also choking of the slit passage. To observe the heat dissipation from the melt into the crucible walls, temperature contours are recorded from the simulations. Figure 15.15(a) shows the temperature contours in the crucible and the slit in a rectangular slit passage and Figure 15.15(b) shows that of no-slit passage nozzle. As the boundaries AP and DQ are at no heat flux condition, no heat transfer is shown from the melt to the crucible inner walls. Slit walls ABFEA and CDHGC are solid domains of quartz with thermal conductivity k = 1.38 W/mK. Even though the thermal conductivity for quartz is very low the slit walls show the temperature gradient in the solid domain.

It is interesting to observe from the temperature contours that there exists a change in temperature even in the melt near to the slit walls. Flow velocity may be affected if the difference in temperature is high. To know the difference in temperature a reference line is taken at half the distance of the slit length. Figure 15.15(a') and 15(b') shows the position of the reference line taken in both the cases. 'MN' represents the slit passage and 'LM' and 'NO' represent the reference line across the slit wall. Figure 15.17 (a) and 17(a') shows the variation of temperature along the reference line MN

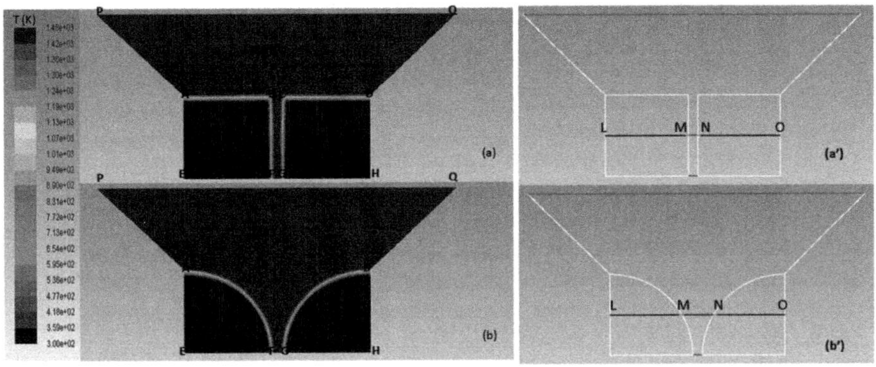

FIGURE 15.15 Temperature contours in the crucible with (a) Straight edge (b) No-slit passage at 8 milliseconds, reference line in (a') straight edge nozzle (b') no slit passage nozzle.

shown in Figure 15.15(a') and 15(b') for rectangular slit passage and no slit passage nozzle respectively. The X-axis shows the reference line and the Y-axis shows the change in temperature (K) along the length of the reference line. The initial temperature of the melt in both the cases is 1473 K. It is interesting to observe that the initial temperature of the melt is slightly reduced in the case of a rectangular slit passage. It is to be noted that the reference line is at 2.5 mm below the slit entrance 'MN.' A significant temperature gradient is also observed within the melt across the slit width of 0.6 mm. Dissipation of heat into the slit wall is less (Figure 15.17(b)) due to the low thermal conductivity of the quartz crucible. Heat is dissipated to a distance of 0.5 mm in the slit wall. The gradient near the wall is much sharp showing a sudden drop in temperature. This is due to the thermal resistance offered by the quartz crucible material. In the no slit passage nozzle, the temperature in the slit passage (Figure 15.16 (a')) is at its initial value at the reference line 'MN.' This is due to the bigger slit passage slowly converging to 0.6 mm at the slit exit FG (Figure 15.15(b)). The influence of wall heat transfer is less in this case as the distance from the wall is more. This clearly indicates that there exists a temperature reduction along the slit passage although it is less in a rectangular slit. Figure 15.16(b') shows the temperature profile along the reference line 'NO' shown in Figure 15.15(b') for the curved slit wall. The profile is similar to the case of a straight slit edge. As the crucible material is same the profile shows a similar gradient in both the cases. The difference in the temperature values is due to the change in the melt temperature in the slit passage. As the melt temperature in the no-slit passage nozzle is higher, the heat dissipation in the nozzle wall is also deeper when compared to that in the straight slit passage nozzle. This infers that to maintain uniform temperature till the slit exit, nozzle with curved slit wall (no slit passage) is preferred.

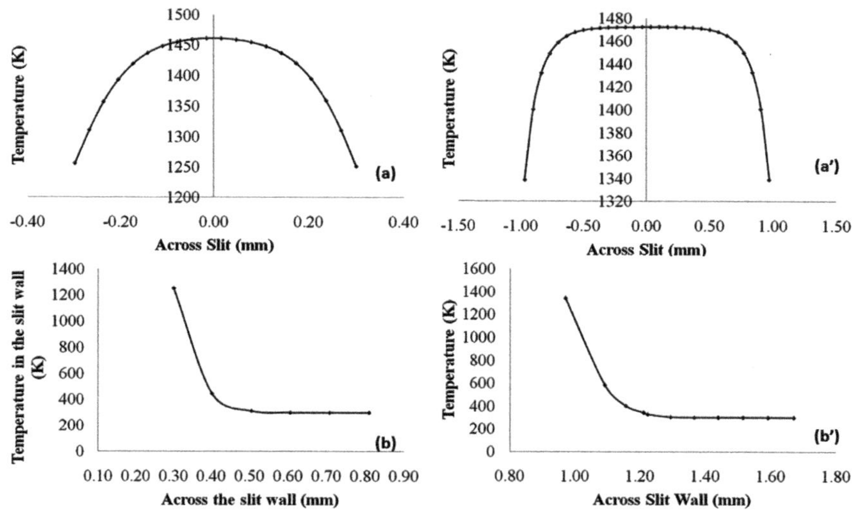

FIGURE 15.16 (a) Temperature plot across the reference line MN for Straight Edge (b) no slit passage nozzle, (a') Temperature plot across reference line 'NO' for Straight Edge and (b') no slit passage nozzle at 8 milliseconds.

15.3.6 PRESSURE DISTRIBUTION IN THE CRUCIBLE

Pressure change/distribution in the nozzle is important for the melt to exit from the constricted slit passage. Applied pressure is the driving force for the melt to flow in the crucible, nozzle, and the slit passage. Hence pressure plots are deduced from the simulations. Figures 15.17 (a) and 15.17(b) shows the pressure along the center line of the slit domain for the straight slit and no slit passage nozzle respectively. It is interesting to observe that the melt pressure in the straight slit passage drops to negative near to the slit entry and increases to zero at the exit. The zero pressure at the exit shows the difference in pressure of the melt and atmosphere at the nozzle exit. This zero difference in pressure indicates that the applied inlet pressure is to be increased to push the melt out of the straight slit exit. In case of no slit passage nozzle, there exist positive pressures throughout the slit length and the difference in pressure at the exit is very near to the applied pressure. Loss of pressure or pressure drop is very negligible in case of the no slit nozzle design.

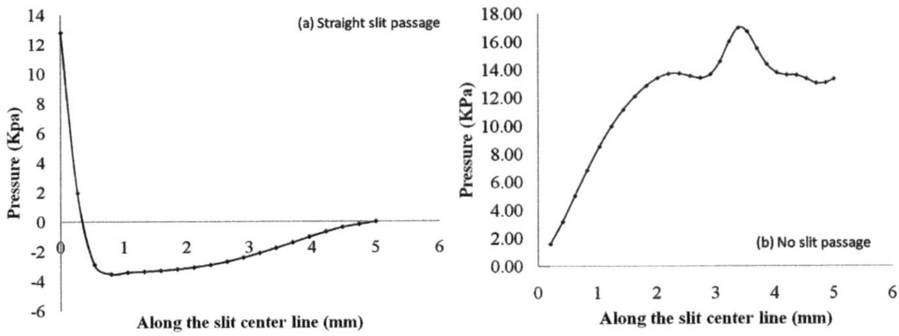

FIGURE 15.17 Pressure along the center line of the slit (a) straight passage (b) Curved/ no slit passage nozzle.

During polymer processing, the flow patterns [17] in the reservoir before the flow attains stability include: secondary flows, re-circulations etc. These patterns show the difference for low-density polyethylene (LDPE) and high-density polyethylene (HDPE) melts. The present model can be used for simulation of flow in an extruder die to study the effect of sudden contraction of the reservoir with a large diameter to the tube or orifice like passage exit of smaller diameter.

15.4 CONCLUSIONS

A two-dimensional transient flow analysis is carried out to simulate the initial melt flow through the nozzle during the planar flow melt spinning process. The inferences drawn are as follows:

1. Flow through the straight slit passage requires more time to turn to fully developed at the nozzle exit.
2. Curved nozzle wall with a straight slit passage has no significant effect on the flow time.
3. Straight nozzle walls with chamfered slit walls reduce the time required by the flow to convert to fully developed, at the exit.

4. Increase in chamfering reduces the time for the formation of fully developed flow.
5. Increase in chamfering increases the size of the air bubble in the melt.
6. Curved nozzle walls with no-slit condition improve the uniform flow conditions at the exit but the flow is haphazard in the passage.
7. Drop in the melt temperature in the nozzle with the no-slit passage is negligible and is significant across rectangular slit passage.
8. The heat dissipation in the nozzle wall is deeper in the nozzle with no slit passage when compared to that in the straight slit passage nozzle.
9. Rectangular slit passage shows negative pressures indicating more ejection pressure required to push the melt through the slit exit.
10. Pressure drop is negligible for the melt in the nozzle with no slit passage or 5 mm chamfered slit edge at its exit.
11. Nozzle with 5 mm chamfering of slit edge with straight incline wall is preferred.
12. The model developed can be used for any material- nozzle design combinations to predict the possible flow through the nozzle exit during the planar flow melt spinning process.

ACKNOWLEDGMENT

The present work is an extension of the research carried out by the author at Defense Metallurgical Research Laboratory (DMRL), Hyderabad. The author thanks the organization for the support during the research (2014).

KEYWORDS

- **crucible flow**
- **crucible-nozzle**
- **nozzle-slit flow**
- **planar-flow melt-spinning**

REFERENCES

1. Byrne, C. J., (2007). Planar flow melts spinning: Process stability and microstructural control, PhD Dissertation, Cornell University.
2. Theisen, E. A., (2007). Transient behavior of the planar-flow melt spinning process with capillary dynamics, PhD Dissertation, Cornell University.
3. Toshio, H., & Shinsuke, S., (2013). Single roll method for foil casting of the aluminum alloy. *J. Materl. Pro. Tech., 137*, 86–91.
4. Thomas, J. P., Jim, S. J. C., & Ampere, A. T., (1994). An experimental study of process behavior in planar flow melt spinning, *Metall. Mater. Trans. B., 26B,* 1199–1208.
5. Majumdar, B., Akhtar, D., & Chandrasekaran, V., (2007). Planar flow melt spinning of soft magnetic amorphous ribbons, *Trans. IIM, 60*(2), 343–347.
6. Carpenter, J. K., & Steen, P. H., (1992). Planar flow spin casting of molten metals: Process behavior, *J. Mater. Sci., 27*, 215–225.
7. Sowjanya, M., (2015). Numerical analysis of planar flow melt spinning process: Issues of puddle stability, heat transfer and ribbon topography of the amorphous ribbons, PhD Dissertation, Jawaharlal Nehru Technological University Hyderabad.
8. Majumdar, B., Sowjanya, M., Srinivas, M., Babu, D. A., Kishen, T., & Reddy, K., (2012). Issues on puddle formation during rapid solidification of Fe–Si–B–Nb–Cu alloy using planar flow melt spinning process, *Trans. Indian Inst. Met., 65*(6), 841–847.
9. Sowjanya, M., & Kishen, K. R. T., (2017). Flow dynamics in the melt puddle during the planar flow melt spinning process, *Materials Today: Proceedings, 4*, 3728–3735.
10. Sowjanya, M., & Kishen, K. R. T., (2017). Obtaining stable puddle and thinner ribbons during the planar flow melt spinning process, *Materials Today: Proceedings, 4*, 890–897.
11. Sowjanya, M., Kishen, K. R. T., Srivathsa, B., & Majumdar, B., (2014). Simulation of initial ribbon formation during the planar flow melt spinning process. *Applied Mechanics and Materials 446–447*, 352–355.
12. Sowjanya, M., & Kishen, K. R. T., (2014). *Wavy Ribbon Formation During Planar Flow Melt Spinning Process – A 3D CFD Analysis* (10th edn.). Int. Conf. HEFAT, Orland, Florida, 1789–1793.
13. Sowjanya, M., & Kishen, K. R. T., (2014). Cooling wheel features and amorphous ribbon formation during planar flow melt spinning process, *J. Mater. Proc. Tech., 214*, 1861–1870.
14. Chen, C. W., & Hwang, W. S., (1995). A three-dimensional fluid flow model for puddle formation in the single-roll rapid solidification process, *App. Math. Modeling, 19*, 704–712.
15. Chiriac, H., Tomut, M., Naum, C., Necula, F., & Nagacevschi, V., (1997). On the measurement of surface tension for liquid FeSiB glass-forming alloys by sessile drop method, *Mat. Sci. Engg. A., 226–228*, 341–343.
16. Liu, H., Chen, W., Qiu, S., & Liu, G., (2009). Numerical simulation of the initial development of fluid flow and heat transfer in planar flow casting process, *Metall. Trans. B., 40B,* 411–429.
17. Evan, M., (2010). Modeling and simulation in polymers–computational polymer processing, John Wiley & Sons, 127–193.

CHAPTER 16

CORROSION PROTECTIVE METHYL SUBSTITUTED POLYANILINE COATINGS ON OXIDIZABLE METALS

VANDANA P. SHINDE and PRADIP PATIL

Department of Physics, North Maharashtra University, Jalgaon – 425001, Maharashtra, India

ABSTRACT

Poly(o-toluidine) (POT) coatings were electrodeposited as corrosion protective coatings on low carbon steel (LCS), 304 stainless steel (SS) and copper (Cu) substrates. POT coatings on metal substrates have been carried out under cyclic voltammetric conditions in an aqueous sodium salicylate solution. The role of substrates during electrochemical polymerization also studied. Then, POT coatings were characterized by UV-visible (UV) and Fourier transforms infrared (FTIR) spectroscopic techniques. Corrosion resistance performances of these coatings were studied in NaCl solution by potentiodynamic polarization measurement and electrochemical impedance spectroscopy (EIS). The results reveal that POT coating on metals prevents corrosion, reduces the corrosion rate.

16.1 INTRODUCTION

Metals and alloys are mechanically very important materials due to high strength and ductility in the industry as well as in marine applications. But, these materials attains thermodynamically stable states. Therefore, corrosion of metals and alloys resulted to a passive thin layer of oxides/chlorides on its top surface, which changes overall its properties. Despite of strongly adherent and chemically stable such thin oxide layer; in specific aggressive environments, especially those containing chlorides, this layer also degrades, which boost initiation and propagation of localized corrosion.

Low carbon steel (LCS), 304 stainless steel (SS) and copper (Cu) are widely used materials in technological and industrial applications, among the huge family of metals and alloys. LCS presents a limited resistance to corrosion; however, seawater and acid solutions are the most aggressive agents for this metal. On the other hand, SS, and Cu has a good resistance to corrosion but is attacked by oxidant acids and salts of heavy metals [1, 2].

Simple and most adopted way, for prevention of corrosion of metals and alloys commercial organic coatings, means paints are preferred. Nevertheless, life and durability of these paints are not good since coatings are electrically inactive and porous [2]. After, the discovery of conducting polymers (CPs) [3] Mengoli et al. [4] and Deberry et al. [5] proved CPs can be used for corrosion protection among wide applications. Generally, polyaniline (PANI), polypyrrole, its substituted derivatives and their composites widely used for corrosion protection of oxidizable metals/ alloy substrates such as copper, aluminum, zinc, iron, mild steel and stainless steel etc. and has been investigated extensively, in last four decades due to its interesting corrosion mechanisms [6, 7].

A number of researchers are studying the actual corrosion mechanism of CPs against different environments. In actual cases, corrosion mechanism offered by CPs on oxidizable metals and alloys varies accordingly synthesis process parameters, experimental conditions, type of corrosive environments, quality of coatings etc. On the beginning of corrosion, oxidation of the substrate changes the oxidation states, i.e., redox conversion of these CPs which forces the doping agent to be released at the affected area. Therefore, self-healing mechanism is possible due to its storing and transportability of charges. Further, local pH in the vicinity of CP coated oxidizable substrate is lowered and diffusion of corrosive anions, as well as the dissolution of metal substrates, are prevented. In additionally, CPs are preferring for corrosion protection because non-toxicity, relatively low costs, and ease of production on oxidizable metals using electrochemical techniques [1, 6–8].

Electrochemical polymerization (ECP) is widely accepted technique for deposition of CPs coatings on oxidizable metals and alloys since it provides in-situ complete growth study of this polymerization if reaction or process controlled through potentiostat and connected with software. Although ECP technique is more informative but metallic substrates dissolute in presence of aqueous electrolyte before oxidation of monomers, hence choosing and controlling appropriate process parameters successful ECP is possible.

CPs acts as a physical barrier between the metal and the electrolytic medium. However, when this CPs has polar functional groups with several heteroatoms (i.e., nitrogen, oxygen, sulfur) and/or p-type bonding of the aromatic ring as well as of the –C=N–double bonds in their structure, they can act as macromolecular inhibitors, shifting the Tafel curves of the coated substrate towards the second quadrant such that the corrosion rate is reduced [6, 8]. These polymers can also induce the formation of a homogeneous oxide layer between the metal and the polymer layer [1, 8, 9]. Among all CPs PANI is considered the most extensively studied polymer, as its redox states maintain a passive film on the oxidizable substrates [8] and facilitate the anodic protection for substrates such as iron and its alloys due to its electro-active. But PANI coating would eventually fail due to porous morphology, brittleness nature and appearance of large defects. Hence substituted derivatives of PANI have been used for corrosion protection on oxidizable substrates. Since substituent groups play important role in case of adhesion properties to metallic substrate increases due to its substituent groups.

Many groups [10–22] reported Tafel curves of both uncoated and PANI-coated metal, immersed in 3.5% NaCl. The polarization curve for PANI-coated steel showed a smaller corrosion current density and a more noble potential than that observed with uncoated steel. On the other hand, Santos et al. [23] studied the corrosion protection for different types of steel coated with PANI in chloride ion-containing solutions. They found that the corrosion potential shifted to more positive values when the steel was coated with the polymer (0.1 V for carbon steel and 0.297 V for stainless steel). Additionally, the inhibition efficiency for the corrosion processes, as evaluated by weight loss, was almost 100% for both types of steel used. Brusic et al. [24] have reported that thin films of PANI, chemically synthesized (emeraldine) and deposited on copper by spin-drying, were able to reduce the corrosion current density in water saturated with air from 10.0 to 0.3 mA cm^{-2}, resulting in a lower corrosion rate, relative to the uncoated metal. These results demonstrated that PANI in the emeraldine oxidation state is able to protect metals such as copper against corrosion. Widera et al. [11] studied the influence of anions on the electrodeposition, properties, and overoxidation of poly(o-anisidine) by means of an electrochemical quartz microbalance in the presence of HCl and $HClO_4$ acids. The effect of anions during polymerization was explained in terms of the interactions of anions with neutral or oxidized monomer and the trapping of hydrated anions in the polymer films. Also, the degradation of the polymer films was considered, taking into account the presence of water molecules in the polymer films. Scanning electron microscopy (SEM) was used to detect the changes in polymer morphology caused by degradation. We also reported the synthesis and corrosion studies of poly(o-toluidine)coatings on low carbon steel, copper in an aqueous medium [25–27]. Very recently, Sinem Ortaboy et al. showed conductive PANI films were deposited on mild steel by an electropolymerization technique in the presence of different types of phosphonium-based ionic liquids, including tetrabutylphosphonium bromide, tetraoctylphosphonium bromide, and methyl tributyl phosphonium diethyl phosphate. Further resulting coatings had considerable corrosion protection efficiency in an aggressive medium of 3.5% NaCl solution [28]. Nautiyal et al. studied the comparative study of electrodeposition of PANI on carbon steel from oxalic and salicylate medium and results are shows that in salicylate medium passivation peak or curve is not observed [29].

The work reported in this paper, we have made an attempt to synthesize strongly adherent POT coatings on LCS, 304 SS and Cu substrates by electrochemical polymerization from aqueous salicylate solution and examined the ability of these coatings to serve as corrosion protective coatings on steel. The objectives of the present study are to synthesize and characterize electrochemically deposited strongly adherent POT coatings on LCS, 304 SS, and Cu substrates. Further to identify the role of the substrate during ECP of POT on LCS, 304 SS, and Cu. To show that sodium salicylate is suitable supporting electrolyte for ECP of the o-toluidine monomer on LCS, 304 SS, and Cu substrates. And again to examine the possibility of utilizing the POT coatings for corrosion inhibition of LCS, 304 SS and Cu substrates in aqueous 3 wt.% NaCl by electrochemical techniques.

16.2 EXPERIMENTAL

16.2.1 SYNTHESIS

16.2.1.1 CHEMICALS

All analytical grade chemicals are used for experimental work. The monomer *o-toluidine* was purchased from S.D. Fine-Chem Ltd. (India) and doubly distilled prior to being used for the synthesis. Sodium salicylate ($C_7H_5O_3Na$) was obtained from Himedia laboratories, (India) and used as received without further purification. Sodium chloride (NaCl) was procured from Qualigens Fine Chemicals, (India) for corrosion studies. Double distilled water was used to prepare all the solutions.

16.2.1.2 SUBSTRATES

The chemical composition (by weight%) of the LCS used in this study was: 0.03% C, 0.026% S, 0.01% P, 0.002% Si, 0.04% Ni, 0.002% Mo, 0.16% Mn, 0.093% Cu and 99.64% Fe.

The chemical composition (wt.%) of the 304 SS used in this study was: 0.066 C, 0.040 S, 0.045 P, 0.030 N, 0.590 Si, 10.040 Ni, 0.320 Mo, 2.490 Mn, 0.320 Cu, 19.010 Cr and 67.049 Fe. The Cu (99.98% purity) substrates used in this study. All the substrates (size ~ 10 × 15 mm^2 and 0.5 mm thick) were polished with a series of emery papers of different grit sizes (400, 600, 800 and 1200), followed by thorough rinsing in acetone and double distilled water and dried in air. Prior to any experiment, the substrates were treated as described and freshly used with no further storage.

16.2.1.3 ELECTROCHEMICAL SYNTHESIS

The POT-sodium salicylate coatings were synthesized by the electrochemical polymerization of *o-toluidine* on LCS, 304 SS and Cu substrates from aqueous sodium salicylate solution using cyclic voltammetry. The aqueous solution containing 0.1M sodium salicylate and 0.1M *o-toluidine* was used as the electrolyte for the synthesis of the coating. The cyclic voltammetric conditions were maintained using a SI 1280B Solartron Electrochemical Measurement System (UK) controlled by corrosion software (CorrWare, Electrochemistry/Corrosion Software, Scribner Associates Inc. supplied by Solartron, UK) [30]. The synthesis was carried out by cycling continuously the electrode potential between −1.0 and 1.8 V vs. SCE at a potential scan rate of 0.02 V/s. After deposition, the working electrode was removed from the electrolyte and rinsed with double distilled water and dried in air.

16.2.1.4 CHARACTERIZATIONS

The FTIR transmission spectra of POT coatings on LCS, Cu, and 304 SS substrates were recorded in the spectral range 4000–400 cm^{-1} in HATR (Horizontal Attenuated Total Reflectance) mode with a Perkin Elmer spectrometer (1600 Series II, U.S.A.). The UV–vis. absorption spectra of POT coatings on LCS, 304 SS and Cu substrates were recorded ex-situ in DMSO solution in the wavelength range 300–1100 nm with a microprocessor controlled double beam UV-vis. spectrophotometer (Model U 2000, Hitachi, Japan). The sample was placed in a quartz cell having the inner dimensions of 10 mm in width (optical path), 10 mm in depth and 45 mm in height. The adhesion of the of POT coatings on LCS, 304 SS and Cu substrates were determined by the standard sellotape test (TESA–4204 BDF) (19 mm wide) which consists of cutting the coating into small squares (~1–2 mm^2), sticking the tape and then stripping it. The percentage of adherence was calculated by taking the ratio of the number of the remaining adherent coating squares to the total number of the squares. The thickness of the coatings was measured by a conventional magnetic induction based microprocessor controlled coating thickness gauge (Minitest 600, Electro Physik, Germany). The error in the thickness measurements was less than 5%.

16.2.1.5 CORROSION TESTS

Corrosion tests were carried out in an aqueous 3 wt.% NaCl solution at room temperature for POT coated LCS, 304 SS and Cu using potentiodynamic polarization technique and EIS. For these tests, a Teflon holder was used to encase the polymer coated metal substrates so as to leave an area of ~ 40 mm^2 exposed to the solution. All the measurements were repeated several times and good reproducibility of the results was observed.

The potentiodynamic polarization measurements were performed by sweeping the potential between – 0.25 V and 0.25 V from OCP at the scan rate of 0.002 V/s. Before polarization the substrates were immersed in the solution near about 15–20 minutes and the OCP was monitored until a constant value was reached. The potentiodynamic polarization curves were analyzed by using Corr-view software from Scribner Associates [30]. This software performed the Tafel fitting and calculated the values of the corrosion potential (E_{corr}), corrosion current density (j_{corr}) and corrosion rate (CR) in mm per year.

The EIS measurements of the bare LCS, 304 SS and Cu and POT coated LCS, 304 SS and Cu were carried out at the OCP in an aqueous 3 wt.% NaCl solution. The frequency was varied from 0.1 Hz to 20 kHz using an ac excitation potential of 0.01 V. The analysis of impedance spectra was done by fitting the experimental results to equivalent circuits using Z-view software from Scribner Associates [31]. The quality of fitting to the equivalent circuit was judged firstly by the chi-square value (χ^2, i.e., the sum of the square of the differences between theoretical and experimental points) and secondly by limiting the relative error in the value of each element in the equivalent circuit to 5%.

16.3 RESULTS AND DISCUSSION

16.3.1 ELECTROCHEMICAL POLYMERIZATION OF O-TOLUIDINE ON LCS, 304 SS, AND CU FROM AN AQUEOUS SALICYLATE MEDIUM

16.3.1.1 ELECTROCHEMICAL POLARIZATION OF LCS, 304 SS AND CUIN AN AQUEOUS SALICYLATE MEDIUM

The LCS, 304 SS and Cu electrodes were first polarized in 0.1 M aqueous sodium salicylate solution (without a monomer) by cycling continuously the electrode potential between –1.0 and 1.8 V versus SCE at a potential scan rate of 0.02 V/sec to understand the different processes occurring at the electrode surface. The cyclic voltammogram of the first scan recorded during the polarization of the LCS, 304 SS and Cu electrode in 0.1 M aqueous sodium salicylate solution is shown in Figure 16.1. In the first positive cycle, a small broad and common anodic peak A_1 at ~ 0.98 V versus SCE for SS and ~ 1.1 V for LCS and Cu substrates are observed indicating oxidation of sodium salicylate. The onset of a sharp peak at 1.2167 V versus SCE is observed which is attributed to the starting oxidation of the salicylate electrolyte and a sharp increase in current density resulted into a peak B indicating oxygen evolution process is observed at 1.799 V versus SCE. During the negative cycle, the anodic current decays very sharply and negligibly small current density is observed till the end of the negative cycle and negative cycle terminated by a small reduction peak in between ~ 0.4 and –0.97 V indicating hydrogen evolution process.

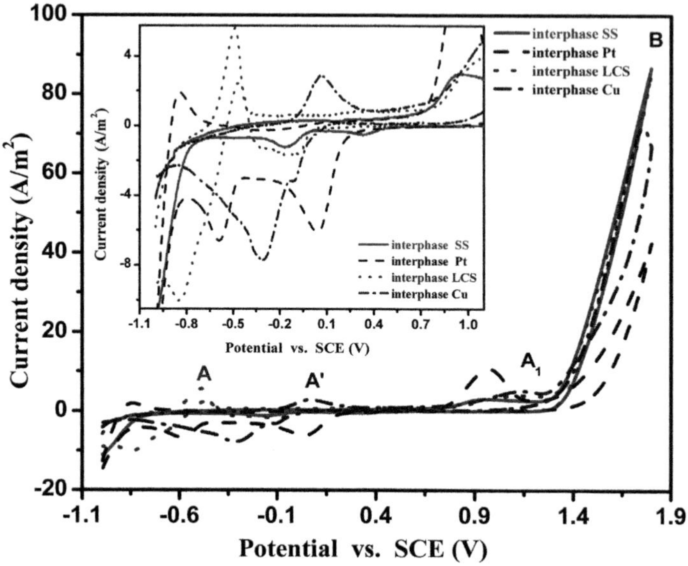

FIGURE 16.1 The first scan of cyclic voltammograms of without monomer in salicylate solution on Pt, LCS, SS, and Cu insect view shows a magnified portion of the figure.

But for LCS and Cu a significant peak between –0.8 V to 0.4 V is observed during positive scan indicating dissolution of substrates. Actually, for LCS a peak *A* ~ –0.72 V and for Cu a peak A' ~ 0.1 V is observed suggesting dissolution of the metal electrodes surface which produces metal cations in its vicinity. These ions interact with the salicylate counter-ions of the electrolyte to form a thin insoluble passive layer at the electrode surface, corresponding to complexation of the metal cations by salicylate counter-ions which inhibits the dissolution of the electrode surface. As a result, just after the peak A and A,' the current density decreases and attains a negligibly small value.

While in the case of SS substrate very small broad peaks is observed in between –0.8 V to 0.5 V and it is found to be increased in scan numbers. On repetitive cycling, the voltammograms identical to that of the first scan are obtained. But current densities corresponding to these peaks are observed to decreases for Cu electrode and increases for LCS and SS, suggesting the dissolution and simultaneous passivation of SS and LCS substrates are continuously observed up to the tenth cycle. While in the case of Cu substrate dissolution is significantly observed during the first cycle only.

To understand the behavior of polarization of oxidizable metals we also polarized the platinum (Pt) electrode under the identical experimental conditions and corresponding cyclic voltammogram of the first cycle is presented in Figure 16.1 (b) for comparison. When the platinum electrode was used, the anodic peak A_1 is observed well defined at 0.971 V versus SCE again indicating oxidation of sodium salicylate. Further, a peak indication oxygen evolution process is attributed at 1.799 V versus SCE. But current density corresponding to this peak B which is observed in case of polarization of Pt electrode is near about half value as compared to the current density of peak B which observed when the polarization of SS electrode. This may be due to the noble nature of Pt electrode as compared to LCS, 304 SS, and Cu electrodes. The repetitive cycling provides identical results. But current densities corresponding to peaks are observed to increases.

Overall magnified view of the first scan of cyclic voltammograms recorded during polarization of LCS, 304 SS, Cu, and platinum (Pt) electrode in 0.1 M sodium salicylate solution is shown in the inset of Figure 16.1. One more difference in these cyclic voltammograms is a broad peak like appearance in between –0.8 to 0.5 V versus SCE is observed during polarization LCS, 304 SS, and Cu electrode. While during polarization of Pt electrode there is no any broad peak is found. This study indicates that on LCS, 304 SS and Cu electrodes metal oxide, metal hydroxide and/or metal salicylate complex passive interphase is deposited. Since, according to Pourbaix diagrams of SS, LCS, and Cu substrates, this potential region and pH of electrolytic bath ~ 6.3 are within the corrosion domain. While on Pt electrode no any such passive layer is deposited. As stated earlier [25,32–33], the formation of passive interphase is always formed on oxidizable metal/alloy electrodes like iron, low carbon steel, copper etc. when they polarized in an aqueous media. Deposition of uniform, quantity/thickness and a strongly adherent passive oxide layer on oxidizable metals and alloys obviously depends on the type of supporting electrolyte, the concentration of supporting electrolyte, pH of the solution and potential range.

16.3.1.2 ELECTROCHEMICAL SYNTHESIS OF POT ON LCS, 304 SS, AND CU

The cyclic voltammogram of the first scan recorded during the synthesis of POT coating on the LCS, 304 SS, Cu, and Pt electrode from an aqueous solution containing 0.1 M *o-toluidine* and 0.1 M sodium salicylate is shown in Figure 16.2.

The first positive cycle is characterized by – (i) an onset of oxidation wave after ~0.4 V followed by oxidation peak (A_2) between 0.9 V to 1.3 V and beyond this potential anodic current decreases. The oxygen evolution process is observed at 1.799 V versus SCE (peak (b). During the reverse cycle, the anodic current density decreases rapidly and a negligibly small current is observed till – 0.077 V. The negative cycle terminates with a reduction peak C between –0.1 V to –0.9 V. The oxidation peak $A_2$is attributed to the oxidation of *o-toluidine,* since a black, uniform film is generated on the steel substrate. The reduction peak C observed during the negative cycle is attributed to the reduction of an anodically formed species. During the second scan, an anodic peak A_2 is again observed and the rest of the features are similar to that of the first scan. On repetitive cycling, the voltammograms identical to that of the second scan are obtained up to the fourth scan. However, the current density corresponding to the anodic peaks decreases gradually and shifted towards at high potential range with the number of scans. After the fourth/fifth scan, the cyclic voltammogram does not show well-defined redox peaks. After tenth scan LCS, 304 SS and Cu electrode are observed completely covered by a uniform, smooth, and black colored POT coating. The thickness of POT coating on LCS, 304 SS, and Cu was measured by using a conventional magnetic induction based microprocessor controlled coating thickness gauge after twenty-five scans are ~ 20μm, 21μm and 24 μm respectively.

The cyclic voltammogram of the first scan recorded during the electrochemical polymerization of o-toluidine on Pt electrode under the identical experimental conditions is shown in Figure 16.2 for comparison. During first scan a small peak A_3 at 0.1230 V versus SCE indicating initiation of the polymerization process and a well-defined peak A_2 at 0.9650 suggesting actual polymerization processes. The position of peak A_2during ECP of o-toluidine on LCS, SS, and Cu follows the sequence as: Pt< Cu < SS < LCS. Since in galvanic series also these oxidizable metal/alloys have a relevant position. During reverse scan a peak C at –0.7680 suggesting a reduction of the coating. On repetitive number, during the electrochemical synthesis of POT coating on Pt electrode, the current density corresponding to anodic peak A_3 is found to be increases and peak A_2 is found to be decreased as the number of scan increases. And oxygen evolution process is found to be lowered, as compared to the polarization of Pt electrode in an aqueous sodium salicylate solution Overall study indicates that on LCS, 304 SS and Cu electrodes slightly, nonuniform passive layer is formed may be due to dissolution of metal electrodes and then this passive layer inhibits further dissolution and then ECP of POT coating is occurs. Therefore, one more thing oxidation of monomer during ECP on LCS, 304 SS and Cu are observed later as compared to ECP on a platinum electrode. Delay of the polymerization process on LCS, 304 SS and Cu as compared to Pt electrode is due to the reactive nature of LCS, 304 SS, and Cu electrodes.

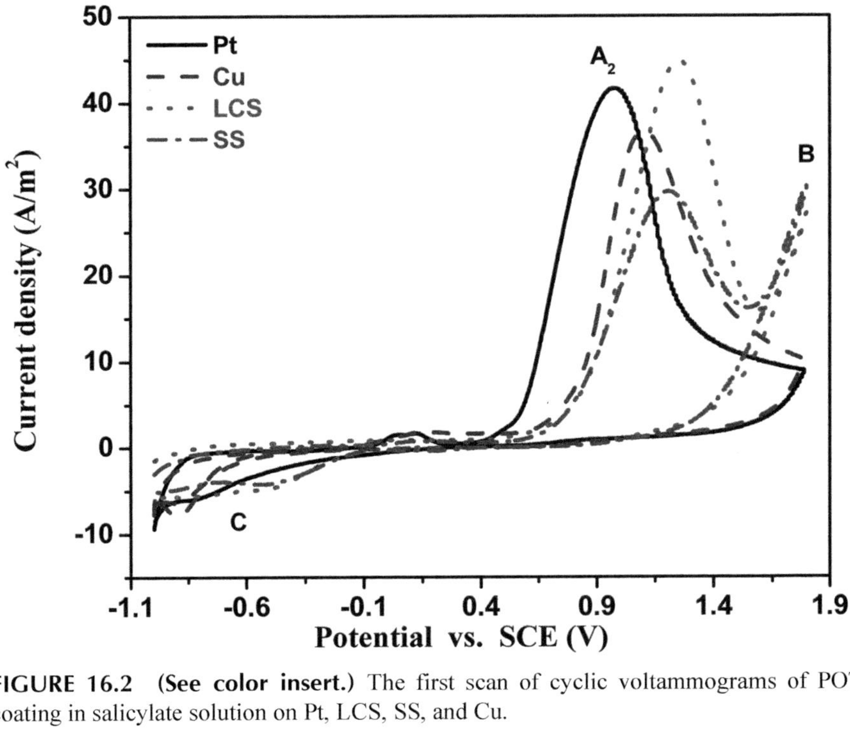

FIGURE 16.2 (See color insert.) The first scan of cyclic voltammograms of POT coating in salicylate solution on Pt, LCS, SS, and Cu.

Patil et al. [33] also revealed that the polymerization process postponed due to the formation of the formation of the passive layer, consisting of copper oxide, zinc oxide, and salicylate complex. This passive layer reduces the further dissolution of brass without preventing the electro-chemical polymerization of o-toluidine and allows the deposition of the shiny and smooth POT film.

The coating thickness was also estimated by using the equation:

$$d = \frac{QM}{2F\rho} \qquad (16.1)$$

where Q is the specific overall charge for the electrochemical polymerization, ρ is the density of the POT (1.34 g/cm^3) and M is the molar mass (107 g/mol) and F is the Faraday constant (96,500 (c). It is interesting to observe that the thickness calculated by using this equation is in fairly agreement with those measured by using a conventional magnetic induction based microprocessor controlled coating thickness gauge.

The FTIR spectrum of POT coatings synthesized on LCS, 304 SS, and Cu electrodes under cyclic voltammetric conditions (10 cycles) recorded in HATR mode is shown in Figure 16.3. These spectra exhibit the following spectral features [25–27] – a broadband at ~ 3375 cm^{-1} due to the N-H stretching vibration, the bands at ~ 3045 and 2929 cm^{-1} due to C–H stretching in methylene group, bands at 1589 and 1481 cm^{-2} due C = N and C = C stretching respectively for the quinoid (Q) and benzoid (B) rings, indicating the presence of imine and amine units in POT coating. The relative intensity (1589/1481) of these bands is greater than unity. Therefore, the coating composed of a mixed phase of the pernigraniline base (PB) along with emarldine salt (ES) is observed. Moreover, the presence of weak bands at ~ 1700 and 1237 cm^{-1} is due to the carboxylic group supports the formation of the oxidized form of POT. Further, the C-N bending band at ~ 1378 cm^{-1}, the bands at ~ 1107, 1016 and 805 cm^{-1} are due to the 1–4 substitution on the benzene ring, the bands around ~ 853 cm^{-1} and between 800–700 cm^{-1} reveals the occurrence of the 1–2 and 1–3 substitutions and the strong band at ~754 cm^{-1} indicates that *ortho*-substituted benzene ring. Overall, FTIR spectra of POT coatings on metal substrates indicates C-H band is very weak and N-H band is shifted towards higher wavelength in case of POT coating on Cu. Further O-H stretching band is also present in POT coating on Cu. FTIR spectra of POT coatings on SS and LCS are closely resembled.

The optical absorption spectrum of POT coating on LCS and 304 SS (Figure 16.4) shows a well-defined peak at ~500 nm and a shoulder at ~ 700 nm. The peak at 500 nm is attributed to the formation of the pernigraniline base (PB), which is the fully oxidized form of POT. The shoulder at 700 nm is the signature of the formation of the emeraldine salt (ES) form of the POT. The ES is the only electrically conducting form of POT. The simultaneous appearance of the 500 nm peak and the shoulder at 700 nm reveals the formation of a mixed phase of PB and ES forms of POT. While POT coating on Cu shows an only a single peak at ~540 nm indicating only the PB form of POT [34]. This may be attributed due to hydroxide content in POT coating on Cu substrate is found to be more as compared to POT coatings on LCS and SS which may be responsible for the formation of PB phase on Cu substrate rather than a mixed phase.

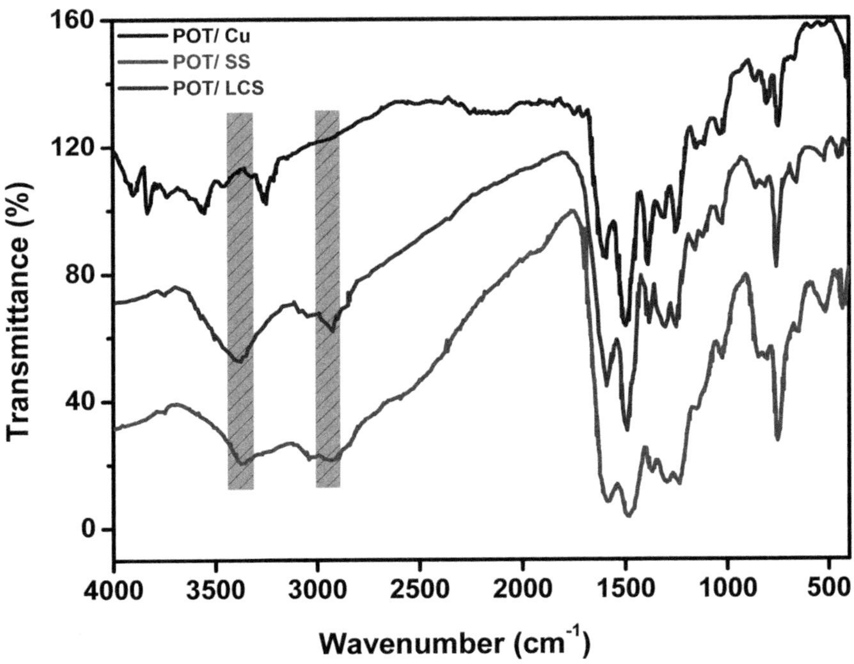

FIGURE 16.3 FTIR spectra of the POT (salicylate) coatings on LCS, SS, and Cu.

SEM image of the POT (10 cycles) deposited on LCS, SS, and Cu is shown in Figure 16.5. It clearly reveals that the coating is uniform, crack-free and featureless. These studies indicate that in an aqueous salicylate medium electrochemical polymerization of *o-toluidine* on the LCS, 304 SS and Cu substrates results in the formation of POT.

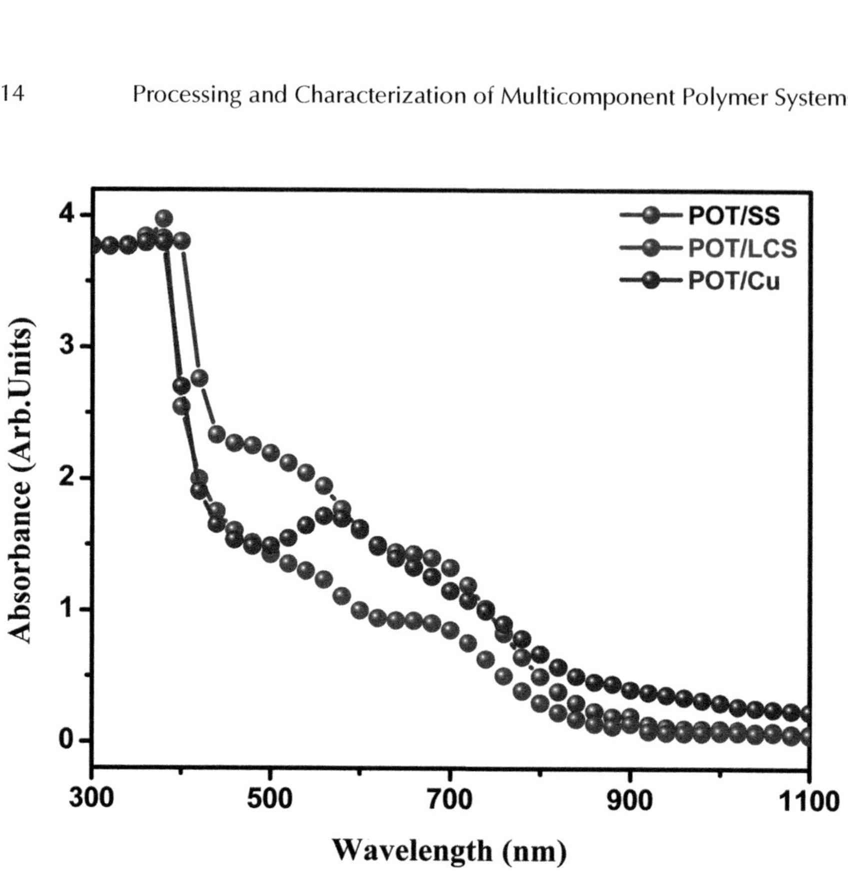

FIGURE 16.4 (See color insert.) UV-Vis. spectra of the POT (salicylate) coatings on LCS, SS, and Cu.

16.3.2 CORROSION PROTECTION PERFORMANCE OF THE POT COATING

16.3.2.1 THE POTENTIODYNAMIC POLARIZATION STUDIES

The potentiodynamic polarization curves for bare metal (LCS, SS, and Cu) and POT coated on metal (25 scans) in aqueous 3% NaCl are shown in Figure 16.6 (a-c), respectively.

The values of the corrosion potential (E_{corr}), corrosion current density (j_{corr}), anodic, and cathodic slopes or Tafel constants (β_a and β_c), polarization resistance (R_{pol}) and corrosion rate were calculated using software [30]. Many researchers explained the procedure to achieve the Tafel parameters and using them for corrosion studies of organic coatings. Potentiodynamic polarization parameters such as E_{corr}, j_{corr}, β_a, and β_c as

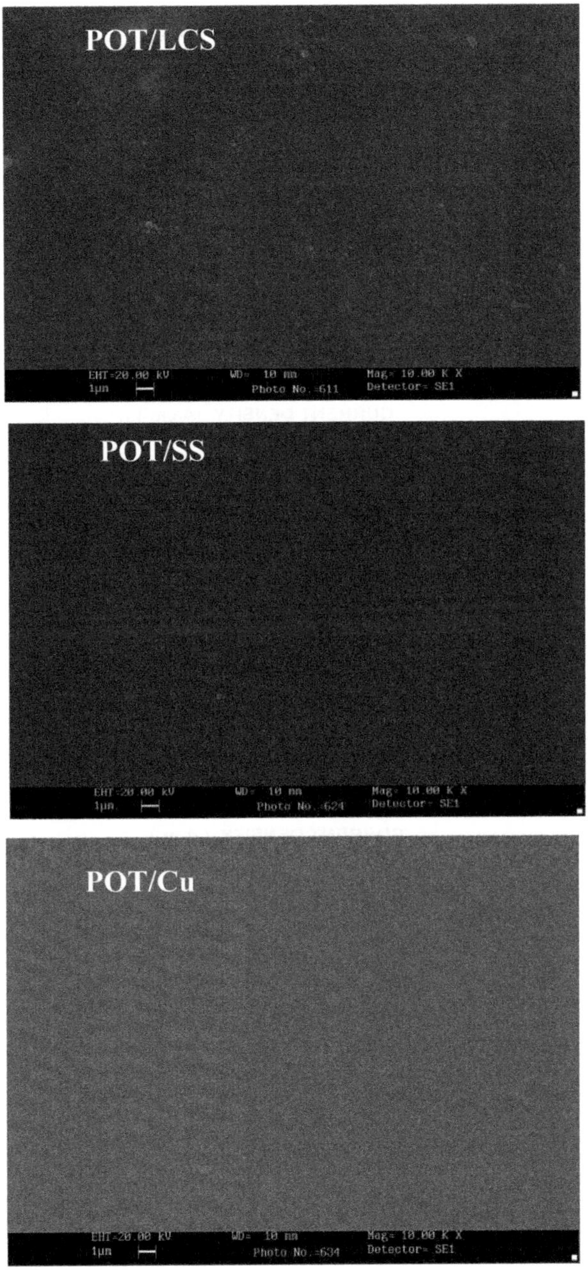

FIGURE 16.5 SEM images of the POT coatings synthesized in salicylate on metal substrates by cyclic voltammetry.

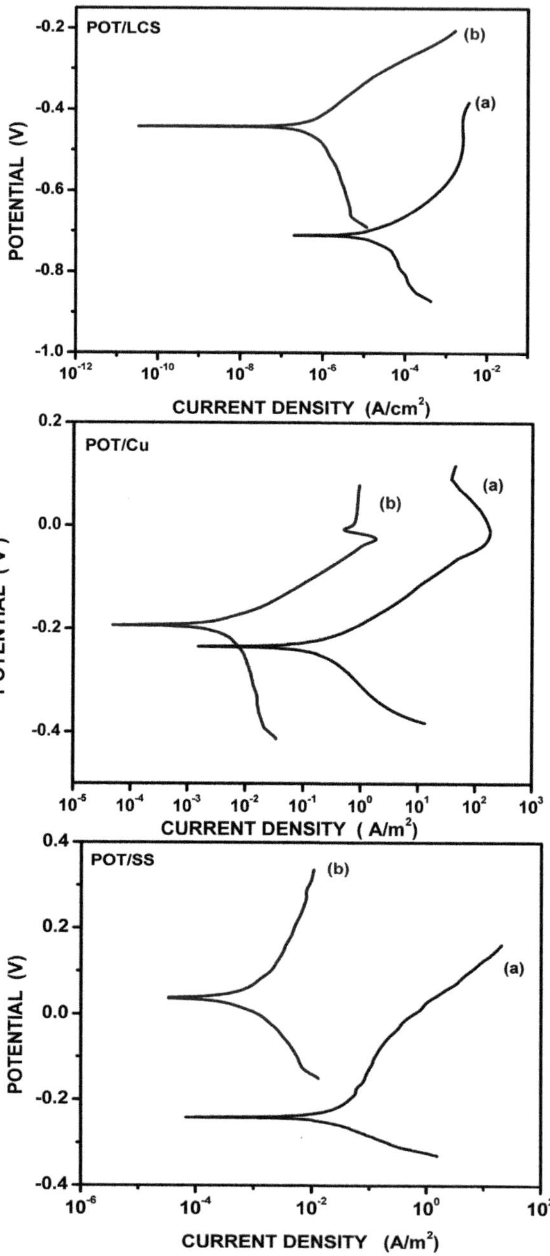

FIGURE 16.6 Potentiodynamic polarization curves recorded for (a) bare metal and (b) POT coated metal substrate in aqueous 3% NaCl solution.

well as the percentage of protection efficiency (*%P.E.*) are obtained and all electrochemical parameters associated with potentiodynamic polarization measurements were calculated and listed in Table 16.1.

It is clearly observed that the j_{corr} decreases and E_{corr} increases for POT coated metal than bare metal. The positive shift in E_{corr} indicates the protection of the metal surface by the POT coating. The corrosion rate (*CR)* is significantly reduced as a result of the reduction in the j_{corr}.

The potentiodynamic polarization curve for POT coated LCS, 304 SS and Cu are shown in Figure 16.6, clearly observed that the j_{corr} decreases from ~ 3.1 × 10^{-6} A/cm^2, 3.2 × 10^{-6} A/cm^2 and 2.46 ×10^{-6} A/cm^2 for bare LCS, 304 SS and Cu to ~ 2.02 × 10^{-8} A/m^2, 7.09 × 10^{-8} A/m^{2}and 0.31 × 10^{-8} A/m^2(*cf.* Table 16.1) for POT coated LCS, 304 SS and Cu respectively.

The E_{corr} for POT coated steel increases from –0.710 V, –0.260 V and –0.234 V (E_{corr} for bare LCS, 304 SS, and Cu) to –0.380 V, 0.055 V and –0.180 V vs. SCE, respectively. The positive shift of 0.33 V. 0.315 V and 0.055 V vs. SCE in E_{corr} and significant reduction of 3.08 × 10^{-6} A/m^2, 3.13 × 10^{-6}A/m^2 and 2.46 × 10^{-6} A/m^2 in j_{corr} indicates the reduction in corrosion rate of the LCS, SS, and Cu surface by the POT coating on LCS, SS, and Cu, respectively.

The CR of POT coated LCS, 304 SS and Cu are found to be ~ 0.02, 0.0008 and 0.003 mm/year which is ~ 18, 45 and 93 times lower than that observed for bare metal. Polarization resistance estimated for bare metal and for POT coated metal for which is also compliment to potentiodynamic polarization result. These results reveal that the POT acts as a protective layer on metal and improves the overall corrosion performance than bare metal.

The SEM images of POT coatings after corrosion testing seen that no apparent change in the surface morphology of the coating occurred after the potentiodynamic polarization measurements shown in Figure 16.7(a–c). Indeed, the visual observation revealed no cracks or defects produced in the coating and the coating remains strongly adherent to the metal substrate. Thus, POT coating has strong adherence to the metal substrate and it is resistant to the corrosion in aqueous 3% NaCl.

The corrosion resistance of the coating mainly depended on the porosity of the coating. The presence of porosity in the coatings can rapidly lead to the electrochemical dissolution of the substrate. The characterization of porosity in a quantitative manner is of great interest to researchers

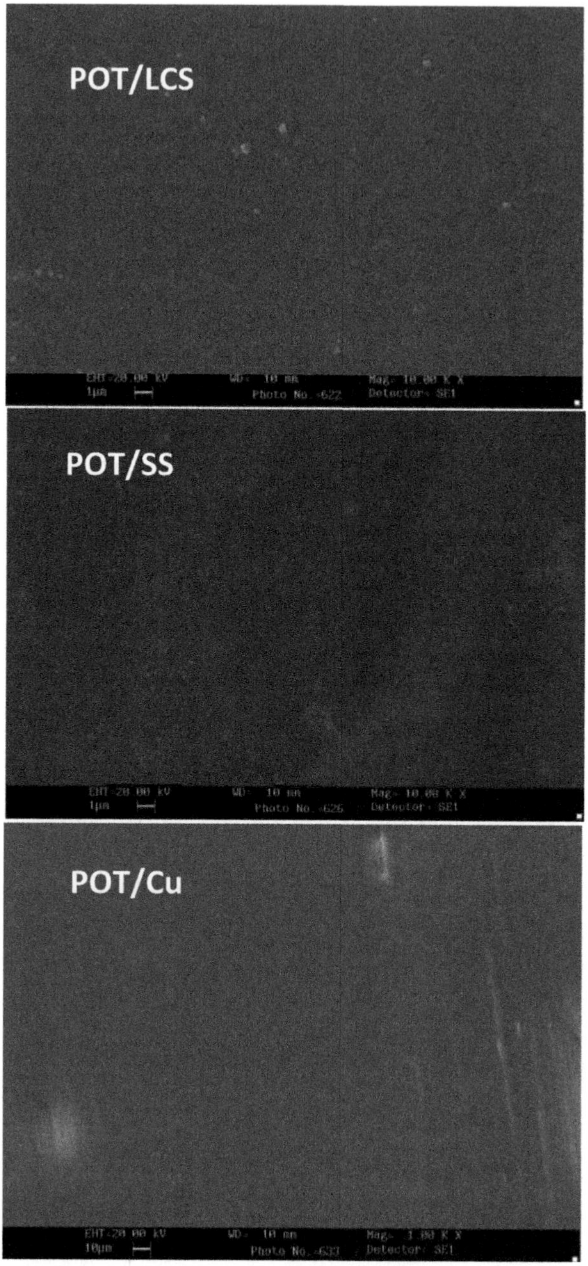

FIGURE 16.7 SEM images of the POT coatings after potentiodynamic polarization study synthesized in salicylate on metal substrates by cyclic voltammetry.

but critical task. Recently electrochemical methods are used for porosity measurement. Due to the electrochemical active nature of POT coating porosity of these coating is unclear but its knowledge gives an idea about its electrochemical nature.

The porosity in POT coating on metal substrates was calculated using the relation [35] given in Table 16.1.

$$\%P = \frac{R_{Pb}}{R_{Pc}}\, 10^{-\left|\frac{\Delta E_{corr}}{\beta_{as}}\right|} \times 100 \qquad (16.2)$$

where P is the total percentage porosity, E_{corr} is the difference between corrosion potentials, R_{pb}, and R_{pc} denotes the polarization resistance for bare and POT coated metal respectively and $_{has}$ the decadic anodic Tafel slope for the bare metal substrate. Percentage porosity of POT coated metal substrates is given in Table 16.1. The lower values of porosity of the POT coating indicating the greater its ability to resist diffusion process. Lowest porosity is observed for POT coated SS. The lower value of porosity suggests that chemical inertness of POT coating due to an increase of the corrosion resistance by hindering the access of the electrolyte to the metal substrates, since as mentioned earlier POT coating composed of mixed phases of conducting ES and insulating PB. The results are fairly agreements with this. Percentage protection efficiency (PE) was calculated by using the expression:

$$\%PE = \left(\frac{j_{corrb} - j_{corrc}}{j_{corrc}}\right) \times 100 \qquad (16.3)$$

where, j_{corrb}, and j_{corrc} denote the current densities for bare and POT coated metal substrate respectively.

16.3.2.2 EIS STUDIES

The corrosion behavior of bare metal and POT coated metal was also investigated by EIS. Figure 16.8–16.11 (a and (b) shows the Nyquist complex plane plots recorded in aqueous 3% NaCl, for bare metals and POT coated metals, respectively. These plots were analyzed using equivalent electrical circuit models shown in Figure 16.9 (a and (b), where R_s represents the electrolyte resistance, CPE_c the constant phase element

TABLE 16.1 Potentiodynamic Polarization Results for POT Coatings on Metal Substrates

Sample	Q (C/cm^2)	Thickness (μm)	β_a (V/decade)	β_c (V/decade)	E_{corr} (V)	j_{corr} (A/cm^2)	R_p (Ω-cm^2)	CR (mmpy)	P Porosity
Bare LCS	NA	NA	0.085	0.186	–0.710	3.1×10^{-6}	825	0.350	NA
POT/LCS (25 scans)	0.049	20	0.058	0.20	–0.380	2.02×10^{-6}	9664	0.020	1.13×10^{-5}
Bare SS	NA	NA	0.265	–0.046	–0.260	3.20×10^{-6}	2717	0.040	NA
POT/SS (25 scans)	0.052	21	0.101	–0.167	0.055	7.09×10^{-8}	27550	0.0008	0.65×10^{-3}
Bare Cu	NA	NA	0.072	0.11	–0.234	2.46×10^{-8}	779	0.280	NA
POT/Cu (25 scans)	0.060	24	0.049	0.31	–0.180	0.31×10^{-8}	56913	0.003	6.40×10^{-4}

referred to as coating capacitance, C_c also connected parallel with R_{po} the coating pore resistance, CPE_{dl} the constant phase element that represents all the frequency dependent electrochemical phenomena, namely double layer capacitance, C_{dl}, and diffusion processes, and R_{ct} the charge transfer resistance connected with corrosion processes in the bottom of the pores. The constant phase element, *CPE*, is introduced in the circuit instead of a pure capacitor to give a more accurate fit and compensate nonhomogeneity of coating [32].

The choice of equivalent circuit is decided from χ^2 value and sum of square values. In addition modeling of the equivalent circuit based on intuition as what kind of impedance is expected to be present in the sample, examination of experimental data to see whether the response is consistent with proposed circuit and inspection of circuit parameters values that are obtained to check that they are realistic and their dependence with each other.

The Nyquist impedance plot of uncoated LCS recorded in 3% NaCl is shown in Figure 16.8(a) and is modeled by an electrical equivalent circuit model depicted insect of Figure 16.8(a), where the impedance plot of the bare LCS can be fitted with a semicircle, which is attributed to the uniform type corrosion processes occurring at the LCS surface, means system undergoing dissolution with the precipitation of a corrosion film at the electrode surface. Bode phase and amplitude diagram of bare LCS is shown in Figure 16.8 (c and d), it clearly shows the presence of only one time constant. The single time constant predicts the occurrence of a single charge transfer in the process. Bode phase diagram also shows a rapid increase in the phase angle and phase angle maxima smaller than −90° (approximately −63°); at ~ 6.3 Hz. After the decrease in phase value is observed. Bode amplitude diagram shows the linear part in between frequency range 0.5–159 Hz and slope in this range is approximately –0.6.

The Nyquist impedance plot of bare steel (Figure 16.10 (a)) indicates an almost arc-type nature. Initially, the Nyquist plot shows a linear nature, but in the latter, the behavior bends towards the real axis, suggesting incomplete half-depressed circle. The electrical equivalent circuit depicted in the inset of Figure 16.10 (a) issued to describe the impedance response of the bare 304 SS.

The Bode amplitude diagram for bare SS shows (Figure 16.10 (c)), the plateau region observed at the highest frequency region is related to the R_s and the straight line with a slope approaching approximately −0.87. The Bode phase plot for bare SS shown in Figure 16.10 (d), a depressed capacitor-like behavior with one time constant is observed. This plot also show phase angle maxima smaller than −90° (approximately −81°); initially, a rapid increase in the phase angle was observed and it attained the value of approximately −81° at ~4 Hz. Thereafter, it remained almost constant at approximately −80° from the frequency of ~20 Hz up to the lower frequency of 1.5 Hz. This constant region corresponds to the linear part of the Nyquist impedance plot. Then again, it is found to slightly decrease towards a lower frequency limit of 0.1 Hz. As this value is slightly lower than that observed for an ideal capacitor, it is attributed to the contribution of some resistive component (like a diffuse layer). The phase angle deviation can be interpreted as a deviation of the double layer from ideal capacitance behavior. Such behavior is termed as frequency dispersion and considered to be due to local in homogeneities in the coating, porosity, mass transport, and relaxation effects [36, 37].

Actually, 304 SS is nobler that LCS in electrochemical series but according to Pourbaix diagram of SS and LCS, it shows an active region for dissolution for SS is more negative than LCS. However, in the case of SS dissolution iron and chromium cations are produced and react with hydroxide anions forming iron hydroxide and chromium hydroxide. While in the case of LCS ferric oxide is produced and which are unstable as compared to metallic hydroxides.

The Nyquist plot of the bare Cu display a depressed broad semicircle at the high frequency and continues up to middle-frequency range, followed by an emerging straight line at low frequency. Such behavior suggests that merging of two-time constants representing components of the corrosive thin layer and the Cu metal Figure 16.11 (a). The high-frequency semicircle may be caused by the charge transfer process, which is related to the relaxation time constant of the charge-transfer resistance (R_{ct}) and the double layer capacitance (C_{dl}) at the copper/electrolyte interface. The high-frequency loop is not perfect semicircles, which can be attributed to the frequency dispersion as a result of the roughness and inhomogeneity of the electrode surface [36]. Further, the emergence of low-frequency diffusion tail is observed which may bend towards the real axis at low frequencies, indicating both a.c. and d.c. diffusion layer thickness is comparable. Resulting starting of the skewed semi-circle and the diffusion is expressed as open Warburg element (shown in Figure 16.9 (c)) which suggest that diffusion layer has finite dimensions and one boundary imposes a fixed concentration for the diffusing species as in the case of oxygen conducting electrodes and corrosion-related diffusion [36, 37]. This indicates that which is attributed to the anodic diffusion process of $CuCl_2$ from the surface of the electrode to the bulk solution and/or to the cathodic diffusion process of dissolved oxygen from the bulk solution to the surface of the electrode. Overall, active type of corrosion is observed in the vicinity of Cu substrate. Basically, in the case of Cu initially cuprous chloride (CuCl) then conversation of it's into cupric chloride ($CuCl_2$) is produced, this corrosive product is porous. The Bode phase diagram (shown in Figure 16.11 (c)) for bare Cu shows broad phase maxima at ~ 63° which is constant in between frequency range 251 to 784 Hz. Later phase decreases gradually from ~52° to 25° during frequency change from 50 Hz to 0.4 Hz respectively and finally attains ~23° at lower frequency limit [38]. Bode amplitude diagram (shown in Figure 16.11 (d)) is not showing exactly straight line it shows the curved in-out type curve, suggesting a merging of two semicircles, hence influx breakpoint frequency points observed at ~ 0.5 Hz (–26°) and ~ 50 Hz (–52°). Further slopes for linear nature for this plot are found to be ~ – 0.27 and – 0.55 in lower and higher frequency regions. Basically, the slope of linear part of Bode amplitude diagram is −1 caused by the capacitor but absolute values of the impedance modulus slope lower than unity in all three bare metal substrates; again, it is in agreement with the use of *CPE* instead of a capacitor.

The Nyquist complex plane plot of POT coated metals shown in Figure 16.8, 16.10, and 16.11(b) is also modeled by using the equivalent circuit model with two semicircles (shown in Figure 9(b)). However, the parameter values of the best fit to the impedance plot are significantly different as compared to those obtained for bare metal. In this case, the first capacitive loop is attributed to the characteristics of the POT/electrolyte interface and it is characterized by the pore resistance (R_{po}) and the coating capacitance (C_c). The second semicircle in the low-frequency region is attributed to the POT/metal interface and it is characterized by the charge transfer resistance (R_{ct}) for the charge transfer reactions occurring at the bottom of the pores in the coating and the double layer capacitance (C_{dl}). Parameters obtained from the EIS study are summarized in Table 16.2. The R_{po} value is significantly higher than that of the bare metal. This indicates that the porosity in the POT coating, as already noted, is considerably lower. The R_{ct} value is found to be higher for the POT coated metal substrates. The higher value of the R_{ct} is attributed to the effective barrier behavior of the POT coating [36]. The lower values of C_c and C_{dl} for the POT coated metal provide further support for the protection of metal by the POT coating. Thus, the higher values of R_{ct} and R_{po} and lower values of C_c and C_{dl} indicate the excellent corrosion performance of the POT coating. The PE and porosity were also calculated by using expression given in V. Shinde et al. [25], from the EIS data. These results reveal that the POT acts as a protective layer on metal and improves the overall corrosion performance.

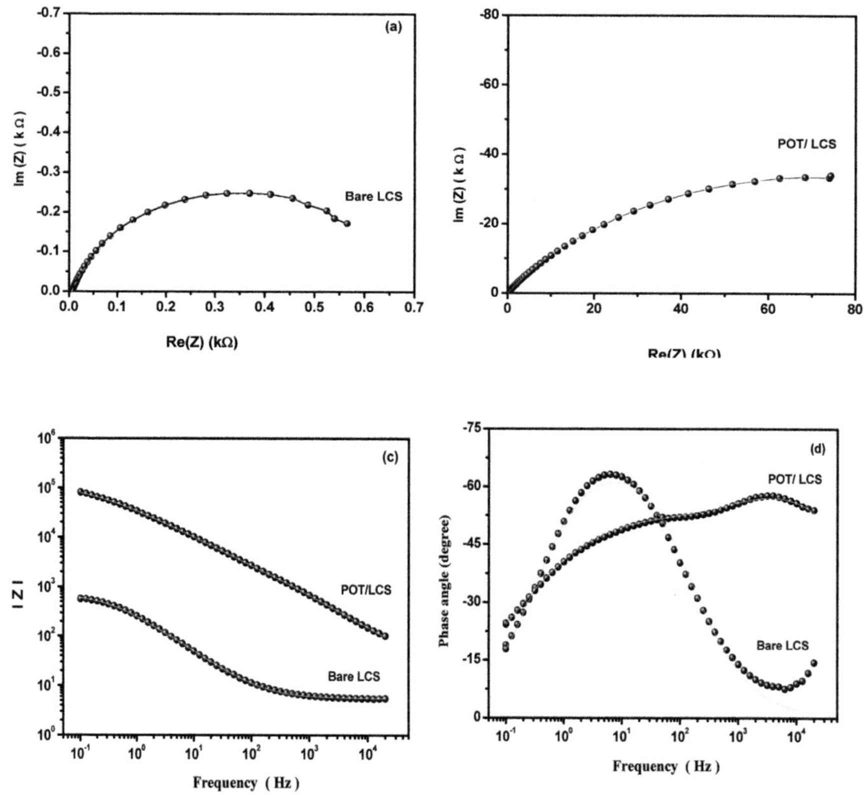

FIGURE 16.8 (a) and (b) Nyquist complex plane plots, (c) Bode amplitude plots and (d) Bode phase plots recorded in aqueous 3% NaCl solution for Bare LCS and POT coated LCS substrates; with fitted curve shown by dotted line. (Inset is equivalent electrical circuits).

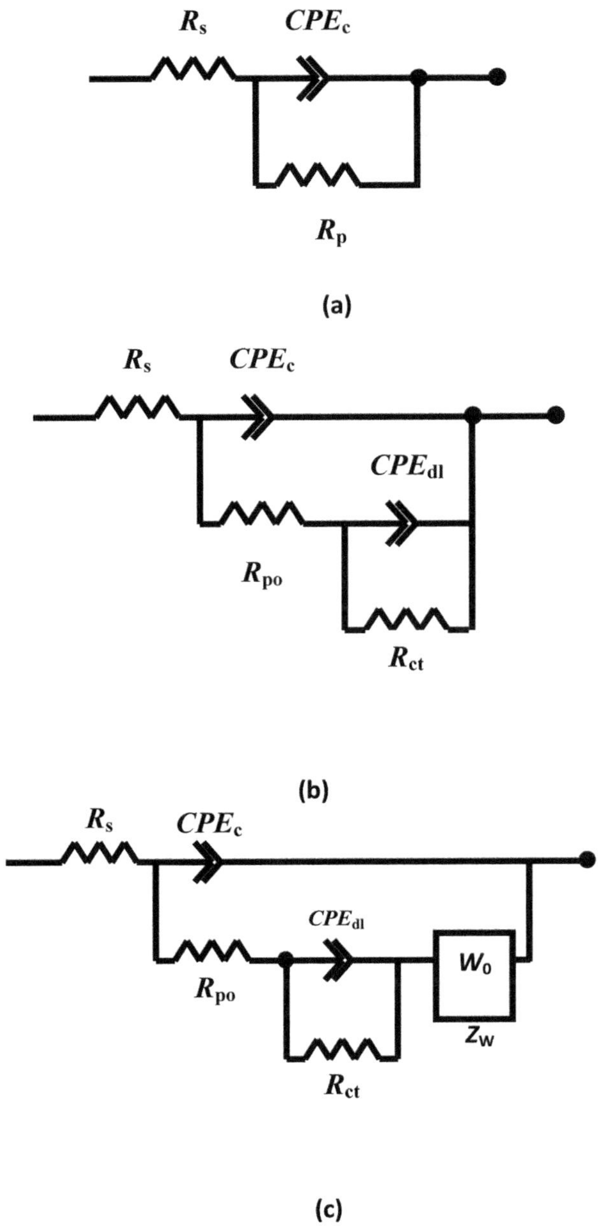

FIGURE 16.9 Equivalent circuits (a), (b) and (c) used for the numerical fitting of the EIS data.

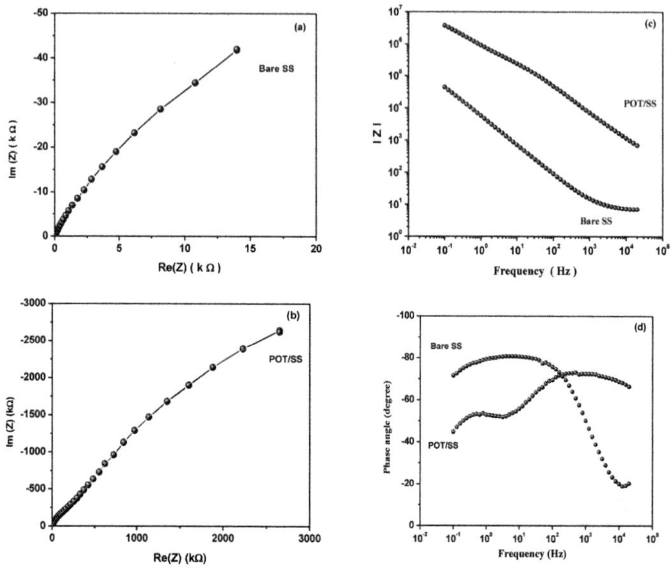

FIGURE 16.10 (a) and (b) Nyquist complex plane plots, (c) Bode amplitude plots and (d) Bode phase plots recorded in aqueous 3% NaCl solution for Bare SS and POT coated SS substrates; with fitted curve shown by dotted line. (Inset is equivalent electrical circuits).

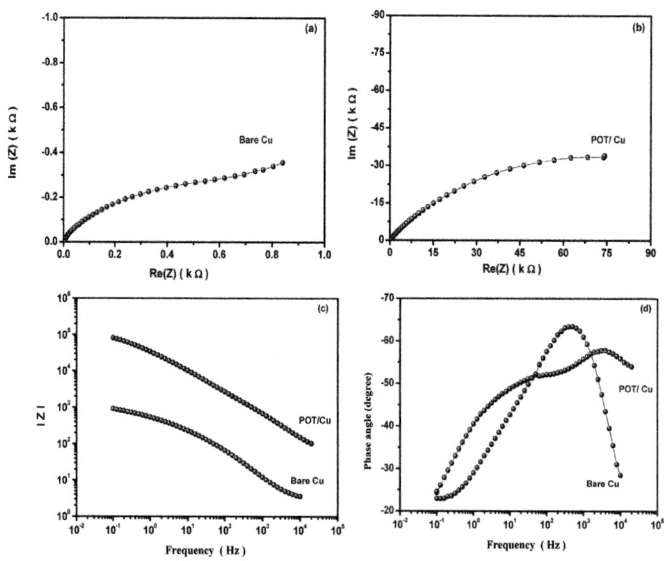

FIGURE 16.11 (a) and (b) Nyquist complex plane plots, (c) Bode amplitude plots and (d) Bode phase plots recorded in aqueous 3% NaCl solution for Bare Cu and POT coated Cu substrates; with fitted curve shown by dotted line. (Inset is equivalent electrical circuits).

TABLE 16.2 EIS Results for Bare and POT Coated Metals

Sample	R_s (Ω/cm^2)	R_{po} (Ω/cm^2)	C_c (F/cm)	R_{ct} (Ω/cm^2)	C_{dl}(F/cm)	Wo (Ω/cm^2)
Bare LCS	5.51	NA	NA	700.5	7.99×10^{-6}	NA
POT/LCS (25 scans)	38.62	0.0021	1.67×10^{-5}	70821	5.43×10^{-5}	NA
Bare SS	6.99	NA	NA	2.33×10^{5}	$3.41\ 10^{-5}$	NA
POT/SS (25 scans)	112.3	55.7×10^{4}	1.00×10^{-7}	915×10^{4}	2.32×10^{-7}	NA
Bare Cu	2.69	128.4	5.03×10^{-5}	820.9	4.75×10^{-5}	1308
POT/Cu (25 scans)	17.71	2698	2.64×10^{-6}	1.36×10^{5}	6.38×10^{-6}	NA

In the present case, POT accepts electrons from the metal and donates them to oxygen. This reaction produces the formation of a passive metal oxide layer at the polymer/metal interface and formation of metal/insulator/semiconductor structure creates an energy barrier indicating the difficulty for the electron from escaping hence a lower probability of oxidation/corrosion or pitting. And finally lowers the corrosion rate and shifts the E_{corr} to more positive values. The result of the potentiodynamic polarization measurements for the POT coated metal substrate shows a shift in the E_{corr} to more positive value and reduce the corrosion rate of the metal substrate. The positive shift of in E_{corr} indicates the protection of the metal surface by the POT coating.

As mention earlier ECP process also observed via formation passive oxide layer; which is necessary for strongly adherent coating and indirectly protection mechanism. As explained in ECP of POT coatings on LCS, SS, and Cu the cyclic voltammogram clearly shows the polymerization process started on Cu substrate as earlier than SS and finally on LCS substrate.

In the present work, the UV-vis absorption spectroscopy revealed the formation of a mixed phase of PB and ES forms of POT (cf. Figure 16.4) on LCS and SS substrates. Emeraldine base form due to redox conversion merely acted like a physical and chemical barrier and doped form emeraldine salt, displayed a remarkable corrosion protection [4, 56]. While in the case of Cu substrate POT coating composed by only PB form. But PB phase in more suitable for corrosion protection [26]. Thus POT coating serves the LCS, 304 SS and Cu against corrosion in 3 wt.% NaCl solution.

Hydrophobic nature of the alkali salicylates and its strong complexation with POT, which having π conjugation backbone; dominate the corrosive tendency of metals and hydrophilic nature of polymer in chloride medium. Overall POT coatings synthesized in salicylate electrolyte hinders the rate of diffusion of corrosive chlorine anion and water towards metal substrates and lowers local pH at defects centers further stops the dissolution of metal cations. Many researchers explained in the literature, corrosion protection is due to the formation of emeraldine salt (ES) which formed through redox conversion from mixed phases of ES along with EB/ PB and provides corrosion protection due to its charge transportability. [1, 8, 39, 40].

16.4 CONCLUSIONS

In summary, the following conclusions have been drawn from the present investigation: The electrochemical polymerization of *o-toluidine* from aqueous salicylate medium generates uniform and strongly adherent POT coatings on LCS, 304 SS and Cu metal substrates. Sodium salicylate is a suitable electrolyte for ECP of POT coatings on LCS, 304 SS, and Cu metal substrates. ECP process is observed via the formation of passive interphase. The FTIR and UV-visible spectroscopic studies clearly reveal that formation of POT coating on metal substrates. The potentiodynamic polarization result and EIS result are compliments to each other and clearly reveals that the POT acts as a corrosion protective layer on metals in 3% NaCl solution and the corrosion protection properties of POT coating on 304 SS and Cu are found to excellent as compared to POT coating on LCS. Insulating PB phase provides good corrosion protection. This study reveals that the POT coating has excellent corrosion protection properties in aqueous 3% NaCl.

ACKNOWLEDGMENTS

The financial support from University Grants Commission (UGC), New Delhi, India under SAP-DRS programme (Project No.: F.530/2/DRS/2010 (SAP-I, Phase -II) and the Board of Research in Nuclear Sciences (BRNS), Department of Atomic Energy (DAE), Mumbai, India are gratefully acknowledged.

KEYWORDS

- **copper**
- **electrochemical impedance spectroscopy**
- **emeraldine salt**
- **low carbon steel**
- **poly(o-toluidine)**
- **stainless steel**

REFERENCES

1. Fontana, M. G., (1987). *Corrosion Engineering* (3rd edn.) McGraw- Hill Book Company, New York.
2. Jones, D. A., (1996). *Principles and Prevention of Corrosion* (2nd edn., p. 223). Prentice Hall, Upper Saddle River, NJ.
3. Skotheim, T. A., (1986). *Handbook of Conducting Polymers* (Vol. I and II). Marcel Dekker Inc., New York.
4. Mengoli, G., Munari, M., Bianco, P., & Misiani, S., (1981). *J. Appl. Polym. Sci., 26*, 4247–4257.
5. Deberry, D. W., (1985). *J. Electrochem. Soc., 132,* 1022–1026.
6. Tallman, D. E., Spinks, G., Dominis, A., & Wallace, G., (2002). *J. Solid State Electrochem.,*673–684.
7. Spinks, G. M., Dominis, A. J., Wallace, G. G., & Tallman, D. E., (2002). *J. Solid State Electrochem, 685*, 100.
8. Wessling, B., (1994). *Adv. Mater., 6*, 226–228.
9. Sazou, D., (2001). *Synthetic Metals, 118*(1–3), 133–147.
10. Bernard, M., Joiret, S., Hugot-Le Goff, A., & Viet Phong, P., (2001). *J. Electrochem. Soc., 148B*, 12–16.
11. Wider, J., Skompska, M., & Jackowska, K., (2001). *Electrochim. Acta., 46,* 4125–4131.
12. Lu, J. L., Liu, N. J., Wang, X. H., Li, J., Jing, X. B., & Wang, F. S., (2003). *Synth. Met., 135–136*, 237–238.
13. Vera, R., Romero, H., & Ahumada, E., (2003). *J. Chil. Chem. Soc.,* 4835–4840.
14. Spinks, G. M., Dominis, A. J., & Wallace, G. G., (2003). *Corrosion,* 5922–5931.
15. Kinlen, P. J., Silverman, D. C., & Jeffreys, C. R., (1997). *Synth. Met., 85*, 1327–1332.
16. Camalet, J. L., Lacroix, J. C., Aeiyach, S., & Lacaze, P. C., (1998). *J. Electroanal. Chem.,* 445117–445124.
17. Fahlman, M., Jasty, S., & Epstein, A. J., (1997). *Synth. Met., 85,* 1323–1326.
18. Epstein, A. J., Smallfield, J., Guan, H., & Fahlman, M., (1999). *Synth. Met., 102,* 1374–1376.

19. Meneguzzi, A., Ferreira, C. A., Pham, M. C., Delamar, M., & Lacaze, P. C., (1999). *Electrochim. Acta., 44*, 2149–2156.
20. Gasparac, R., & Martin, C., (2001). *J. Electrochem. Soc., 148B*, 138–145.
21. Choi, S., & Park, S., (2002). *J. Electrochem. Soc., 149*, 26–34.
22. Tallman, D. E., Pae, Y., & Blerwagen, G. P., (1999). *Corrosion, 55*, 779–786.
23. Santos, J. R., Mattoso, L H., & Motheo, A. J., (1998). *Electrochem. Acta., 43*, 309–313.
24. Brusic, V., & Angelopoulus, M. G., (1997). *J. Electrochem. Soc., 144*, 436–441.
25. Shinde, V., & Patil, P. P., (2013). *J. Solid State Electrochem.*, 1729–1741.
26. Shinde, V., Sainkar, S. R., & Patil, P. P., (2005). *J. Appl. Polym. Sci., 96*, 685–695.
27. Shinde, V., Mandale, A. B., Patil, K. R., Gaikwad, A. B., & Patil, P. P., (2006). *Surf. Coat. Technol., 200*, 5094–5101.
28. Sinem, O., (2016). *J. Appl. Polym. Sci., 133*(38). doi: 10.1002/APP.43923.
29. Amit, N., & Smrutiranjan, P., (2016). *Prog. Org. Coat., 94*, 28–33.
30. *Electrochemical Corrosion Software-CorrWare and CorrView*, Scribner Associates Inc., Southern Pines, NC.
31. *Electrochemical Impedance Software-Z-Plot and Z-View*, Scribner Associates Inc., Southern Pines, NC.
32. Chaudhari, S., Sainkar, S. R., & Patil, P. P., (2007). *J. Phys. D: Appl. Phys., 40*, 520–533.
33. Patil, D., & Patil, P. P., (2008). *J. Appl. Polym. Sci., 118*, 2084–2091.
34. Feast, W. J., (1996). *Polymer, 37*(22), 5017–5047.
35. Porosity, C. J., Mazille, H., & Idrissi, H., (2000). *Surf. Coat. Technol., 130*(2–3), 224–232.
36. Walter, G. W., (1986). *Corr. Sci., 26*(9), 681–703.
37. Macdonald, J. R., (1987). *Impedance Spectroscopy*, New York, NY, John Wiley & Sons.
38. Yang, Z., Shenying, X., Lei, G., Shengtao, Z., Hao, L., Yulong, G., & Fang, G., (2015). *RSC Adv., 5*, 14804–14813.
39. Rammelt, U., Nguyen, P. T., & Plieth, W., (2003). *Electrochim. Acta., 48*(9), 1257–1262.
40. Rohwerder, M., Isik-Uppenkamp, S., & Amarnath, C. A., (2011). *Electrochim. Acta., 56*(4), 1889–1893.

INDEX

1

1,4-diazabicyclo[2,2,2]-octane (DABCO), 46
1-ethyl-3-(3-dimethyl aminopropyl) carbodiimide (EDAC), 46

2

2D
  crystal, 103
  electron gas, 105

3

3D, 61, 102, 282
  assembled gold nanostructures, 35

A

Absorption
  bands, 122, 125, 126, 198
  spectra, 133, 199, 306
Acetic acid, 47, 48, 69, 73
Acetone, 30, 33, 34, 305
Aerospace elements, 78
Agro waste/biowaste, 63
Air
  domain, 285
  exposure, 232
  melt interface, 289
Alfrey's approximation, 253, 255
Alkaline phosphatase, 61
Alkyl
  ammonium, 11, 12
  chains, 12
  phosphonium actions, 11
Alloy
  electrodes, 309
  substrates, 302
Aluminum, 182, 278, 302
Amide I, 125, 126
Amide II, 125, 126
Amino acids, 69
Aminolysis, 43, 45, 47, 50, 61
  hydrolysis, 47
Aminolyzed
  hydrolyzed films, 47
  multilayered surfaces, 44, 55, 60
Amorphous
  condensed matter, 259
  materials, 258
  phase, 8, 248
    polymers, 237
  ribbon formation, 287
  ribbons, 277, 278
  structure, 278
Amplitude diagram, 321, 323
Amyloid-derived diffusible ligands, 39, 41
Analytical
  grade, 69, 119, 304
  solution, 244
Anatase, 215
Angular frequency, 168, 244, 264, 267, 272
Anionic sodium dodecyl benzene sulphonate (SDBS), 191, 193
Anions, 302–304, 322
Anisotropic property, 68, 72
Anisotropy, 163
Annealing process, 29, 41
Antibacterial
  activity, 117
  properties, 118
Antigen, 26, 31, 32, 39, 41
Antimicrobials, 118
  photocatalytic, 118
Antistatic property, 66
Aqueous
  medium, 104, 120, 304
  salicylate medium, 313, 328
  solution, 29, 40, 133, 305

Aromatics, 66
Asymmetric stretching vibration, 198
Atomic
force microscopy (AFM), 119, 120, 128
mass, 265, 267
Autofluorescence, 52, 54
Automobile industry, 136
Automotive
components, 79
industry, 64, 65
parts, 78
Azobisisobutyronitrile (AIBN), 43, 179, 182–186, 188, 189

B

Bacteria, 117, 215
Banana
fiber, 63, 79, 81–83, 86, 91, 93, 97
jute fiber-reinforced polyester, 81
Benzoyl peroxide (BPO), 179, 182–189
Binary
layer, 169
mixture, 173
Binomial distribution, 245
Biocidal
capacities, 214
effects, 118
Biocompatibility, 4–6, 8, 13–15, 18, 19, 26, 28, 102, 111, 118
Biodegradable
nature, 68
rate, 21
Bioelastomer films, 46
Biofibers, 64
Biological
entities, 26, 35
implants, 118
labeling, 28
molecules, 28, 105
properties, 118, 121
safety cabinet, 47
Biomass material, 63
Biomaterials, 3–6, 21
Biomedical
applications, 4, 6, 8, 13, 14, 41, 102, 110, 118
device application, 6
implant applications, 6
industry, 19
material, 3, 5, 11, 19
plastics, 19
polymer, 19
Biomimetic, 6
Bio-molecules, 34, 35, 37
Biosensing applications, 25
Biosensors, 6
Biostability, 4–8, 12–14, 19, 20
Biostabilizing effects, 4
Bode
amplitude diagram, 321
phase
diagram, 321, 323
plots, 324, 326
Bonded silicon clusters, 262, 270
Bone
marrow, 44
plates, 6
prostheses, 118
Bovine
growth hormone, 31, 32, 40
serum albumin (BSA), 46, 49
somatotropin (bST), 31, 40, 42
Bradford assay, 49, 59, 61
Bragg's angle, 124
Bromodeoxyuridine, 60
Bruggeman model, 173

C

Calcium silicate, 231
Calibration curve, 31
Calomel electrode, 195
Carbon
black (CB), 132, 138
electrode, 192, 194
nano fiber (CNF), 108, 109
nanotube (CNT), 101–103, 107–109, 135, 138, 139, 145–147, 150
steel, 301–304, 309, 329
Carboxyl groups, 104, 312
Cardiovascular repair, 44
Carpet backing cloth (CBC), 68
Cartesian system, 263

Cartilage tissue engineering, 44
Catalysis, 118, 192
Catalyst, 82, 83, 131, 132, 180
Cathodic
 cetyltrimethylammonium bromide (CTAB), 191–194, 196–209
  CTAB-CuS sample, 206
 charges, 206
 slopes, 314
Cell
 assay, 59
 culture, 43, 47
 growth, 60, 61
 locomotion, 45
 membrane, 19, 48, 49
 motility, 52
Cellophane tape, 102
Cellular processes, 58, 59–61
Cellulose, 67, 68, 71, 93
Cement mortar, 213, 214, 216, 228, 233
Cetyltrimethylammonium bromide, 191, 193, 209
Chemical
 admixture, 217
 attacks, 214, 216
 barrier, 327
 bonding, 115
 bonds, 180
 exfoliation, 103
 functionalization, 111
 gas, 215
 properties, 74, 192
 reaction, 82, 240, 242
 resistance, 66, 181
 stability, 79, 102, 106
 structure, 8, 111
 terminology, 180
 testing, 64, 70
 treatment, 71, 73
 vapor deposition (CVD), 27, 103
Chemisorption, 106
Chitosan (Cts), 43, 46, 47, 116, 118, 119, 121–126, 128–133
Chloride, 33, 64, 109, 214, 216, 224, 231, 305, 328
Chloroauric acid, 29, 30, 40
Chromium
 cations, 322
 hydroxide, 322
Cluster
 distributions, 257
 relative weights, 254
 sizes, 240, 247, 250–252, 255, 256, 258
 weight distributions, 253
Clustering, 237, 240, 243, 244, 246, 254–256, 258, 259
 probabilities, 256
 process, 256
 theory, 244, 246, 254, 255
Cobalt naphthenate, 82
Coconut sheath (CS), 80
Coffee, 63, 68–71
 grounds, 74
 production, 68
Composite materials, 63, 64, 78–80, 106, 107, 110, 150, 156, 169
Compression molding process, 63, 70
Compression testing machine (CTM), 221, 225
Computational domain, 279, 280, 282
Computational fluid dynamics (CFD), 280, 287
Conducting polymer composites (CPCs), 138
Conductivity, 106, 110, 117, 138, 156, 167, 171, 193, 287, 294
Conventional industrial method, 8, 13
Cooperative/clustering segmental theory of materials dynamics (CSTMD), 258, 259
Copolymer, 4, 7–10, 12–15, 18–20, 21, 25, 132
Copolymerization, 7, 9
Copper (Cu), 120, 159, 184, 193–198, 200, 202, 204, 206, 301–314, 317, 320, 322, 323, 326–329
Corona discharge treatments, 29
Corrosion, 6, 64, 301–306, 309, 314, 317, 319, 321–324, 327, 328
 mechanism, 302
 protection, 302, 303, 327, 328
  efficiency, 304
 protective coatings, 301, 304
 rate, 301, 303, 314, 317, 327
 resistance performances, 301

tests, 306
Corr-view software, 306
Cotton, 66, 67, 77
Coulombic interaction, 206
Coupling
agent, 71
model (CM), 195, 245, 259
Covalent
bond, 262
counterparts, 270
silicon cluster, 262
Cross-linked polymers, 180
Crosslinking agent, 180
Crucible-nozzle, 277–279, 284, 291, 298, 299
Crystalline, 66, 117, 123, 124, 136, 151, 191, 196, 197, 203, 220, 248, 250, 282
Crystallization, 7, 278
Cultivation, 67, 81
Culture
conditions, 44
medium, 47
plate, 47
Cuprous chloride (CuCl), 323
CuS
electrode, 192–194, 205–209
nanoparticles, 208
nanostructure, 191, 194, 197, 198, 203, 208
stretching modes, 198
Cyclic
voltammetric conditions, 301, 305, 312
voltammetry (CV), 191, 192, 195, 205,206, 209, 305, 315, 318
voltammogram (CVs), 205, 327
Cyto Viva's hyperspectral microscope, 38
Cytotoxicity
data, 19
level, 4, 19
study, 18
test, 18

D

Damping
behavior, 149, 181, 186
properties, 149, 179, 181, 188
Decimal digit mantissa, 259
Defectfree ribbon, 278
Degenerated tissues/organs, 6
Degradation, 18, 66, 120, 130, 304
De-ionized water, 30, 119, 194
Demagnetization, 164
Density functional theory (DFT), 261, 262, 267, 270
Dental implants, 6
Deposition process, 30
Dicumyl peroxide (DCP), 182, 183
Dielectric, 26, 34, 156, 157, 167–169, 172–174, 238
constant, 26, 172, 173
data, 157, 167
function, 167
materials, 156
measurements, 155
permittivity, 167, 168, 173, 174
properties, 156, 157, 167, 169, 174
Diffraction
intensity, 198
pattern, 123, 159
Diffusion
kinetics, 207
process, 319, 321, 323
Dimethyl sulfoxide (DMSO), 46, 306
Dimethylformamide (DMF), 46, 48
Dimethylhydrazine, 104
Dirac
equation, 105
fermion, 105, 112
Discontinuous
phases, 78
spectra, 252, 257
Dispersed phase, 137, 138, 141
compatibilizer, 138
dispersing technique, 12, 15, 21
geometry, 137
morphology, 137
Dissolution, 302, 308, 311, 321, 322, 328
Distilled water, 15, 119, 195, 220, 305
Distribution transformer, 278
Divinyl benzene (DVB), 183
DNA
assay, 43, 44, 49, 55, 61
concentrations, 61

contents, 43, 49
Dodecyl sulfatepolyacrylamide gel electrophoresis (SDS-PAGE), 60
Domain inlet, 287, 288
Doping agent, 302
Double
coupled clusters method, 267
distilled water, 305
Drug delivery system, 13, 118
Dulbecco's Modified Eagle Medium (DMEM), 46, 47
Dye solutions, 120
Dynamic
equation, 244
heterogeneity, 256, 258
light scattering (DLS), 116, 119, 120, 128
mechanical, 137, 138, 140, 148, 151, 181, 184, 185, 238, 258
analysis (DMA), 140, 148, 181, 184
thermal analysis (DMTA), 184
thermal analyzer, 140, 151
physical properties, 258, 259
properties, 238

E

Effect of surfactants, 200
Effective-medium theory (EMT), 173
Elastic
energies, 243
modulus, 10, 87, 185
Elasticity, 40, 41, 67, 83, 102
Elastomer, 4, 6–8, 156, 181
Elastomeric materials, 3, 9
Electric
batteries, 106
charge, 105, 215
double layer capacitors (EDLCs), 192
vehicles, 107
Electrical
circuit models, 319
conductivity, 108, 110, 138, 193
energy storage, 107
properties, 108, 193
Electrochemical, 104, 192–195, 206, 207, 209, 301, 303, 310–313, 317, 322, 328
active nature, 319
behavior, 194
capacitors, 107, 192
characterization, 195
dissolution, 317
double layer capacitors, 107
impedance spectroscopy (EIS), 195, 207, 301, 306, 307, 319, 323–325, 327–329
EIS studies, 195, 319, 323
measurement, 192
methods, 319
nature, 319
phenomena, 321
polymerization, 304, 305
power, 107
properties, 191–193
studies, 205
synthesis, 305, 310
techniques, 302, 304
workstation, 195
Electrode, 6, 104, 107, 110, 192–195, 205–207, 209, 305, 307–311, 321–323
electrolyte contact area, 207
materials, 104, 107, 110, 192–194, 207, 209
position, 193, 303, 304
preparation, 195
surface, 307, 308
Electrolyte, 192, 193, 195, 205–207, 302, 304, 305, 307–309, 319, 322, 328
Electromagnetic field, 36, 105
Electron
absorbent, 132
hole, 116, 131
recombination process, 131
transfer, 193
Electronic
components, 105
devices, 6, 104, 107
properties, 104
Electrons, 104, 105, 121, 132, 169, 214, 215, 327
Electropolymerization technique, 304
Electrostatic
repulsions, 243
self-assembly, 45

Emeraldine, 303, 312, 327- 329
salt (ES), 312, 313, 319, 327, 328
oxidation, 303
Energy
conversion, 193
dispersive
spectroscopic (EDS), 115, 119, 126, 191, 195, 201, 202, 225–227, 231
X-ray analysis (EDXA), 225
X-ray microanalysis (EDXMA), 225
dissipation mechanism, 90
equation, 285
quantization, 242
storage, 101, 102, 107, 110, 192, 193, 209
devices, 101, 192
Epoxy
composites, 81
resins, 64
Equilibrated gas, 242
Equilibrium
distribution, 182
position, 263–265, 271
Equivalent
circuit, 307, 321, 323
electrical circuits, 324, 326
Ethanol, 40, 47, 157, 193–195
Ethanol solution, 40, 47
Ethylene, 4, 7–9, 21, 65, 109, 118
glycol, 104
vinyl acetate (EVA), 3–5, 7–16, 18–21, 23, 109
carbon monoxide system, 7
chains, 16, 18
copolymer, 7, 8, 12, 14–16, 20, 21
matrix, 12, 14, 16, 20
nanoclay nanocomposites, 10
Ethylenediaminetetraacetic acid (EDTA), 46, 49
EulerLagrange equations, 266, 268
Exfoliation, 5, 12, 14, 16, 20, 103, 112
Expanded graphite (EG), 109, 157, 194, 196
Experimental
behavior, 237, 238, 258
conditions, 302, 309, 310
data, 20, 166, 240, 248, 250, 253, 258, 262, 321
methods, 238
points, 307
results, 281, 282, 307
techniques, 259
tests, 77
time scale, 237
values, 173, 174, 248, 282
verification, 270
work, 304
Extracellular matrix proteins, 52
Extrusion methods, 63

F

Fabrication, 35, 41, 65, 82, 96, 105, 136, 179, 184, 194
Facing center cubic (fcc), 115
F-actin, 48, 50, 51, 53, 54
Faradaic reactions, 107, 206
Faraday constant, 312
Fatty acids, 69
Ferrimagnetism, 156
Ferrite
fillers, 161–163
loading, 155, 166
particles, 155, 165, 166
Fetal bovine serum (FBS), 46, 47, 60
Fiber
aspect ratio, 79, 88
breakage, 90, 94, 96
content, 65, 68, 80, 81, 89, 90
dispersion, 79
geometry, 79
loading, 65, 77–81, 86–92, 94, 96, 97, 149
matrix adhesion, 88
orientation, 79, 86
reinforced
composites, 84, 136
hybrid composites, 79, 80
polymer composites, 78, 79
Fiber-to-fiber interactions, 88
Fibril
breakage, 137
diameter, 136, 141, 143, 145
morphology, 137, 139, 141
Fibrils, 136–138, 140–146, 149, 150
Fibrinogen, 39, 41

Field theory, 172
Filler
  concentration, 166
  loading, 146, 150, 156, 163, 170, 174
  surface, 170
  system, 172
Flammability, 7
Flax, 77, 78
Flexural
  properties, 80, 84, 88, 89
  strength, 66, 78, 80, 84, 88, 89, 96
  test, 84
Flow
  pattern, 277, 278, 282, 289, 291, 293, 297
  velocities, 281
Fluctuation-dissipation theorem, 245
Fluid
  structure interaction, 277, 286
  technique, 277
Fluorescence, 43, 48, 52–55, 60, 105
  images, 50–52, 55
  microscope, 48
  microscopy, 43, 48, 50–54, 60
    images, 52, 54
Foreign Body Reaction (FBR), 6
Formation mechanism, 191
Fourier transforms infrared (FTIR), 115, 119, 120, 124–126, 155, 157, 160, 161, 191, 194, 198, 301, 306, 312, 313, 328
  FTIR analysis, 160
Fructose, 104
Fuel-efficient aircraft, 106
Full width at half maximum (FWHM), 124
Functionalization, 29, 101, 102

G

Geometric reconstruction scheme, 285
Geometry, 280, 282, 293
Geotextiles, 68
Glacial acetic acid, 70, 73
Glass
  fiber, 80, 81
  substrates, 30, 39–41
  transition, 28, 34, 148, 149, 184, 185, 188, 238, 258, 259
    temperature, 28, 181, 188
Glassy
  carbon electrode (GCE), 195
  polymer, 181
Global environmental problem, 64
Glucose, 104
Glycerol, 45, 46
Gold
  colloid solutions, 30
  colloids, 36, 39, 41
  ions, 26
  multilayers, 30
  nano-island, 37
    structures, 35, 40
  nanostructure, 41
Gold-poly (dimethyl siloxoxane) (Au-PDMS), 40, 42
Götze's mode coupling theory, 259
Graphene, 101, 102, 104–106, 108, 109, 111, 113
  based
    electrode material, 107
    materials, 107
    nanocomposites, 111
    polymer composite materials, 106
    polymer nanocomposites, 101, 108–110
  derivative, 108, 109
  graphene derivatives, 103, 106
  layer, 102, 104, 106, 108, 109
  monolayers, 103
  nanosheets, 104, 110
  oxide (GO), 104
  polymer nanocomposites, 107, 109
  research, 111
  sheets, 110
Graphite, 101–103, 108, 109, 112, 279
  flake, 102
  oxide, 103
Graphitic
  carbon-based nanomaterials, 102
  materials, 101, 102

H

Harmonic
  frequencies, 261–263, 267, 269, 270, 272
  oscillators, 266, 268

Heat
dissipation, 294, 296, 298
flux, 281, 286
calculations, 281
condition, 294
transfer, 280, 282, 294, 295
model, 280
Heating-cooling sequences, 37
Hectorite, 4, 8, 11
Heteroatoms, 303
Heterocyclic dye, 130
Heterogeneous, 238
Hexagonal
arrangement, 257
network, 104
system, 196
High-density polyethylene (HDPE), 109, 136, 297
High-speed images, 281
Homogeneous, 33, 126, 136, 182, 303
mixture, 180
nucleation, 33
Homopolymer, 7
Homo-polymer grade polypropylene, 69
Homopolymers, 186
Honeycomb, 101, 102, 104, 106
crystal plane, 102
Horizontal displacements, 252
Hot-melt adhesion, 9
Household and industrial applications, 65
Human
body, 5, 6
placenta, 46, 47
Hummers method, 103
Hyaluronic acid (HA), 43, 46–48, 51–54
chitosan
layers, 60
multilayer, 50
solution, 48
Hybrid
composite, 77–82, 84–97
electric vehicles, 107
Hybridization, 79
Hydrated
anions, 304
products, 231
Hydraulic press, 70, 158, 182
Hydrazine, 104
hydrogen, 104
solution, 104
Hydrochloric acid (HCl), 69, 73, 303
Hydrodynamic
diameter, 120
radius, 128
Hydrogen
evolution process, 307
gas, 106
peroxide, 132
storage, 106
Hydrolysis, 43, 45, 50, 61
Hydrolyzed surfaces, 44, 52, 55, 60, 61
Hydrophilic, 9, 11, 12, 20, 21, 28, 55, 56, 65, 104, 201, 328
hydrophobic, 11
terminals, 201
vinyl acetate, 9
Hydrophobic
characteristic, 12
ethylene, 9
terminals, 201
Hydrophobicity, 28, 29
Hydroquinone, 104
Hydrotalcite, 10, 11
Hydrothermal
method, 208
process, 191, 194
technique, 104, 193
Hydroxyl
epoxy functional groups, 104
ions, 116
radical, 116, 131, 132, 215
Hyperspectral image, 38
Hysteresis loop, 164, 165

I

Impact test, 85
In vitro mechanical performance, 14
In vivo application, 21
Incubator, 48
Indian Synthetic Rubber Limited (ISRL), 182
Indium-tin-oxide (ITO), 104
Industrial techniques, 4

Infrared
band (IR), 199, 261, 267
spectral range, 261, 270
Injection molding, 4, 13, 63, 69, 109
Inorganic
In-situ methods, 27, 29
materials, 121
nanomaterials, 108
Interfacial
bonding, 87, 88, 94, 108, 162
phenomena, 170
polarization, 169, 174
reaction, 71
Intermolecular
interactions, 10, 125, 126
packing, 238
Interpenetrating polymer networks (IPNs), 179–189
formation, 180, 185, 186
manufactures, 181
material, 181
samples, 183–185
synthesis, 180, 181, 188
system, 181, 185, 188
Interplanar spacing, 196
Intraocular lenses, 6
Intrinsic property, 20
Ion
adsorption, 107
implantation, 27, 45
mobility, 171
Ionic
nature, 192, 198
radii, 123
Ionization energy, 118, 267
Iron hydroxide, 322
Isolated gold nanoparticles, 35
Isotherms, 203

J

Joint replacements, 6
Jute, 63, 66–74, 77–83, 86–96
butts, 68
fiber, 63, 66, 67, 71, 72, 80, 82, 86–97
application, 67
properties, 66

K

Kapok/glass composites, 81
Kenaf, 77–79

L

Lagrange's formalism, 261, 262, 266
classical mechanics, 261, 266, 272
Laminar/turbulent, 281
Lattice
parameter, 196, 197
sites, 197
Laughlin-Uhlmann-Angell's fragility index, 246, 248
Layer-by-layer deposition, 45
techniques, 45
Layered double hydroxide (LDH), 108, 109
L-cysteine, 200
Lennard-Jones potential, 262, 263, 272
L-glutamine, 46
Light
compression molding technique, 82
emitting diodes (LEDs), 110
Lignocellulose fiber, 92
Lignocellulosic biomass, 63
Lineage specification, 44
Linear
domain, 241
homogeneous equations, 269
viscoelastic domain, 241
Liquid
crystal displays (LCDs), 104, 106
crystalline polymers (LCPs), 136
medium, 5, 12
Lithium borohydride, 104
Localized Surface Plasmon Resonance (LSPR), 26, 28, 31, 33, 36, 38, 39, 42
Low carbon steel (LCS), 301, 302, 304–314, 317, 320–322, 324, 327, 328
Low density polyethylene (LDPE), 136, 137, 297
Luminescence, 116
Lung fibrosis, 44

M

Macromolecular inhibitors, 303

Macroscopic dynamics, 237
Magnetic
  behavior, 156
  characteristics, 156, 157
  induction, 306, 310, 312
  materials, 163
  measurements, 165
  properties, 155, 156, 163, 174
Magnetism, 156
Magnetization, 156, 157, 163–166
Magnetometer, 157
Manufacturing process, 13, 278
Marine
  applications, 73, 301
  structures, 78
Material
  nozzle design combinations, 277
  properties, 111, 214, 285, 286
Maxwell
  model, 173, 174
  Wagner effect, 171
Maxwell's molecular velocities, 242
Mechanical
  barrier properties, 4, 8
  characteristics, 77, 96
  degradative properties, 45
  electrical properties, 108
  exfoliation method, 103
  performance, 78, 79, 135, 136, 141, 146, 181
  properties, 4–6, 13, 14, 41, 63, 69, 74, 77–80, 83, 93, 94, 96, 97, 109, 116, 118, 136–141, 146, 148, 150, 151, 181
  response, 238
  vibration, 181
Medical
  application, 118
  device, 3–6, 10, 13, 19
  diagnostic technologies, 13
  implantable devices, 4
  industry, 3, 6, 13
  pharmaceutical devices, 19
Melt
  compounding process, 5
  ejection, 289
  flow path, 292
  intercalation, 108, 109
    method, 109
  spinning process, 277, 278, 281, 297, 298
Melting
  compound, 20
  point, 65, 136
Meniscus formation, 292
Mercury-vapor lamp, 120
Mesenchymal
  lineages, 44
  stem cells (MSC), 43, 44, 50, 52, 53, 55, 58, 59, 61
  tissue regeneration, 44
Mesoporous sulfide system, 194
Metal
  cations, 308, 328
  ceramic materials, 13
  insulator/semiconductor structure, 327
  ions, 33, 116, 121, 156
  nanoparticles, 25, 26, 117
  oxide semiconductor, 131
  precursor, 26
  substrates, 301, 302, 306, 312, 315, 318, 319, 323, 324, 328
Metallic
  hydroxides, 322
  substrate, 303
Methane (CH), 106
Methanol, 120, 199
Methyl
  ethyl ketone peroxide, 82
  methacrylate (MMA), 179, 182, 183, 185, 186, 188
Methylene
  blue, 130, 133
  group, 312
Mica fluoride, 11
Microchannel, 26, 29, 30
Microchip bonding, 29
Microelectronics, 118
Microfibers, 143
Microfibril
  composites, 137
  diameter, 143
Microfibrillar
  angle, 91

blends (MFBs), 139
composites (MFCs), 135–141, 145, 146, 148, 150
technology, 136
in-situ composites, 135
Microfibrils, 135, 139, 141, 143, 145
Microfluidic
biosensing, 25, 28
detection, 31
device, 30, 31, 40
environment, 30, 31, 42
sensing, 31
experiments, 38
synthesis, 31, 32
Micrographs, 53, 86, 93–95, 135, 137, 145
Micro-hardness, 78, 81, 85, 91, 92, 96
hybrid composites, 91
properties, 77
Microheterogeneous morphology, 180
Micromechanical cleavage, 102, 103, 112
Microphase separation, 181
Microprocessor, 306, 310, 312
Microscopic
analysis, 48, 184
transitions, 242
Microstructural
behavior, 78
studies, 233
Microstructure, 93, 133, 151, 156, 225, 226
Microwave (MW), 33, 34, 41, 46
assisted reduction reaction, 40
MW-assisted reduction, 33
synthesized composite, 34
Minerals, 11, 69, 108, 215, 216
Miniaturization, 102
Model boundaries, 286
Modulus curves, 254, 256
Moisture content, 139, 158
Molar mass, 312
Molding
complex, 70
machine, 135, 139
Molecular
arrangement, 156
dynamics simulations, 259
interactions, 21
mobility, 148, 258
oxygen, 132
parameters, 148
scale, 180
weight, 9, 238
Molten
alloy, 278
metal flow, 277
state, 109
Momentum equation, 285
Monodispersed clusters, 129
Monodispersive mesoporous distribution, 209
Monolayer, 39, 41, 102, 105
Monomer, 4, 7–9, 21, 26, 32, 46, 65, 108, 180, 182, 183, 302, 304, 307, 308, 311
Monomer o-toluidine, 304
Monomers/monomer combination, 7
Montmorillonite (MMT),3–5, 8, 11, 12, 14–16, 18–20
nanoplatelets, 11, 12
surface modification, 11
Morpholino (ethane-sulfonic acid) (MES), 46–48
Morphological
changes, 29, 44
characterization, 15
control, 193
cross-section, 93, 95
features, 128
properties, 69, 119, 187
structure, 94
studies, 60
Morphology, 20, 26, 29, 30, 36, 37, 40, 43, 44, 50, 52, 60, 86, 126, 132, 135–138, 140–142, 145, 150, 151, 179, 181, 182, 184, 187, 188, 191–196, 202, 303, 304, 317
Mortar cubes, 228
Multilayered surfaces, 52, 53, 55, 57, 60, 61
Multiple sclerosis (MS), 44, 156, 165, 166, 216
Multipotent progenitor cells, 44
Multiwall carbon nanotube (MWCNT), 7, 138, 145–147
Murine embryonic fibroblasts, 18

N

Nanoadditives, 217
Nanoconcrete samples, 224
Nanotitanium di-oxide, 220
Nanoclay, 3–5, 8, 10–12, 14, 21
  structure, 5
  type, 12
Nanocomposite, 3–5, 7, 8, 10–12, 14–16, 18–21, 25–29, 32–42, 79, 101, 102, 107–111, 116–119, 121–123, 125, 126, 128–130, 132, 133, 158, 169, 173, 223
  platform, 41
  properties, 12
  system, 5, 11, 20
Nanoconcrete, 222, 228, 231–233
Nanofiller, 3–5, 7, 10–12, 14–16, 18, 20, 21, 108–110, 138, 146, 169–171
  exfoliation and dispersion, 14, 16
  matrix interaction, 12
Nanohole structures, 39–41
Nanolayers, 170
Nanomaterials, 101, 102, 112, 115, 117, 128, 193, 207, 214, 233
Nanometer
  scale spatial resolution, 128
  sized
    holes, 41
    particles, 10
Nanoparticles (NP), 5, 11, 25–30, 33–42, 117–123, 125, 126, 128, 129, 132, 156, 161, 169, 171, 174, 195, 201, 220
  hole structures, 39
Nanoplatelets, 12, 16, 18
Nanoscale
  interactions, 5, 11
  reinforcement, 5
Nanoscience, 102
Nanosilver lined refrigerators, 118
Nano-sized inclusions, 138
Nanospheres, 129, 201, 202
  lithography (NSL), 39, 41, 42
Nanostructure, 26, 35, 40, 41, 117, 128, 191–194, 197–204, 207
Nanotechnology, 102, 117
  applications, 214
Nano-titanium oxide, 217, 233
  mortar cubes, 217
Nanotubes, 103, 147
Nanowires, 201
Naphtha cracking, 65
Natural
  fiber, 63–66, 77–79, 81, 86, 91, 97
    reinforced biodegradable polymer composites, 64
  fibrous polymers, 65
  layered materials, 108
  materials, 13
  synthetic origin, 5
  rubber, 155–157, 161, 163, 165, 173, 174
    composites, 155
  sunlight-induced reduction, 133
  synthetic polymer composites, 65
Neutral
  atoms, 263, 272
  silicon atoms, 267
Neutrinos, 105
Ngai's
  coupling model, 259
  model, 247
N-hydroxysuccinimide (NHS), 46, 48
Nickel
  ferrite, 155–160, 162–167, 170, 173, 174
    fillers, 157
  nitrate, 157
Nitrogen adsorption, 203
Nonhomogeneity, 321
Nonionic (Triton X–100) surfactants, 191, 193
Non-linear
  solid, 244
  stress relaxation, 237, 259
Non-uniform activation energies, 258
Nozzle
  bottom, 281
  design, 277, 279, 282, 298
  domain, 285, 287
  forms, 291
  slit, 277, 282, 291
    domains, 286
    flow, 298
  walls, 285, 289, 291, 293, 294, 297, 298

wheel gap, 283, 284
Numerical
model, 277, 279–282, 287
operation, 257
processor's memory, 259
scheme, 281
Nylon fibril, 143, 145
Nyquist
complex plane plots, 319, 324, 326
impedance, 321, 322
plot, 207, 321, 322

O

Octahedral
tetrahedral lattice, 156
stretching, 160
Octasilicate, 11
O-H stretching band, 312
Oil palm, 64, 77, 79
Olefin monomers, 7
Open-minded analysis and falsification, 259
Ophthalmic instrument, 6
Optical Bio Microsystems Laboratory, 25, 42
Ordinary Portland Cement (OPC), 213, 218, 229, 231, 233
Organic
cations, 12
chemicals, 131
coatings, 302, 314
compound, 69, 132, 215
degradation, 132
inorganic compounds, 116
peroxide, 132
pollutant degradation, 116
scavenger, 132
solvent, 65, 118, 158
surfactant, 11, 12, 15, 191
synthesis, 103
Organoclay, 4, 5, 8, 14
Organo-MMT
dispersion and exfoliation, 14
exfoliation and dispersion, 14
nanofiller, 14, 15
tactoid exfoliation, 15
Orthopedics, 6
Oscillations, 261–263, 266–268, 272
Osteoblast
differentiation, 60
phenotype, 52, 55, 56
Osteogenic differentiation, 45, 52, 57
O-toluidine, 301, 304, 305, 310, 311, 313, 328, 329
Oxidant acids, 302
Oxidation, 66, 116, 117, 130–132, 206, 302, 307, 309–311, 327
Oxidative stability, 28
Oxide layer, 301, 303, 309, 327
Oxides/chlorides, 301
Oxidizable
metal, 302, 309, 310
substrate, 302, 303
Oxygen
conducting electrodes, 323
containing functional groups, 104
evolution process, 307, 309, 310
Ozone resistance, 13

P

Paraformaldehyde (PFA), 46, 48
Passivation, 304, 308
Pendulum hammer, 85
Penicillin, 46, 47
Perkin Elmer spectrometer, 195, 306
Permittivity, 155, 156, 167–171, 174
Pernigraniline base (PB), 312, 313, 319, 327, 328
Perturbative triple corrections, 267
Petroleum resources, 64
Phalloidin, 46, 48
stained F-actin, 50
Pharmaceutical applications, 13
Phenolic, 64
Phenotype, 52, 53, 57, 60, 61
differentiation, 61
Phenotypes, 58, 60, 61
Phosphate buffered saline(PBS), 46, 47
Phosphonium-based ionic liquids, 304
methyl tributyl phosphonium diethyl phosphate, 304
tetrabutylphosphonium bromides, 304

Photoactivity, 130
Photocatalysis, 130
Photocatalyst, 116, 120
Photocatalytic
  activity, 116–118, 120, 130–133
  behavior, 116
  degradation, 130
  efficiency, 116, 130
  measurement, 120, 130
  process, 116
  reaction, 132
Photocatalyzed organic compound, 131
Photochemically, 26, 33
Photodegradation, 116, 119
Photoelectric effect, 242
Photoexcitation, 131
Photogenerated electrons, 132
Photoinduced electron transfer, 105
Photomicrographs, 86
Photon hypotheses, 242
Photons, 104, 106
Photoreduction, 116
Photovoltaic devices, 101
Physical
  barrier, 303
  chemical properties, 44, 117
  mechanical performance, 10
  methods, 27
  response, 237, 239
Physicochemical properties, 102, 104, 108, 110, 156, 166, 174
Physisorption, 106
Piezoelectricity, 116
Pineapple
  leaf, 64, 78
    fibers, 77
Planar flow melt spinning (PFMS), 278, 279, 298
Planck's constant, 105
  Boltzmann's constants, 243
Plasma treatment, 45
Plasmaor, 29
Plasmon
  band, 25–28, 30, 31, 40, 122
  oscillation, 35
  resonance, 35, 38, 39, 42
  vibrations, 122
Plasmonic
  detection, 26, 40
  sensing performance, 37
  sensitivity, 36, 38
Plastic
  containers, 106
  material, 65
  molding method, 70
Plasticizers, 10
Plastics recycling industry, 136
Platinum (Pt), 308–311
  electrode, 309, 311
Ploy (ethylene glycol) (PEG), 115, 116, 118, 119, 121–126, 128–133
Polar
  functional groups, 303
  groups, 71, 121
Polarization, 156, 157, 167–169, 303, 306, 307, 309, 310, 314, 317–319, 327, 328
Poly (butylene terephthalate) (PBT), 138
Poly (ethylene terephthalate) (PET), 136–138, 250
Poly (vinyl alcohol) (PVA), 12, 20, 21, 26, 36, 38, 39, 109
Poly [methyl methacrylate] (PMMA), 26, 32–34, 109, 179, 181–189, 239–241, 245, 246, 248, 249, 251, 253, 255–257
Poly carboxylate ether (PCE), 217, 220, 221, 228
Poly(carbonate), 259
Poly(dimethylsiloxane) (PDMS), 25, 26, 28, 29, 31, 36–39, 41
Poly(ethylene terephthalate), 248, 259
Poly(glycerol-sebacate)(PGS), 43, 61
  films, 43–51, 56, 60
  substrates, 43–45
  surfaces, 44, 51, 52, 54, 55, 57, 58, 60, 61
Poly(methyl methacrylate), 34, 239, 259
Poly(o-anisidine), 303
Poly(o-toluidine)(POT), 301, 304–307, 309–314, 317–320, 323, 324, 326–328
  coated metals, 319, 323
    substrate, 316, 327
  coating, 301, 304, 306, 310, 312, 313, 315, 319, 327, 328
Poly(o-toluidine)coatings, 304

Poly(styrene) (PS), 26, 36, 39, 41, 56, 59, 109
Poly(vinyl chloride) (PVC), 10, 64, 109
containers, 10
products, 10
Polyaniline (PANI), 302–304
coating, 303
Polyatomic molecules, 266
Polycarbonate (PC), 136, 239
Polycarboxylate Ether (PCE), 217
Polyester, 64, 77–82, 86–96
Polyester composites, 77–81, 94, 96
Polyesterbased hybrid composites, 81
Polyethylene, 7, 8, 12, 64, 136, 151, 297
Polyethylene (PE), 7, 8, 12, 15, 20, 21, 136–138, 297, 319, 324
grafted maleic anhydride (PE-g-MA), 109
Polyethyleneimine (PEI), 46
Polygonal, 45, 52, 53, 55, 56, 60, 61
Polymer
blends, 118, 136, 137, 151, 180, 181, 185
chains, 11, 34, 37, 38, 109, 123, 148, 150, 170
mobility, 18
composite system, 172
conventional graphite-based composites, 109
film, 36, 40, 304
graphene nanocomposites, 101, 108, 110
inorganic composites, 108
matrix, 5, 11, 12, 25, 27, 29, 41, 78, 87, 107–110, 133, 169, 170
nanocomposites, 64, 79, 81, 107–109, 112, 151
network, 37, 41
non-polymer, 238
poly carboxylate, 214
reinforcement, 81
solution, 26
surface, 36, 45
ubstrates, 35
Polymeric
material, 4, 6, 13, 108, 167
science, 180
matrices, 107, 148
substrates, 45
systems, 180
Polymerization, 7–9, 108, 109, 180–182, 184–188, 301, 302, 304, 310, 312, 313, 327, 328
process, 9, 310, 311
technique, 179, 180
Polymer-on-polymer stamping, 45
Polymer-organoclay nanocomposites, 8
Polyolefin's based materials, 7
Polypeptide, 31, 41
Polyphenols, 69
Polypropylene (PP), 7, 64–66, 69, 71–74, 81, 109, 135–151
Polypyrrole, 302
Polysaccharides, 69
Polystyrene, 39, 41, 42, 47, 56, 59, 60
microspheres, 39, 41
monolayer mask, 41
Polyvinylpyrrolidone (PVP), 157, 161
solution, 157
Porosimetry, 225
Porous CuS, 198, 209
Post-synthesis
heat treatment, 35
thermal treatment, 29
Potentiodynamic polarization
curves, 306, 316
measurement, 301, 306, 317
parameters, 314
Pourbaix diagram, 322
Pre-dispersing process, 5, 14–16, 18, 20
Primitive relaxor and ensuing analysis, 238
Primitive relaxors, 238, 245, 248, 254, 256, 257
Proliferation, 43, 44, 50, 52, 55, 59–61
Propylene, 65
Propylene monomer, 65
Protection mechanism, 327
Protein
assay, 44, 48, 58, 60
concentration, 58–60
quantification, 58
results, 58, 60
Protons, 129
Pseudocapacitor, 107, 192, 193
applications, 192
Puddle formation, 281, 287

Q

Quadratic form, 264
Qualigens fine chemicals, 305
Quantitative justification, 256
Quantum
  Hall effect (QHE), 105, 106, 112
  mechanical
    effect, 105
    origin, 263
  relativistic phenomena, 111
Quartz
  crucible material, 295
  microbalance, 303

R

Raw
  cotton, 67
  material, 13, 64, 67
    utilization, 64
Reactive
  ions, 207
  oxygen species (ROS), 19
Redox
  based faradaic reactions, 192
  capacitors, 107
  conversion, 302, 327, 328
  reaction, 107, 207
Reduction
  process, 41
  reaction, 41
Refractive index sensitivity, 27, 31, 38
Reinforcing filler, 5, 69, 163
Relaxation
  effects, 322
  moduli, 248–250, 253
  process, 185
  spectra, 248, 250, 253, 254
  time constant, 322
  times, 241, 244, 245, 254
Relaxor, 237–240, 243–245, 253–257
Renewable
  energies, 111
  resources, 64
Reservoir, 278, 297
Residual catalyst, 120
Resistivity, 105, 156
Reverse cycle, 310
Rheological
  complexity, 250, 252, 256
  properties, 137, 151
Rheometer, 182
Rheumatoid arthritis (RA), 44
Root means square (RMS), 128
Round bottom flask (RB), 157
Rubber ferrite composites(RFCs), 156, 157, 163, 166

S

Salicylate, 304, 305, 307–309, 311, 313–315, 318
  complex, 309, 311
  counterions, 308
  electrolyte, 328
  medium, 304
  solution, 301, 304, 307–309, 311
Sandwich composite, 63
Satellite structure, 204
Saturation magnetization, 156
Scanning
  electron microscope, 78, 81, 86, 93, 97, 119, 135, 151, 195
  electron microscopy (SEM), 29, 33, 40, 86, 94, 115, 119, 120, 122, 126, 132, 133, 135, 137, 140, 141, 143, 145, 151, 187, 191, 195, 200–202, 225, 226, 231, 304, 313, 317, 318
Scherer equation, 123, 159, 196
Sebacic acid, 43, 45, 58
Selectedarea electron diffraction (SAED), 202, 203
Self-cleaning properties, 213
Self-healing mechanism, 302
Semi-crystalline
  material, 248
  nature, 66
Sensing
  abilities, 26
  applications, 26, 33, 34
  experiment, 30, 31, 37, 39, 40
  platform, 31, 35, 40
  process, 31
Silicate

compounds/synthetic clay, 108
delamination, 12
Silicon, 6, 102, 262, 263, 267, 270
atom, 267
carbide (SiC), 103
elastomer, 4, 6, 21
semiconductor technology, 105
Silver
cubic nanoparticles, 126
nanoparticles, 116–118, 129
nitrate solution, 119
Sisal (S), 63, 77–80, 87, 115, 135, 139, 151, 155, 159, 183, 191, 195, 198, 202, 204, 206, 237, 304–306
fiber, 79, 80
Slit
nozzle design, 297
passage, 277–279, 285, 287, 288, 290–298
nozzle, 294–296, 298
Sodium
bicarbonate, 46
borohydride, 32, 104
chloride, 46
citrate, 30, 46
hyaluronate (SA), 47
SA fibers, 47
hydroxide (NaOH), 46, 47, 69, 70, 73, 157
salicylate, 301, 304, 305, 307, 309–311, 328
solution, 305, 307
sulfide, 200
Soft matter theories, 242
Solar
cells, 106, 110, 117
energy, 111, 117, 118, 193
wind energy, 111
Solartron Electrochemical Measurement System, 305
Solid domain, 287, 294
Solidification process, 278
Solvothermal, 103, 193
Sonochemical, 104
Spectroscopic techniques, 301
Spent coffee ground (SCG), 69, 71, 73
Sphere-shaped filler, 169
Spinal cord injury, 44
Spinel
ferrites, 156
structure, 156
Sputtering technique, 225
Stable phase equilibrium, 231
Stainless steel (SS), 182, 301–314, 317, 319–322, 326–329
Statistical
analysis, 44, 56, 58–60
mechanical development, 242
weights, 248, 256
Steam-cracking process, 65
Steric hindrance, 208
Sterility, 47
Sterivex sterile filter unit, 47
Stoichiometric composition, 193
Streptomycin, 46
Stress relaxation, 239, 241, 243, 244, 248, 254, 256–258
cracking resistance, 9
experiments, 241
moduli, 251, 257
strain curves, 84
Stretch ratios, 137, 139, 140, 143, 144
Styrene butadiene rubber (SBR), 179, 181–186, 188, 189
SBR phase, 186
Substrates, 30, 35, 43, 45, 301–308, 313, 319, 324, 326–328
Sucrose, 104
Sunlight
illumination, 132
irradiation, 116, 130, 132
Super plasticizer, 221
Supercapacitors/ultracapacitors, 107, 110, 192–194, 206
Supercapacitor application, 107, 207
Superoxide anion, 132
Superplasticizer type, 217
Surface
morphological analysis, 126
AFM studies, 128
SEM/EDS studies, 126
entrapment modification, 45
gold–polymer nanocomposites, 36
modification, 11, 12, 45
morphology, 93

treatment, 60
methods, 43
Surfactants, 191, 193, 194, 196–198, 200, 201, 205
Synergistic effect, 161, 206
Synthetic polymers, 13
Systemic lupus erythematosus (SLE), 44

T

Tactoid bonding, 5
Tafel constants, 314
Tan delta value, 155
Technical textiles, 66, 68
Technological and industrial applications, 302
Temperature gradient, 294, 295
Tendon, 44
Tensile
fracture, 94
fractured specimens, 96
modulus, 10, 77, 79, 81, 83, 84, 86–88, 92, 96
performance, 145
properties, 65, 79, 83, 84, 143
strains, 239
strength, 6, 10, 12, 14–16, 20, 63, 71–73, 77, 79, 80, 83, 84, 86, 87, 96, 137, 140–147, 162
test, 83, 84
toughness, 16
Testing
machine, 83–85, 140
process, 83
Tetrabutylphosphonium bromide, 304
Tetrahedral site, 160
Textile fibers, 67, 68
Thermal
conductivity, 66, 86, 106, 108, 287, 294, 295
convection, 35, 36, 39
method, 35, 36
degradation process, 18
evaporation/sputtering equipment, 39
expansion coefficient, 66
fluctuations, 103
insulation, 78
manipulation, 38
measurements, 258
mixing method (melt intercalation), 109
procedures, 156
properties, 14, 19, 37, 110, 111
radiation, 240
reduction, 26, 33, 34
method, 26, 33
resistance, 295
stability, 4, 5, 7, 8, 15, 18, 20, 109, 181
Thermochromic
glass transition, 258
materials, 25
state, 262
Thermodynamically stable, 301
Thermoformed trays, 19
Thermogravimetric curves, 18
Thermolysis, 27
Thermoplastic, 3, 4, 7–9, 13, 63–65, 70, 136, 150, 181
characteristics, 4, 8, 13
nanocomposites, 109
polymer, 109
polyurethane (TPU), 3
Thermorheological complexity, 258
Thermoset polymers, 70
Thermosetting matrices, 64
Thioacetamide, 200
Thiourea molecules, 195
Thoracic and lumbar spine, 44
Three dimensional (3D) graphite, 102
Three-atom cluster, 267
Tissue engineering, 43, 44, 61
applications, 43
approaches, 44
Titanium Dioxide,169, 213, 215, 216, 224–227, 231
Toluene and acetone, 46
Toxicity, 26, 28, 117, 118, 302
Transformer core applications, 278
Transition state theory (TST), 240, 242, 246
Transmission
electron micrographs, 188
electron microscope, 184, 195
electron microscopy (TEM), 15, 103, 120, 155, 161, 162, 179, 187, 188, 191, 195, 201–203, 219

TEM analysis, 161
Tunable
bandgap, 102
fragility, 246, 258
properties, 102
Two-atom cluster, 266
Two-state model, 243, 245, 246, 250

U

Ultracut (UCT), 184
Ultramicrotome, 184
Ultrasonic
probe, 15
reduction, 32
waves, 220
Ultrasonication, 15, 16, 18, 19, 220
Ultrathin sections, 184
Ultraviolet (UV), 7, 9, 26, 33–35, 40, 45, 49, 66, 106, 115, 117, 119–121, 130, 131, 133, 191, 195, 199, 301, 306, 314, 327, 328
UV absorption spectra, 131
UV irradiation, 33, 34, 120, 131
UV light, 120
UV radiation, 9
UV resistance, 7
UV-vis spectrophotometer, 35, 306
UV-visible spectroscopy, 26, 115, 120, 195, 199
Umbilical cord blood, 44
Uniaxial tensile test, 83
Unimodal distribution, 129

V

Vacuum, 46, 86, 105, 182, 194, 195, 225, 292
Valorization, 69
Van der Waals-London
bond, 262
bonding, 262
cluster, 261, 262
force, 263
interaction, 261, 263
silicon clusters, 270
Vapor phase, 104
Vascular grafts, 6
Velocity, 85, 139, 267, 269, 288, 294
Vertical deposition method, 39, 41
Vicat
mold, 221
needle, 221
Vicinity, 26, 302, 308, 323
Vinyl acetate (VA), 3–5, 7, 9, 14, 15, 21
Viscoelastic, 149, 179, 181, 182, 184, 186, 245
behavior, 179, 182, 184
domain, 245, 258
material, 148, 149
measurements, 184
properties, 179, 181, 184
Viscosity
equation, 277
ratio, 137
Voltammogram, 205, 307–311
Volume
fluid technique (VOF), 285, 287, 293
fraction, 140, 141, 162, 172, 286
fraction average method, 172

W

Water
absorption, 218
molecules, 14, 116, 304
proof properties, 10
purifier, 116
soluble polymers, 129
Waveguides, 25
Wavelength, 26, 28, 35, 49, 120, 122, 124, 159, 306, 312
Wavenumber, 125
Weak bonding, 262
Weakly bound silicon cluster, 272
Weight
fractions, 166
percentages, 82, 88, 182
ratio, 78–82, 86–97
Western blot, 60
Wheel
domain, 282, 284
surface, 284
Wooden mold, 83
Wound dressings, 118

Wrappers, 68

X

X-ray, 115, 119, 120, 123, 124, 159, 194, 202, 220, 225
  diffraction (XRD), 15, 16, 115, 119, 120, 123, 124, 126, 133, 155, 157, 160, 161, 191, 194, 196–198, 203, 220
    data, 15
    pattern, 16, 120, 191, 194, 197
    spectrum, 115
  diffractometer, 119, 120, 194, 220
  photoelectron spectroscopy (XPS), 119, 195, 203, 204
    analysis, 203, 204
    spectra, 203
    spectrum, 204
  powder diffraction, 159
  spectrum, 126

Y

Y-functions, 265
Young's moduli, 51

Z

Zero
  difference, 296
  dimensional (0D), 102
  pressure, 296
Zeta
  potential, 129
  sizer nano ZS90, 120
  values, 129
Zig-zag flow, 293
Zinc, 121, 123, 302, 311
  oxide (ZnO), 115–126, 128–133, 159, 311
    absorption, 122
    concentration, 119, 128, 133
    content, 115, 126
    doped Ag, 123, 125, 128, 130, 133
    doped Cts, 116, 119, 121, 126, 130, 132, 133
    line, 126
    nanoparticles, 118, 126, 132
    nanostructure material, 117
    particles, 117, 123, 126, 131
    stretching band, 125

9781771887243

9781771887243